Radiolaria

O. Roger Anderson

Radiolaria

With 64 Figures

Springer-Verlag
New York Berlin Heidelberg Tokyo

O. Roger Anderson
Biological Oceanography
Lamont-Doherty Geological Observatory
Columbia University
Palisades, New York 10964
U.S.A.

Cover: Projection of radiolarian skeletons on a transmission electron micrograph of the cytoplasm of a living radiolarian

Library of Congress Cataloging in Publication Data
Anderson, O. Roger, 1937–
    Radiolaria.
    Bibliography: p.
    Includes index.
    1. Radiolaria. I. Title. [DNLM: 1. Protozoa.
    QX 50 A548r]
    QL368.R2A5 1983   593.1'4   83-4266

Media conversion by WorldComp, New York, New York.
Printed and bound by Halliday Lithograph, West Hanover, Massachusetts.
Printed in the United States of America.

9 8 7 6 5 4 3 2 1

ISBN 0-387-90832-3 Springer-Verlag New York Berlin Heidelberg Tokyo
ISBN 3-540-90832-3 Springer-Verlag Berlin Heidelberg New York Tokyo

Preface

The study of marine plankton has traditionally focused on those organisms that appeared to have obvious ecological significance in understanding the major patterns of biological productivity, trophic relations, community structure, and the dynamic interaction of living things with the physical environment. Not infrequently, this thrust has centered on the apparently most abundant and/or larger members of the plankton community, including significant primary producers such as the diatoms, nonthecate algae, and flagellates, or the major consumers—copepods, gelatinous metazoa, and other abundant metazoan invertebrates. Consequently, some of the less well recognized but also abundant microzooplankton have been given less attention. The radiolaria, although widely studied as fossils by micropaleontologists, have been relatively neglected by biologists in modern times. This is lamentable given their widespread distribution in the oceans, remarkably complex form, and not infrequently localized abundance. Their diversity of form, encompassing solitary species of microscopic dimensions and colonial species as large as several centimeters or more, challenges us to explain their evolutionary origins, explore their structural-functional correlates, and comprehend the ecological basis for their widespread occurrence in all oceans of the world from the greatest depth to the surface of the sea. Their intricate and aesthetically pleasing skeletons of enormous variety and fine-detailed design formed from amorphous silica (opaline glass) offer a unique biomineralized product that defies immediate biological explanation. The elucidation of the biogenesis and morphogenesis of these delicate, ornate, glassy structures promises to yield insight into broader principles of biological design and the control of form and function in living systems.

These unique organisms, therefore, are of potential interest to a wide audience of scientists, extending beyond the traditional community of marine scientists, including cell biologists, developmental biologists, biochemists, and molecular biologists. This book is addressed to this wider audience with the hope of encouraging interdisciplinary research with these remarkably complex organisms. The organization of the book has been designed to proceed from fundamental topics on structure and taxonomy toward organismic biology and ultimately to global issues of ecology, distribution in space and time, and the dynamics of interactions between radiolaria, other biota, and the physical environment. Where possible, allusions are made to the pressing unsolved problems, likely heuristic theoretical models, and multidisciplinary questions that currently prevail. An interdisciplinary theme is intended throughout the book by offering cross references to other mineral secreting microplankton and establishing a context for multidisciplinary discussions, particularly in chapters focusing on fine structure, physiology, ecology, and evolution. Particular attention is given to drawing closer ties between biological and micropaleontological research with the hope of encouraging increased cooperation between these two disciplines. Although much of the information reported here is derived from widely diverse published reports, some new information, not published elsewhere, on fine structure, chemical composition, and skeletal morphogenesis of radiolaria is also presented. I hope the assembly of information from diverse sources in the biological and earth sciences and its synthesis within a theme of radiolarian biology will provide some novel insights and more comprehensive perspectives not otherwise available to specialists familiar with one or more aspects. In many respects this book must be considered a prologue rather than an epilogue as so much remains unknown.

During the course of the several years that I have pursued research with radiolaria, and during my intensive effort to reduce a diverse and substantial body of literature into a relatively concise treatise, many colleagues have given encouragement and advice. There are clearly many more people deserving mention than can be acknowledged in a brief statement, but among those who are most immediately concerned, I would like to thank Allan Bé, James Hays, Barbara Hecker, Dave Lazarus, Joseph Morley, and Neil Swanberg, who have read parts of the manuscript and made helpful suggestions. On occasion I have had the pleasure of working with colleagues at other institutions, including Christoph Hemleben and Michael Spindler at Tübingen University, West Germany, and many friends and colleagues at the Bermuda Biological Station for Research Inc.; Bellairs Research Institute at Barbados; Caribbean Marine Biological Institute in the Dutch West Indies; and the Spanish Research Institute at Tenerife, Canary Islands. On pleasant occasions, other colleagues have encouraged my endeavors in radiolarian research, including Jean and Monique Cachon, whose pioneering work with the fine structure and taxonomy of radiolaria is widely known, and W.R. Riedel, who in collaboration with A. Sanfilippo, has contributed significantly to our understanding of the geological history of radiolaria and their taxonomy.

Many younger colleagues have assisted in the laboratory during the various times research was pursued at field stations. I express sincere appreciation for their interest and dedication to the work—John Hacunda, David Caron, Howard Spero, Walter Faber, Martyn Botfield, and Paul Bennett. I gratefully acknowledge financial support from the Biological Oceanography Division of the National Science Foundation, which has funded much of my research summarized in this book. Finally, I am very pleased to acknowledge Gabriella Oldham, who efficiently rendered my handwritten manuscript into legible typewritten form. Philip Manor and the staff of Springer-Verlag contributed much to making this task more efficient and enjoyable.

Columbia University   O. R. Anderson
New York
May, 1983

Contents

Chapter One
Morphology and Systematics

Introduction

Radiolaria are holoplanktonic protozoa widely distributed in the oceans, including arctic, subtropical, and tropical waters. They occur throughout the water column from the near surface to hundreds of meters depth. As with many planktonic organisms, their abundance in a geographical region is related to the quality of the water mass, including such variables as temperature, salinity, productivity, and available nutrients. They are largely nonmotile organisms, and their general morphology clearly reflects an adaptation for a floating existence. Various structures in the cytoplasm enhance buoyancy, including bubble-like alveoli in the peripheral rhizopodia of some species (e.g., *Thalassicolla nucleata*, Fig. 1-1A) and lipid globules localized in the dense central cytoplasm or dispersed within the surrounding complex cytoplasmic network. The skeleton, when present, is composed of amorphous silica, and is probably the most widely recognized morphological feature of the radiolaria (Figs. 1-1B–D and 1-2). It is also a major feature distinguishing the radiolaria from their close relatives, the Acantharia, possessing strontium sulfate skeletons (Fig. 1-3). Acantharian skeletons, moreover, are often much more massive with intersecting rods joined at the center of the skeletal array arranged as a set of Cartesian axes. Some Acantharia, however, possess very delicate spines and great care must be taken to distinguish them from the elaborate glassy skeletons of the radiolaria. The complexity and architectural diversity of these glassy structures have long been a point of curiosity as to their mode of formation and function, in addition to amazement at their pleasing aesthetic properties. The delicateness of form and diversity of space-enclosing structures, including perforated spheres (Fig. 1-1C), ornate geodesic-like polyhedral lattices (Fig. 1-1D), and seemingly endless variations of combinations of solid geometric designs, bears clear witness to the adaptive plasticity and sophisticated phylogenetic devel-

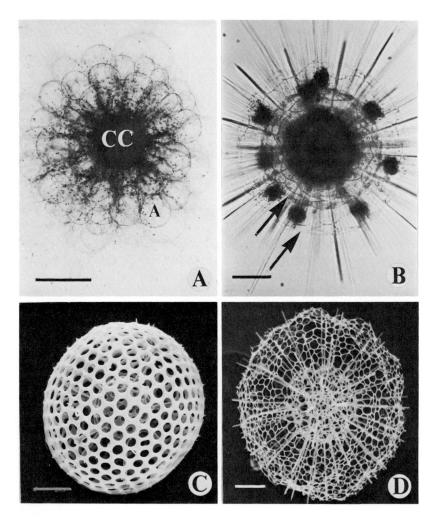

Fig. 1-1. Spumellaria. **(A)** A living *Thalassicolla nucleata* (Collodaria) exhibits the major morphological features of a large skeletonless spumellarian radiolarian possessing a dense spherical central capsule (CC) and frothy extracapsulum consisting of bubble-like alveoli (A) connected to fine rhizopodia that penetrate outward toward the surrounding environment. Scale = 1 mm (Anderson, 1978b). **(B)** A living sphaerellarian species is characteristic of skeleton-bearing species with delicate, concentric, lattice shells (arrow) connected to radially arranged spines. Scale = 50 μm. **(C)** A scanning electron micrograph of a siliceous, robust lattice shell with oval pores (*Theocosphaera inermis*). Scale = 25 μm. **(D)** A delicate arachnoidal lattice skeleton characteristic of some larger sphaerellarian species (*Actinomma arcadophorum*). Scale = 50 μm. **(C** and **D** courtesy of Kozo Takahashi, 1981, 1983)

opment of these single-celled organisms. Although they are protista, some species are large enough to be seen without magnification.

The solitary forms vary considerably in size from ca. 30μm diameter to 2 mm diameter. Some groups of radiolaria (e.g., Spumellaria) form macroscopic colonies consisting of hundreds of radiolarian cells inter- connected by rhizopodial strands and enclosed within a translucent, gelatinous envelope (Figs. 1-4 and 1-5). The colonies may be spherical, varying in diameter from several millimeters to a centimeter or larger. Others are cylindrical to filiform, with lengths varying from under a centimeter to several meters! Some of the colonial Phaeodaria form more complex aggregates of several shells joined by their spines or otherwise held together by simple siliceous networks.

The living substance of radiolarian cells is divided into two major regions (Haeckel, 1887): (1) a central mass of cytoplasm known as the central capsule and (2) a peripheral layer of cytoplasm surrounding the central capsule known as the extracapsulum (Fig. 1-6). The latter is composed of a variety of cytoplasmic structures including a frothy envelope of bubble-like alveoli, usually proximal to the central capsule, and a corona of ray-like axopodia and web-like rhizopodia radiating around the central capsule. The alveoli, when present, are attached to and interspersed among the axopodia and rhizopodia. In some species, particularly among the Spumellaria, the extracapsulum is further aug- mented by a gelatinous coat that may extend outward many times the diameter of the central capsule. It is often so hyaline that it cannot be detected with the light microscope, unless a dark contrasting substance such as India ink is added to the surrounding fluid.

The central capsule usually appears in the light microscope as a more dense mass than the extracapuslum. Indeed, in large solitary species it may be so voluminous as to appear nearly opaque. The cytoplasm within the central capsule (intracapsulum) consists of one or more nuclei, vacuoles of varying size and chemical composition, food reserve substances including lipid (fats and oils), and granular deposits that may be carbohydrates. Vital organelles within the intracapsulum in- clude mitochondria (the respiratory centers of the cell), Golgi bodies (secretory organelles), primary lysosomes (vesicles containing digestive enzymes to be transported to sites of digestion usually in the extracap- sulum), endoplasmic reticulum (intracellular membranous network), and ribosomes (particles mediating protein synthesis). Large digestive vacuoles and vesicles containing digested food products occur in the proximal layer of the extracapsulum. Algal symbionts (yellow-green cells) when present occur in the extracapsulum; however, none have been reported in the intracapsulum.

The central capsule is separated from the extracapsulum by a per- forated barrier known as the central capsular wall (e.g., Fig. 1-6B), varying in composition and thickness among species. It sometimes

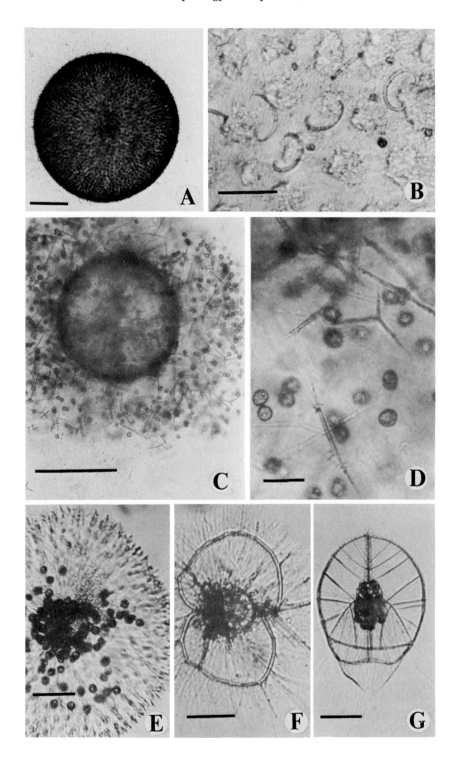

consists of a delicate envelope of living membranes perforated by fine strands of cytoplasm connecting the intracapsulum with the extracapsulum. In other cases, the wall is augmented by a porous organic barrier apparently secreted by the living envelope of membranes and varying in thickness among species, but always perforated by the thin cytoplasmic strands of the membranous envelope. These strands known as fusules (Cachon and Cachon, 1965) provide structural and physiological continuity between intracapsulum and extracapsulum. They are a unique property of radiolaria and their close allies, the Acantharia. Details of fusule structure and their relationship to the central capsular wall are presented in Chapter 2.

The silicate skeleton, when present, is surrounded by cytoplasm, and large segments of it may be found within the intracapsulum, particularly when there is a small central microsphere or internal spongiose shell. In most cases, however, a considerable amount of the peripheral skeleton is located in the extracapsulum. Skeletal morphology in conjunction with the microanatomy of the central capsule and its pattern of perforations is the main characteristic used to identify species of living specimens (Fig. 1-6). Skeletal morphology, moreover, is the major attribute used to identify species of fossil radiolaria where only the nonliving hard parts remain. The skeleton in a living specimen is enclosed within a cytoplasmic sheath called a cytokalymma (Anderson, 1976a, 1980) which is thickened in regions of active silica deposition, or it is thin and partially intact in older regions. The cytokalymma is attached to and is probably produced by the elaborate rhizopodial network surrounding the cell.

The complexity of the extracapsular cytoplasmic network should not be overlooked in developing a critical understanding of morphological

---

Fig. 1-2. Comparative light micrographs of some large living Spumellaria (A–D) and some Nassellaria (E–G). **(A)** *Physematium mulleri* exhibits a large, spherical central capsule surrounded by a nearly invisible jelly envelope. Scale = 1 mm. **(B)** The spicules of *P mulleri*, embedded in the jelly matrix, are C shaped or sometimes S shaped. Scale = 200 μm. **(C)** *Lampoxanthium pandora* exhibits a large opalescent central capsule surrounded by a thick envelope of alveolated extracapsulum containing algal symbionts and numerous siliceous spicules of simple and branched types. Scale = 1 mm. **(D)** Spicules of *L. pandora*. Scale = 100 μm. **(E)** A skeletonless nassellarian with numerous algal symbionts in the robust extracapsular cytoplasm. Scale = 100 μm. **(F)** A coronida nassellarian (*Eucornis*(?) sp.) exhibits the central capsule suspended within a delicate skeleton composed of intersecting rings. Scale = 50 μm. **(G)** *Callimitra* sp. possesses a complex lattice skeleton (also shown in Figs. 2-21–2-24) surrounding the centrally-located cephalis containing the central capsule. A web-like network of fine rhizopodia extends from the basal opening and rhizopodial threads emanate from the lattice. Scale = 100 μm.

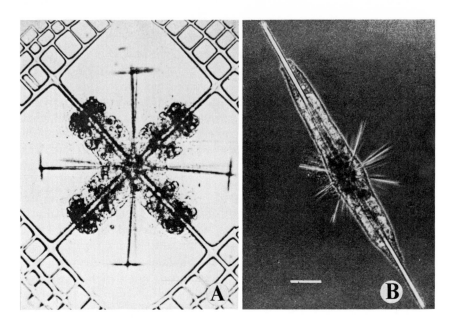

Fig. 1-3. Living acantharia. **(A)** An elaborate species, *Lithoptera mulleri*, exhibits a delicate snowflake design. (Photograph by Cachon and Cachon Villefranche sur Mer) **(B)** A more robust species, *Amphilonche elongata*, with a major axial spine and smaller lateral spines arranged along Cartesian axes. Scale = 20 μm. (Grell, 1973)

diversity among radiolaria and their taxonomic position relative to other protozoa. In some species, the halo of axopodia is the dominant feature of the extracapsulum (e.g., Figs. 1-1B, 2-3B, and 2-7A). These cytoplasmic processes are stiffened by a central shaft of parallel microtubules called the axoneme. The ensemble of axopodia is augmented by a network of web-like rhizopodia that may be particularly well developed proximal to the central capsule and interconnects the numerous axopodia within a continuous cytoplasmic network. Among some of the colonial radiolaria, the rhizopodial network is very elaborate and gives rise to a thin halo of axopodia and/or filopodia radiating from the surface of the colony. Filopodia are fine cytoplasmic strands, rather straight, but without an axoneme. The cytoplasmic network is very mobile and exhibits cytoplasmic streaming with varying velocities, depending on the position of the strand within the network and apparently the physiological state of the organism. The flowing motion is accompanied by fusion of the strands (coalescence) to produce nodes or thickened segments at some places, while simultaneously at other places, strands bifurcate or thin into a sheet and become reticulate as the cy-

Fig. 1-4. Light micrographs of living colonial radiolaria showing **(A)** A disc-shaped colony *Solenosphaera* sp. (Swanberg, 1979) and **(B)** A cylindriform colony of *Collozoum caudatum*. Scale = 1 cm. (Swanberg and Anderson, 1981)

toplasm segregates into fine interconnected strands. In some species, the axopodia are contractile and can be withdrawn upon contact. Occasionally, fine axopodia radiating from the pericapsular layer of cytoplasm exhibit rhythmic waves of contraction. This rapid contractile movement may aid in carrying prey into the proximal layer of extracapsular cytoplasm where digestion occurs. In some solitary species, axopodia may become coalesced into a whisker-like projection of specialized cytoplasm called an axoflagellum (Haeckel, 1887; Hollande and Enjumet, 1960). It is occasionally much longer than the halo of axopodia and originates from a pore field on the capsular wall. On the whole, however, the ensemble of rhizopodia and axopodia is very fine in contrast to the alveolate and somewhat thickened, vacuolated cytoplasmic layer proximal to the capsular wall. Radiolaria, moreover, are clearly contrasted to other amoeboid protozoa by the fineness of their extracapsular structures. They very seldom, if ever, possess thickened pseudopodia as observed in amoebas and close relatives, nor do they exhibit quite the degree of branching and anastomosing of rhizopodia observed in foraminifera (e.g., Sheehan and Banner, 1972; Adshead, 1980). In many respects, their morphology is more closely related to the straight, fine axopodia observed in Heliozoa (Hausmann and Patterson, 1982).

Comparative Morphology and Taxonomy

Based on the structure of their delicate pseudopodial network, radiolaria have been classically assigned to either the Sarcodina, which includes all amoeboid and rhizopod-bearing organisms including foraminifera, or the Actinopoda-bearing axopodia. The Actinopoda include the Heliozoa (sun animals), Acantharia, and radiolaria. In a modern classification (Levine *et al.*, 1980), the radiolaria are assigned to the Kingdom Protista, Phylum Sarcomastigophora (including amoebas, flagellates, and the rhizopod- and actinopod-bearing protozoa), Subphylum Sarcodina (pseudopod-bearing protozoa), and the Superclass Actinopoda. This scheme assumes a five-kingdom classification (Monera, Protista, Plantae, Fungi, Animalia). The Protozoa are considered to be a subkingdom of the Kingdom Protista (see Appendix). Within the Superclass Actinopoda are included the classical axopod-bearing protozoa: Acantharia (Acantharea), Radiolaria (Polycystinea and Phaeodarea), and the Heliozoa (Heliozoea). Levine *et al.* do not recognize radiolaria as a natural group apparently based on the substantial differences among the species in the two subgroups—Polycystinea and Phaeodarea. Some of these differences will become apparent as their comparative morphology is presented. However, the term radiolaria will be used here as a convenient label for polycystina and phaeodaria. Many researchers in the field still recognize radiolaria as a taxonomic category, including Petrushevskaya (1977) and Riedel and Sanfilippo (1977).

It is of historical interest to note that early taxonomic schemes (e.g., Haeckel, 1887) included the Acantharia among the radiolaria. This was partially attributed to similarities in cytoplasmic morphology, including the presence in both groups of a perforated capsular wall. The clear differences, however, in chemical composition of the skeleton and differences in the skeletal morphology offer compelling evidence to separate them. Consequently, the Acantharia are not considered in detail in this treatise; however, where appropriate, comparisons with radiolaria are presented to elucidate the unique characteristics of these closely allied organisms.

Only one putative radiolarian is motile—*Sticholanche zanclea*; however, not all biologists agree that it belongs with the radiolaria. It bears

---

Fig. 1-5. Colonial radiolaria. **(A)** A living *Collozoum* sp. exhibits a tip of a cylindriform colony with internal alveoli (A), nearly opaque central capsules (CC) and a peripheral distribution of algal symbionts (Sy). Scale = 500 μm. **(B)** Central capsules of *Collozoum* sp. interconnected by a rhizopodial network containing symbionts (Sy). Scale = 5 μm. **(C and D)** Central capsules of *Sphaerozoum punctatum* containing a centrally located lipid inclusion (LI) surrounded by the intracapsulum. Siliceous symmetrical-triradiate spicules (Sp) and dinoflagellate symbionts (Sy) occur in the extracapsulum. Scales = 50 μm. **(E)** Edge of a spherical colony illustrative of *Collosphaera globularis*. Scale = 200 μm. **(F)** Detailed view of a central capsule of *Collosphaera globularis* surrounded by a perforated siliceous shell and containing a refractile lipid droplet at the center of the intracapsulum. Scale = 25 μm. (Anderson, 1976c, 1978a)

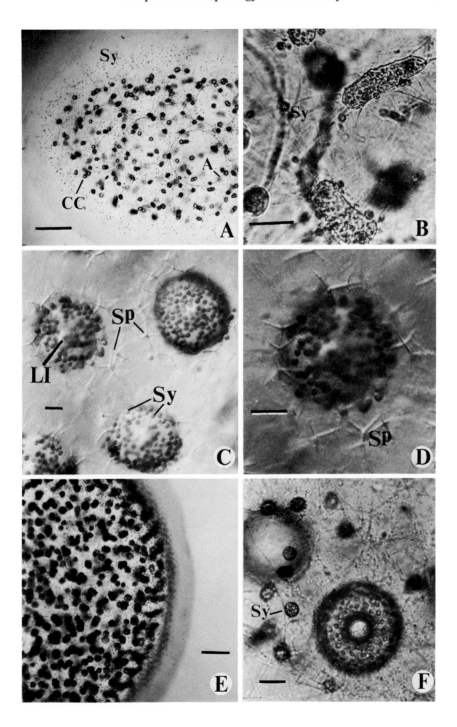

modified axopods that behave as minute oar-like appendages propelling
it feebly through the water. This unusual adaptation of the axopodia
and other morphological features (Fol, 1883; Hollande and Enjumet,
1954; Hollande et al., 1967) suggest that it may be more closely related
to the Heliozoa (Cachon et al., 1977). It lacks a typical well-organized,
peripheral, perforated capsular membrane as occurs in the major three
groups of radiolaria to be described hereafter.

Among the radiolaria, most modern taxonomists recognize three divisions: Spumellaria (Spumellarida), Nassellaria (Nasselarida), and
Phaeodaria (Phaeodarea). It is essential to recognize at the outset that
little consensus exists among taxonomists on the classification of radiolaria. This can be attributed in part to our fragmentary knowledge
about essential aspects of radiolarian biology, including mode of reproduction, physiology, ontogeny, genetics, and phylogeny. The diverse applications of radiolarian taxonomy within the disciplines of
micropaleontology, paleoecology, and marine biology often impose contrasting criteria as to the most scientifically useful scheme. Ideally, this
disparity should subside as more interdisciplinary knowledge becomes
available. The lack of modern biological knowledge about radiolaria
has been a hinderance to micropaleontologists, who have developed
many of the current taxonomies based largely on skeletons from the
fossil record. Nonetheless, they have made remarkable strides in elucidating natural affinities based on phylogenetic models derived from
fossil evidence. These are based largely on spumellarian and nasselarian specimens, as the fragile phaeodarian skeletons are not well
preserved in the fossil record. As background for a more detailed taxonomic analysis, the fundamental distinguishing characteristics of the
three groups are presented.

Spumellaria (Spumellarida)

Spumellaria are considered to be among the most primitive radiolaria.
They possess a spherical to spheroidal central capsule perforated by
nearly isodiametric pores distributed uniformly over the surface of the
capsular wall (Fig. 1-6A). The fusules penetrating the pores are connected to a dense cytoplasmic layer immediately surrounding the central capsule. Ray-like axopodia and/or fine filopodia, continuous with
the fusules, emerge through the cytoplasmic layer surrounding the central capsule. The siliceous skeleton, when present, consists of a variety
of forms: (1) The simplest forms are individual spicules varying in
complexity from simple needles distributed throughout the extracapsulum to complex, symmetrically arranged, triradiate spines distributed
in the extracapsulum or clustered around the central capsule. (2) More
elaborate skeletons are composed of spheroidal to spherical shells that
are either solitary or multiple-concentric, enclosing the central capsule.
These shells bear perforations often with complex or ornate embellish-
ments. Sometimes the concentric shells are interconnected by radial

beams, or the shells may be spongiose or formed by a thin, solid wall. (3) Complex polyhedral skeletons resembling lattices or geodesic structures are reinforced in some groups by radial beams extending beyond the shell as long spicules, sometimes bearing teeth, fine spines, or other forms of ornamentation. Some examples of spumellarian skeletons are presented in Figs. 1-7 and 1-9A–D. In species with concentric shells, the innermost is known as a medullary shell (or microsphere or macrosphere in some systems) and the outermost shells are cortical shells. The range in shape and number of concentric shells may vary considerably. When there is extreme axial growth, the skeleton may be ellipsoidal, composed of a lattice structure or spongy network. Growth along two axes produces a flattened or discoidal shape usually with a spongy-textured skeleton bearing either simple or forked projections forming arms that are connected by an interbrachial spongy veil called a patagium (Fig. 1-9B). Other variations include forms with peculiar girdle structures of simple or triradiate form surrounding the shell (Fig. 1-9C). These may also have slits or large perforations termed gates between the girdles. Further variations include large, vaulted, dome-like structures affixed to the shell. In some species, there is a proliferation of chambers to form a planispiral (coil within a plane, Fig. 1-7L), heliocoidal (snail-like coil), or irregular aggregate shell without definite geometric order.

Among skeletonless Spumellaria, the species are identified by (1) the presence or absence of alveoli either within the intracapsulum or in the extracapsulum, (2) the size and shape of the central capsule, and (3) the color of pigment when present. The latter characteristic is of questionable value, as it may reflect the age or physiological condition of the specimen rather than a species difference. Hence considerable caution is advisable when using color as a taxonomic criterion.

Solitary and colonial species occur among the Spumellaria and exhibit some of the most complex development and largest sizes among the radiolaria. The large solitary species of *Thalassicolla* (Fig. 1-1A), *Thalassolampe*, *Physematium* (Fig. 1-2A), or some *Lampoxanthium* sp. (Fig. 1-2C) may reach diameters of 1 to several mm. Spherical colonies achieve diameters of several centimeters, whereas filiform species occur up to several meters in length and several millimeters in diameter. Algal symbionts are commonly observed in solitary and colonial Spumellaria. They are most frequently dinoflagellates or prasinophytes (Anderson, 1980). Little research, however, has been done on the diversity of algal symbionts, and further investigations may yield additional kinds.

Nassellaria (Nassellareda)

The Nassellaria possess an egg-shaped or prolate central capsule with a single pore field called a porochora (Haeckel, 1887) located at one pole of the central capsule. Cytoplasmic strands (fusules) pass through the pores in the pore plate and provide continuity between intracap-

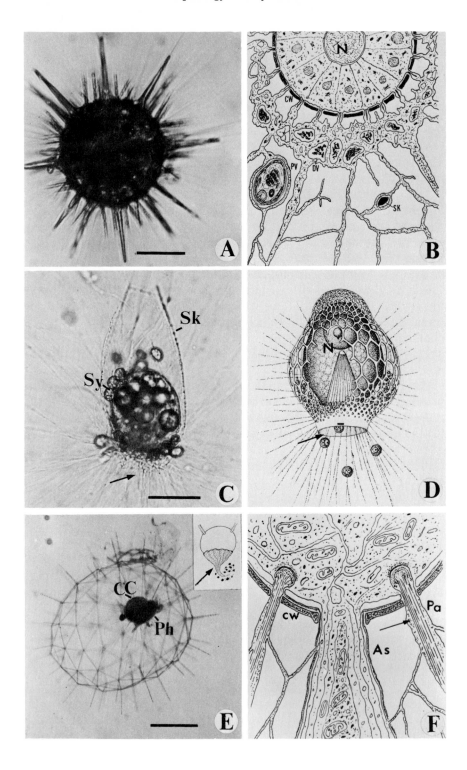

sulum and extracapsulum. The pole of the central capsule containing the pore plate is taken as the base, and the opposite pole is the apex. This corresponds also to the orientation of the central capsule within the more complex skeletons exhibiting a helmet-shaped structure (Figs. 1-6C, 1-8, and 1-9E–J), where the broad, basal opening is coincident with the pore field, and the closed, apical part of the skeleton covers the upper nonperforated portion of the central capsule. Axopodia and a reticulated system of cytoplasmic strands emerge from the basal opening (arrow, Fig. 1-6C, D) and radiate around the skeleton.

Three fundamental types of skeletons are found in the Nassellaria: (1) The simplest form is a tripod located near the base of the central capsule (Fig. 1-9E). It is composed of three divergent bar-like elements united at a common central point and oriented in such a way that one leg is toward the posterior direction and the other two are forward-right and -left anterolateral, labeled p, r, and l, respectively, in Fig. 1-9E. (2) A more complex form consists of a conical or helmet-shaped, complexly perforated shell (cephalis) sometimes surrounding the tripod or situated at an elevated position relative to the center of the tripod, but usually fixed at the common center of the tripod (Fig. 1-9F). (3) A sagittal ring that reinforces the latticed shell in the medial, sagittal plane (Fig. 1-9G).

Fig. 1-6. Comparative views of Spumellaria **(A and B)**, Nassellaria (**(C and D)**, and Phaeodaria **(E and F)**. Spumellaria **(A)** frequently exhibit a spherical architectural plan clearly exemplified by the central capsule **(B)** surrounded by a spherical capsular wall (CW) penetrated by cytoplasmic strands (fusules) producing the elaborate extracapsular network of rhizopodia and ray-like axopodia. Digestive vacuoles (DV), symbiotic algae enclosed within perialgal vacuoles (PV) and newly deposited skeletal structures (SK) contained within the secretory sheath (cytokalymma) occur in the extracapsulum. Scale = 50 μm. Nassellaria **(C)** possess a spheroidal central capsule enclosed, in some species, by a siliceous lattice shell (Sk) called a cephalis. Axopodia (arrow) emerge from the operculum in the base of the shell. Dinoflagellate symbionts (Sy) are often abundant. Scale = 50 μm. The central capsule **(D)** contains a cone-like array of microtubules, called a podoconus, that gives rise to the fusules penetrating the capsular wall. Phaeodaria **(E)** sometimes possess an elaborate geodesic-like lattice shell surrounding a central capsule (CC) suspended within a network of rhizopodia containing a dense granular mass known as a phaeodium (Ph). The central capsule (inset) has a major emergent strand of cytoplasm called an astropyle (arrow) and two or more finer strands parapylae sometimes situated opposite the astropyle. Scale = 0.5 mm. A diagramatic view **(F)** shows the emergent astropyle (As), two laterally adjacent parapylae (Pa) with rods of microtubules (arrow) within them. (B adapted from Anderson, 1983; D from Grell, 1973)

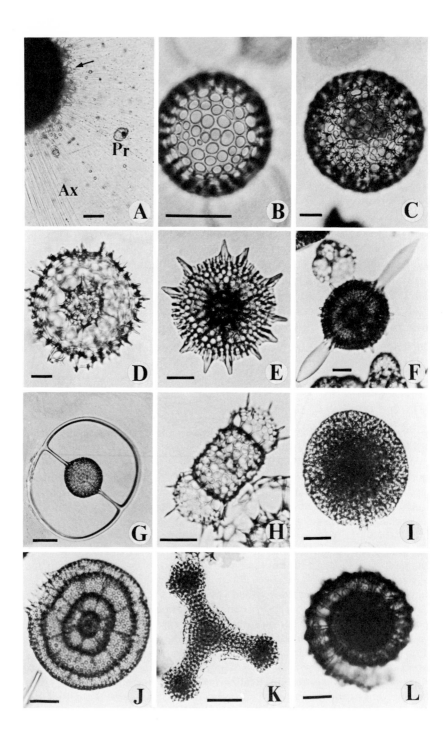

The three skeletal components are not consistently joined in a regular arrangement and may occur singly or in combinations thus increasing the complexity of the skeletal types. Seven categories of skeletons have been determined based on the composition and arrangement of the skeletal elements: (1) The skeleton consists of the tripod alone. (2) The basal tripod may be alone or with a vertical apical spine rising from its center commonly elaborated with an irregular framework arising from the bars of the tripod, but without trace of a cephalis or sagittal ring. (3) The skeleton is composed only of a latticed shell (cephalis) or a single chamber and lacks a basal tripod or sagittal ring. (4) A sagittal ring and basal tripod occur without a latticed cephalis. (5) The basal tripod is absent, and only a latticed cephalis and sagittal ring are present. (6) The skeleton is composed of a basal tripod and latticed cephalis which may possess an apical cupola or dome but lacks a sagittal ring. (7) The skeleton is composed of all three components—a basal tripod, a sagittal ring, and a latticed cephalis.

In some groups of Nassellaria, the latticed shell is further complicated by the presence of one or more transverse strictures that segment it into joints (e.g., Fig. 1-8I). Among the morphological features of taxonomic importance are the number of radial apophyses protruding from the shell and the architecture of the basal shell orifice, which may be open as in most species or closed off and fenestrated. In some species, the base of the shell may be completely closed by a spherical, fenestrated segment resembling a round-bottom flask (Figs. 1-8O and 1-9I).

The intracapsular cytoplasm is clearly dominated by a large conical-shaped structure called the podoconus (Haeckel, 1887) consisting of a cone of cytoplasmic tubules oriented with the apex toward the top of the central capsule and the base intersecting with the pores in the pore field (Fig. 1-6D). A large nucleus lies in close proximity to the apex of the podoconus. Numerous dinoflagellate symbionts (Figs. 1-2E and 1-6C) are attached to the rhizopodia and axopodia radiating from the orifice and/or pores in the skeleton. The skeletons of Nassellaria often

---

Fig. 1-7. Light micrographs of some representative Spumellaria. (A) Periphery of a living specimen showing axopodia (Ax), protozoan prey (Pr), and skeletal spines (arrow) projecting from the cytoplasm Scales = 200 μm. Cleaned skeletons (B–L): (B) Cenosphaera sp., (C) Spongoplegma antarcticum = Actinomma antarcticum, (D) Actinomma sp. exhibiting central medullary shell, (E) Heliodiscus asteriscus, (F) Stylatractus sp. = Xiphotractus sp., (G) Saturnalis circularis, (H) Ommatartus tetrathalamus, (I) Spongopyle osculosa, (K) Euchitonia sp., and (L) Lithelius nautiloides. Scales = 50 μm. (B and F), courtesy of David Lazarus; C–E and G–L courtesy of James Hays and Grace Irving, Columbia University)

exhibit remarkable complexity consisting of elaborately reinforced lattices exhibiting exquisite principles of architectural design on the microscopic scale (Figs. 1-8 and 1-9G–J).

Phaeodaria (Phaeodarea)

The Phaeodaria are in many respects distinctively different from the Spumellaria and the Nassellaria. The central capsule is a large oblate spheroid slightly depressed in the direction of the main axis. A major perforation occurs at one locus on the capsular wall and forms an inverted, cone-like opening—the astropyle. A limited number, usually two, additional pore fields called parapylae may occur adjacent to the astropyle or at variable locations in the capsular wall. The astropyle contains a small cylindrical central extrusion termed the proboscis and a conical collar known as the operculum shown in Figs. 1-6E, F.

A dense mass of dark-pigmented granular substance (phaeodium) is suspended in the cytoplasm usually in the oral region of the extracapsular cytoplasm. The phaeodium sometimes contains reddish, granular deposits dispersed throughout the dark-pigmented mass. The function of the phaeodium is unknown, although it has been described as metabolic wastes or sometimes hypothesized to possess light sensing properties or to contain zooxanthellae (algal symbionts). Some phaeodarian skeletons consist of an organic and siliceous matrix, and thus are much less resistant to diagenesis than polycystine skeletons. The skeletons of some phaeodaria are constructed of hollow tubes containing living cytoplasm, and many contain substantial amounts of organic material at the places where they are joined to one another. These joints are thus very fragile, and the skeletons frequently become disarticulated before they are completely fossilized. In some species, the skeletons form remarkably elegant geodesic-lattice structures (Figs. 1-6E and 1-10G–I) composed of bars, and may be reinforced at their nodes by rosette-like connectors. At these axes, radial spines variously ornamented with tubercles or minute knobs may protrude from the surface of the lattice. These structures are elegantly portrayed in some of the detailed drawings presented by Haeckel (1887) (Fig. 1-10). Other species possess shells of very fine reticulate construction resembling diatom frustules, whereas others exhibit a porcelaneous texture and are composed of a silica cement with numerous fine needles within the matrix. Some shells are bivalved (clam shaped), whereas others possess a distinctly pouch-shaped architecture with broad, oral apertures sometimes bearing teeth on the rim (Figs. 1-9K–M). Most of the species are solitary; however, occasional complex assemblies occur of several porcelaneous pouch-like shells joined together with an elaborate anastomosing set of appendages uniting the ensemble into a net-like framework. The pouch-like shells occur at the periphery of the framework with their mouths oriented toward the center of the array. Phaeodaria usually occur at great depths in the ocean; however, some species are also found in surface water.

Given the unique properties of the central capsule and the clearly amorphous quality of the skeleton composed of hollow tubes, it is very plausible to conclude that the Phaeodaria are distinctly different from the polycystine radiolaria, which possess much more solidly constructed skeletons and exhibit similarities in organization of the central capsule. Hence the revised classification of Levine *et al.* (1980) and the detailed taxonomic treatments of Riedel (1967a–c), Riedel and Sanfilippo (1977), and Petrushevskaya (1977) that recognize the polycystine radiolaria as separate taxa from the Phaeodaria are much more likely to represent a natural classificatory scheme than some of the older systems that placed them in taxa of equivalent level.

Historical Perspectives on Radiolarian Taxonomy

The taxonomic classification of radiolaria has not been comprehensively investigated in recent times, although several excellent treatises have appeared on specialized topics (e.g. Polycystine radiolaria: Riedel, 1967a–c; Petrushevskaya, 1971a; Foreman and Riedel, 1972; Goll, 1972a,b; Pessagno, 1977; Dumitrica, 1970; Riedel and Sanfilippo, 1977; Nigrini and Moore, 1979; and Collosphaerid radiolaria: Strelkov and Reshetnyak, 1971). A complete treatise on the systematics of the polycystine and phaeodarian radiolaria would be beyond the objectives and scope of this book; therefore, a survey of historical perspectives and current knowledge is presented as a conceptual framework for taxonomic references made in the remainder of the book.

The earliest substantial treatise on radiolaria was the comprehensive report presented by Haeckel (1887), based on samples obtained from the Challenger Expedition 1873–1876. Although his system of classification is not regarded as a natural one, based on clear phylogenetic lineages, it was one of the more thorough treatments to appear during the 19th century. It has remained until very recently the major source of information on radiolarian diversity and taxonomy. His system, though elegant in scope and profusion of species, is often difficult to apply in species identification. Some of these problems have been set forth in modern treatises (e.g., Hollande and Enjumet, 1960; Riedel, 1967a–c). Nonetheless, Haeckel's thorough analysis provided a conceptual framework that has influenced thought well into the 20th century. He treated "Radiolaria" as a class with four legions: (1) Spumellaria, (2) Acantharia, (3) Nassellaria, and (4) Phaeodaria. He defended these divisions as natural groups stating:

The four principal groups of Radiolaria to which we have given the name 'legions,' are natural units, since the most important peculiarities in the struc-

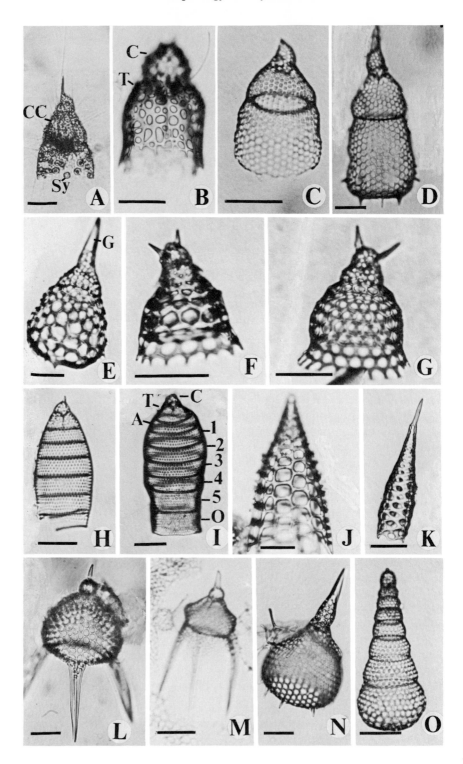

ture of the central capsule are quite constant within the limits of the same legion, and since all the forms in the same legion may be traced without violence to the same phylogenetic stem. (Haeckel, 1887, p. iii)

The Spumellaria and Acantharia were grouped in a subclass, "Porulosa," and the Nassellaria and Phaeodaria were placed in a subclass, "Osculosa," based on the distribution and size of the pores in the central capsular wall. Each legion was divided into two sublegions, the total subsuming 20 orders, 85 families, and thousands of species. Seventeen hundred species were described for the Spumellaria alone. With reference to species, Haeckel was clearly aware that categories, even at this level of analysis, are relative and not fixed by absolute criteria of categorization—decisions about the inclusion or exclusion of varying types are as much a matter of logical choice as finding true joints of Nature. Consequently, categories may be grouped together or split apart depending on the fineness of defining characteristics chosen by the taxonomist. Moreover, given the limited knowledge of radiolarian ontogeny, there is the clear possibility that developmental stages may be misidentified as separate species. Haeckel rationalizes his approach to species definition with the following logic.

According to the individual views of the systematist and the general survey which he has attained of the smaller and larger systematic groups, the conception of a species adopted in his practical work will be wider or narrower. In the present systematic arrangement, a medium extent has been adopted. (Haeckel, 1887, p. ciii)

Notwithstanding a medium extent of analysis, Haeckel identified no less than 3,508 new species and described an impressive array of 4,318 species in total. The authenticity of some of Haeckel's species has been challenged more than once on varying grounds, from sheer incredulity at the overwhelming number and bewildering array of geometric variations exhibiting seemingly every possible combination of geometric

---

Fig. 1-8. Light micrographs of some representative species of Nassellaria. (A) Living specimen (*Theoconus* sp.) with algal symbionts (Sy), central capsule (CC) enclosed within the shell, and peripheral halo of rhizopodia. Cleaned skeletons (B–O). Symbols: A, abdomen; C, cephalis; G, galea; O, operculum; T, thorax; segments, 1–5. (B) *Antarctissa* sp., (C) *Theoconus zancleus* = *Pterocorys zancleus*, (D) *Theocorythium dianae*, (E) *Androcyclas gamphonycha*, (F) *Cycladophora davisiana* = *Theocalyptra davisiana*, (G) *Theocalyptra bicornis*, (H) *Eucyrtidium acuminatum*, (I) *Eucyrtidium* sp., (J) *Peripyramis circumtexta*, (K) *Cornutella profunda*, (L) *Lychnocanium grande*, (M) *Pterocanium praetextum*, (N) *Anthocyrtidium ophirense*, and (O) *Cyrtopera languncula* (lost apical spine). Scale = 50 μm. (B, L, and M courtesy of David Lazarus; C–K, N, and O courtesy of James Hays and Grace Irving)

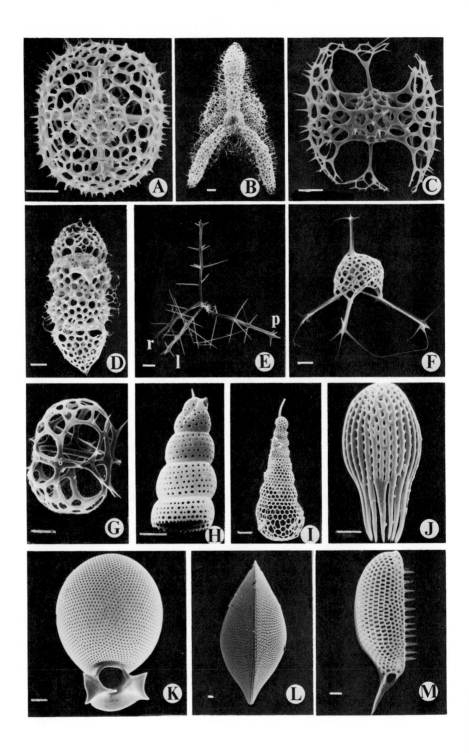

shapes to reservations about the pertinence of erecting species based on a single specimen, often a skeleton alone. Haeckel admits some reservation in his species designations based on single specimens, and candidly acknowledges that some forms may be only different stages of ontogenetic development. This is particularly suspect when species divisions are made on the number of concentric shells of similar morphology but varying only in diameter as occurs in the Spumellaria. Some Nassellaria, moreover, possess a tripod element at an early stage of growth and later it is supplemented with a complete lattice skeleton. Although Haeckel recognized these predicaments, he chose to be exhaustive in his categorization of skeletons rather than permit even a single potential species to pass unnoticed. The question of the natural affinities of Haeckel's species and indeed their accuracy remains to be answered. Much additional research on fossil assemblages coupled with more intensive biological investigations on radiolarian ontogeny will be required to elucidate these issues.

As an aid in comprehending the major systematic groups in Haeckel's treatment, a brief synopsis of families among the Spumellaria, Nassellaria, and Phaeodaria are presented and illustrated with modern micrographs and reproductions of his figures for some representative species.

Spumellaria

Table 1-1 presents the basic morphological features of the families in the Spumellaria. The major distinguishing feature in the two orders of the Spumellaria (Collodaria and Sphaerellaria) is skeletal morphology. Collodaria lack skeletons or possess only spicules, the Sphaerellaria possess complete shells composed of a lattice work or spongy texture. The families of the Collodaria are identified jointly by the construction of the skeleton and the solitary versus colonial form of the organisms. The Thalassicollida of the suborder Colloidea, for example, are solitary and skeletonless as exemplified by *Thalassicolla* sp. (Fig. 1-1A). The Collozoida are skeletonless, colonial forms as, for example, with *Col-*

Fig. 1-9. Scanning electron micrographs of some Spumellaria **(A–C)**, Nassellaria **(D–I)**, and Phaeodaria **(J–L)**. **(A)** *Stylosphaera?* sp., **(B)** *Euchitonia elegans*, **(C)** *Octopyle stenozona*, **(D)** *Didymocyrtis tetrathalamus tetrathalamus*, **(E)** *Tetraplecta pinigera* showing the three tripodal feet: anterolateral right (r) and left (l), and posterior (p), **(F)** *Cladoscenium ancoratum*, **(G)** *Lophospyris pentagona hyperborea*, **(H)** *Spirocyrtis subscalaris*, **(I)** *Cyrtopera languncula*, **(J)** *Carpocanistrum acutidentatum*, **(K)** *Protocystis murrayi*, **(L)** *Conchopsis compressa*, and **(M)** *Conchidium caudatum*. Scale = 25 μm. (Courtesy of Kozo Takahashi, 1981, 1983)

lozoum sp. (Figs. 1-5A, B). The Suborder Beloidea possessing spicules is represented by the solitary forms (Fig. 1-2C) in the family Thalassosphaerida and colonial forms in the family Sphaerozoida. An example of the colonial form is *S. punctatum* displayed as Figs. 1-5C, D.

Among the Sphaerellaria, the suborders are separated according to the shape of the shell ranging from the Sphaeroidea with one or more spherical shells to the Larcoidea with an ellipsoidal shell bearing three axes (triaxon–ellipsoid). The major five families are discriminated by the number of spines on the shell. In Haeckel's terminology, a spine is a major rod-like protrusion often tapered toward the tip, sometimes triangular in cross section or twisted in a spiral near the distal end. The Liosphaerida are spineless. The remaining families are characterized by increasing number of spines. Stylosphaerida (two radial spines), Staurosphaerida (four radial spines), Cubosphaerida (six radial spines), and Astrosphaerida with numerous spines (eight to many). The families are further subdivided according to the number of concentric shells present and the extent to which they are spongy in texture as displayed in Table 1-1. These five major families are further subdivided according to the six levels of concentric shell development thus yielding 30 different subfamilies.

Nassellaria

The Nassellaria encompasses two orders, Plectellaria without complete lattice shell and Cyrtellaria with complete shell (Table 1-2), and embraces 6 suborders based on the complexity of the skeletal elements. There are 26 families including Nassellida representing the skeletonless genera (*Cystidium* and *Nassella*) and 25 encompassing the skeleton-bearing genera. The descriptions of the families are summarized in Table 1-2.

The Nassellida are considered the most primitive Nassellaria as they possess no skeleton. The spheroidal to spherical central capsule exhibits a typical porochora (pore field) and internal cone-like podoconus. The capsule is surrounded by a jelly envelope penetrated by numerous pseudopodia emitted from the pores of the porochora. Two genera are distinguished by the presence or absence of alveoli in the extracapsulum. The simplest representative of this family is *Cystidium princeps*, a skeletonless species. Once more, caution is warranted in accepting this as a naturally occurring species. It may be an early developmental stage of a skeleton-bearing species that has not yet deposited a skeleton.

The two families of the suborder Plectoidea are distinguished according to the absence of a wicker work (Plagonida, Fig. 1-10D) or its presence (Plectanida, Fig. 1-10E) on the radial spines that support the central capsule. The radial spines are peculiarly united at a central point or at two poles of a common central rod. In Plagonida, the distal ends are free and never united as occurs in the Plectanida, which are

Table 1-1. Orders and families of the Spumellaria[a]

I. COLLODARIA: Skeleton wanting or quite imperfect, not latticed	
Skeleton entirely wanting	1. Colloidea
Solitary cells, living as isolated individuals (Fig. 1-1A)	A. Thalassicollida
Associated cells, living in colonies or coenobia (Fig. 1-5A)	B. Collozoida
Skeleton represented by numerous scattered spicules	2. Beloidea
Solitary cells, living as isolated individuals (Fig. 1-2C)	A. Thalassosphaerida
Associated cells, living in colonies or coenobia (Fig. 1-5C)	B. Sphaerozoida
II. SPHAERELLARIA: Skeleton a perfect shell of lattice work, or spongy and resembling wicker-work.[b]	
Lattice shell spherical or composed of concentric spheres (Figs. 1-6A and 1-7A–C)	3. Sphaeroidea
Shell lacking spines (Fig. 1-7B)	A. Liosphaerida
Spines opposite on one axis of the shell (Figs 1-7G and 1-9A)	B. Stylosphaerida
Four radial spines arranged perpendicular to each other	C. Staurosphaerida
Six radial spines arranged as Cartesian axes in space (Fig. 1-14C)	D. Cubosphaerida
Numerous spines: 8 to 20 or more in a radial array (Fig. 1-7D)	E. Astrosphaerida
Lattice shell ellipsoidal or prolonged in one axis	4. Prunoidea
A. MONOPRUNIDA: Shell without transverse stricture.	
Shell simple, latticed (not spongy), without enclosed internal shells	A. Ellipsida
Shell composed of two or more concentric latticed shells (not spongy) (Fig. 1-7F)	B. Druppulida
Shell partially or wholly composed of an irregular Spongy framework	C. Spongurida
B. DYOPRUNIDA. Shell bilocular, divided by an equatorial stricture into two communicating hemiellipsoidal shells	
Shell simple, without enclosed internal shells	A. Artiscida
Shell composed of two or more concentric shells	B. Cyphinida
C. POLYPRUNIDA. Shell multilocular, divided by three or more parallel transverse strictures into four or more serial camerae	
Shell with three parallel strictures and therefore four camerae	A. Panartida
Shell with five or more parallel strictures and therefore six or more camerae (Fig. 1-7H)	B. Zygartida
Shell a biconvex lens or a flat disk	5. Discoidea

Table 1-1. (continued)

III. PHACODISCARIA: Discoidea with external phacoid shell (or lenticular latticed cortical shell)	
Phacoid shell simple, without enclosed medullary shell	A. Cenodiscida
Phacoid shell with simple or double-enclosed medullary shell.	
Margin without chambered girdles (Fig. 1-7E)	B. Phacodiscida
Margin surrounded by chambered girdles	C. Coccodiscida
IV. CYCLODISCARIA: discoidea without external phacoid shell (no lenticular latticed cortical shell)	
Surface of the shell covered by convex or even porous sieve plates (not spongy).	
Concentric rings around the central chamber complete (without open spaces) (Fig. 1-7K)	D. Porodiscida
Concentric rings around the central chamber interrupted by three open spaces	E. Pylodiscida
Surface of the shell spongy, not covered by peculiar porous sieve plates	F. Spongodiscida
Shell a triaxon-ellipsoid, with three different axes	6. Larcoidea
Larcoidea with a regular or symmetrical shell, the growth of which is determined by the three dimensive axes	
Cortical shell completely latticed, without external gates (or interzonal fissures), without annular constrictions and domes	
Medullary shell absent or simple (spherical or lentelliptical)	A. Larcarida
Medullary shell trizonal or *Larnacilla*-shaped (composed of three dimensive girdles)	B. Larnacida
Cortical shall incompletely latticed, with two to four or more symmetrically disposed gates or fissures remaining between latticed dimensive girdles (Fig. 1-9C)	C. Pylonida
Cortical shell completely latticed, without external gates (or interzonal fissures), with two to four or more annular constrictions, which separate three to six or more dome-shaped protuberances	
Constrictions of the cortical shell in diagonal planes; domes in dimensive axes	D. Tholonida
Constrictions of the cortical shell in dimensive places; domes in diagonal axes	E. Zonarida

Table 1-1. (continued)

Larcoidea with a symmetrical or irregular shell, either with spiral growth or with quite irregular growth		
Cortical shell with spiral growth		
Spiral cortical shell bilateral (with plane spiral) (Fig. 1-7L)	F.	Lithelida
Spiral cortical shell asymmetrical (with ascending spiral)	G.	Streblonida
Cortical shell with quite irregular growth		
Cortical shell simple with one single chamber	H.	Phorticida
Cortical shell composed of a number of heaped up or aggregated chambers	I.	Soreumida

[a] After Haeckel (1887). Family names are designated by capital letters.
[b] Shell-bearing colonial radiolaria are included here in the family Collosphaerida.

further characterized by an elaborate wicker work formed by the branching of the radial spines. Four families in the Suborder Stephoidea are characterized by the presence of skeletal rings, varying from a simple vertical sagittal ring only (Stephanida, Fig. 1-10A) to a complex arrangement of crossed rings forming in the simplest set a "signet ring configuration" with one ring vertical to the plane of the second ring (Semantida), or a set of two crossed vertical rings intersecting each other's planes. There may also be a third basal ring lying below the crossed rings (Coronida, Fig. 1-10B), or in the most complex case a pair of horizontal rings separated by one ring, or a pair of crossed vertical rings (Tympanida, Fig. 1-10C). Among the Stephoidea, the skeleton may be embellished with numerous irregular spines. In some species, the rings may not be complete or they may be embellished with a loose irregular lattice or framework as observed in some Tympanida or Coronida.

The four families of the Suborder Spyroidea all possess a complete lattice shell known generally in the Nassellaria as a "cephalis" resembling a helmet. But in the Spyroidea the cephalis is constricted by a vertical ring producing a bilocular shell. In the majority of Spyroidea, the three essential elements of the nassellarian skeleton can be detected; namely, (1) the vertical sagittal ring of the Stephoidea, (2) a basal tripod as observed in the Plectoidea, and (3) the latticed cephalis of the Cyrtellaria. The vertical ring, however, is so incorporated within the structure of the cephalis that it appears as an external, longitudinal more or less distinct constriction, separating the lateral halves of the inflated bilocular cephalis (e.g., Fig. 1-9G). The three bars of the tripod emerge from the basal plate of the cephalis as three divergent feet. The sagittal

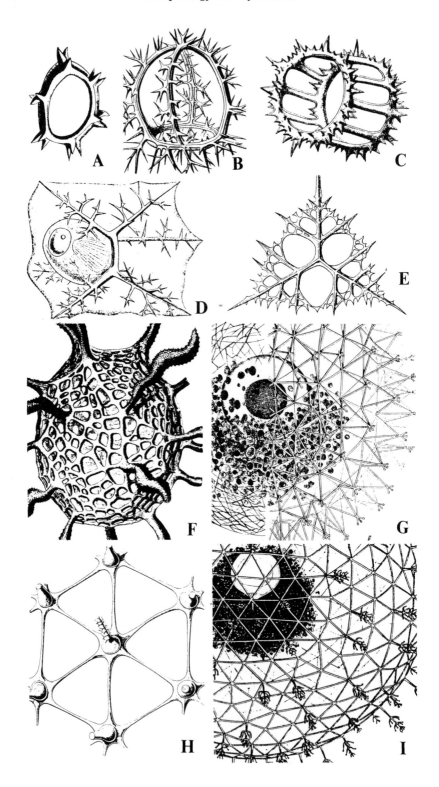

ring, although constantly present, may be obscured by its incorporation with the cephalis.

The bilocular construction of the latticed cephalis with its sagittal constriction is characteristic of all Spyroidea and distinguishes them from the Botryodea with a multilobate cephalis, and the Cyrtoidea with a simple, entire cephalis. In the Spyroidea, moreover, the patterns of fenestration on the several faces of the bilocular cephalis are usually quite variable differing in the size, number, and/or distribution of pores. The four families of Spyroidea are distinguished by the combination of two attributes, the presence or absence of a second shell joint on the base of the cephalis (e.g., Fig. 1-8B) and the presence or absence of a galea (apical dome or cupola atop the cephalis, e.g., Fig. 1-8E). Many subtle variations of these features occur among the species represented in the four families thus making precise taxonomic identifications difficult. Other subtle features such as the number, disposition, and arrangement of the tripodal feet are important in distinguishing subfamilies of the Zygospyrida according to Haeckel.

There are three families included in the Suborder Botryodea characterized by a complete lattice shell possessing a lobate and multilocular cephalis, with three to five or more separated lobes, and two to three or more constrictions. The construction and position of the lobes as well as the organization of the constrictions on the shell surface are quite variable giving rise to complex aggregates of lobose segments in the cephalis (e.g., Fig. 1-8I). The three families are distinguished according to the number of principal segments of the cephalis. The family Cannobotryida is characterized by a monothalamus shell consisting of only a lobate cephalis. The Lithobotryida possess a dithalamous shell composed of a lobate cephalis supplemented by a second bell-like segment (simple thorax) attached at its base. The Pylobotryida possess a trithalamous shell, composed of a lobate cephalis, a thorax, and a third segment called an abdomen (e.g., Fig. 1-8C). The abdomen may be open

---

Fig. 1-10. Some representative illustrations from *The Challenger Expedition Report* (Haeckel, 1887) showing Nassellaria **(A–E)** and Phaeodaria **(F–I)** are represented for historical perspective. **(A)** *Zygocircus triquetrus* (Stephanida) with only a single ring skeleton. **(B)** *Coronidium acacia* (Coronida), **(C)** *Eutympanium militare* (Tympanida), **(D)** *Plagonium sphaerozoum* (Plagonida), **(E)** *Triplecta triactis* (Plectanida). **(F)** *Orosphaera serpentina* (Orosphaerida), **(G)** *Sagmarium spongodictyum* (left half) and *Sagoscena castra* (right half) (Sagosphaerida). **(H)** Triangular mesh in skeleton of *Aulosphaera sceptrophora*. **(I)** Segment of *Aulosphaera dendrophora* (Aulosphaerida). The Orosphaerida have been assigned to the Spumellaria in modern treatises (e.g., Campbell, 1954; Riedel and Sanfilippo, 1977).

with a basal mouth or closed by a perforated wall yielding a porous flask-like structure. The radial apophyses (bar-like arms) found in the majority of Botryodea seem to correspond in position on the cephalis and in their spatial relationships to those typically observed in other Nassellaria; namely, three descending basal feet arising from the base of the cephalis and an ascending vertical horn arising from the apex of the cephalis.

The sixth and last surborder, Cyrtoidea, is characterized by a complete lattice shell, exhibiting a simple or reduced cephalis, which is nonlobate and without sagittal constriction. The shell, however, may be transversely constricted to form two or more joints, the number being an important discriminating feature for the families in this suborder. There are 12 families distinguished by the conjunctive characteristics of the number of joints in the shell (row headings in Table 1-2) and the number of apophyses (projections) arising from the shell (column headings in Table 1-2).

There are three major family groups based on the number of apophyses. The Pilocyrtida with three radial apophyses (these comprise the majority of species in the Cyrtoidea), the Astrocyrtida with four to nine or more apophyses, and the Corocyrtida with no radial apophyses. There are four categories of shell configuration; namely, monothalamic (one segment), dithalamic (two segments), trithalamic (three segments), or polythalamic (more than three segments). These four categories taken in conjunction with the three major family groups generate the 12 families cited in the matrix of Table 1-2. Haeckel further subdivided each family into two subfamilies based on whether the mouth of the shell was open, or closed by a lattice wall. The radial apophyses are undoubtedly modifications of the tripodal elements present in most Nassellaria.

The size of the thorax or second shell joint in Cyrtoidea is generally in inverse proportion to that of the cephalis (first shell segment). The more the cephalis is reduced, the more the thorax is developed. Its form is variable, usually three-sided pyramidal in the species bearing three apophyses, polyhedral in the multiradiate species, and conical or cylindrical in species lacking apophyses. The abdomen, or third shell joint of the Cyrtoidea when present may be a simple, large chamber as in the Tricyrtida, but may form an annulated body composed of a variable number of successive joints in the Stichocyrtida. The lattice structure of the shell in the Cyrtoidea exhibits remarkable variety and was employed by Haeckel to discriminate species. The cephalis has usually very small and simple pores, whereas the lattice work of the thorax is ornamented with radial structures. The abdomen usually exhibits numerous, regular pores. The numerous joints in the annulated abdomen of the Stichocyrtida commonly exhibit little variety (e.g., Fig. 1-8I).

Table 1-2. Orders and families of the Nassellaria[a]

I. PLECTELLARIA: Nassellaria without complete lattice shell	
No skeleton (Fig. 1-2E)	1. Nassoidea
Body composed only of central capsule and extracapsulum,	A. Nassellida
Skeleton a basal tripod without ring	2. Plectoidea
Skeleton composed of radial spines united in a common center and supporting the central capsule but lacking a wicker-work, (Figs. 1-9E and 1-10D)	A. Plagonida
Skeleton wattled with wicker-work composed of the united radial spines enclosing the central capsule, (Fig. 1-10E)	B. Plectanida
Skeleton with a sagittal ring	3. Stephoidea
Skeleton composed of a single vertical sagittal ring only, (Fig. 1-10A)	A. Stephanida
Skeleton composed of two crossed rings (vertical and basal)	B. Semantida
Skeleton composed of two crossed meridional rings (sometimes also a basal ring) (Fig. 1-10B)	C. Coronida
Skeleton composed of two parallel horizontal rings (Fig. 1-10C)	D. Tympanida
II. CYRTELLARIA: Nassellaria with a complete lattice shell	
Cephalis bilocular, with a sagittal constriction (Fig. 1-9G)	4. Spyroidea
Cephalis without thorax joint or galea (no apical cupola)	A. Zygospyrida
Cephalis without thorax joint but with a galea or cupola	B. Tholospyrida
Cephalis with thorax but without a galea or cupola	C. Phormospyrida
Cephalis with a thorax, galea, or cupola	D. Androspyrida
Cephalis multilocular, with two or more constrictions and lobes	5. Botryodea
Shell monothalamous, one chamber (Fig. 1-6C and D)	A. Cannobotryida
Shell dithalamous, cephalis and thorax (Fig. 1-8B)	B. Lithobotryida
Shell trithalamous, cephalis, thorax, abdomen, (Fig. 1-8C)	C. Pylobotryida
Cephalis simple, without constriction and lobes	6. Cyrtoidea

Twelve families characterized by the number of radial apophyses and the number of segments in the shell

Shell type	Three apophyses	Four apophyses	No apophyses
Monothalamic	A. Tripocalpida	E. Phaenocalpida	I. Cyrtocalpida
Dithalamic	B. Tripocyrtida	F. Anthocyrtida	J. Sethocyrtida
Trithalamic	C. Podocyrtida	G. Phormocyrtida	K. Theocyrtida
Polythalamic	D. Podocampida	H. Phormocampida	L. Lithocampida

[a] Based on Haeckel (1887). Family names are designated by capital letters.

Phaeodaria

There are four orders assigned to the Phaeodaria based on shell morphology (Table 1-3). They subsume 15 families largely distinguished
by the variations in fine lattice work of the shell, its surface texture,
physical composition, and the kinds of surface ornaments. The diversity of surface ornaments has been elegantly described by Haeckel:

> "The hollow or solid spines, which arise from the shell of the Phaeodaria,
> exhibit an extraordinary variety and elegance in the production of different
> branches, bristles, hairs, secondary spine, and thorns, hooks, anchor-threads,
> pencils, spathillae. . . . They are organs partly for protection, partly for retention
> of food. They are much more interesting than in other Radiolaria. (Haeckel,
> 1887, p. 1,541).

The first order, Phaeocystina, includes species lacking a skeleton, or
composed of numerous single, scattered pieces. The central capsule
occurs in the center of the gelatinous envelope. The three families
encompassed here are (1) Phaeodinida without a skeleton, (2) Cannor-
raphida with skeleton composed of numerous scattered pieces, which
are either hollow tangential spicula or cup-shaped dishes, or fenestrated
rings scattered loosely in the extracapsulum; and (3) Aulacanthida characterized by skeletons composed of numerous hollow radial tubes, the
proximal ends in contact with the central capsule, and the distal ends
projecting outward, penetrating the gelatinous envelope surrounding
the central capsule. In all but one genus of the Aulacanthida, the surface
of the extracapsular gelatinous envelope is covered by an arachnoidal
veil or mantle, composed of thousands of very fine, hollow, tangential
needles. The radial tubes are always cylindrical, usually straight but
sometimes slightly curved and occasionally tapered toward both ends,
sometimes spindle shaped. The siliceous wall of the cylindrial tubules
is usually very thin, fragile, and perfectly structureless; however, in a
few species, it may consist of thick walls composed of concentric cylindrical layers.

The second order is the Phaeosphaeria, consisting of species with
skeletons composed of a single or a double usually spherical lattice
shell. There are four families: (1) Orosphaerida, (2) Sagosphaerida, (3)
Aulosphaerida, and (4) Cannosphaerida. The Orosphaerida and Sagosphaerida (Fig. 1-6E) form a set based on shell morphology. They
possess shells composed of a simple, nonarticulated lattice framework
(lacking joints where the skeletal rods are united at nodes in the framework). The nodes, moreover, are *without astral septa*, star-shaped complex plates, that reinforce the union of the skeletal elements at the
nodes. The Orosphaerida (e.g., Fig. 1-10F) possess a robust and coarse
skeleton with irregular polygonal meshes. The bars are very thick and
partly hollow, whereas the Sagosphaerida exhibit a very delicate and
fragile skeleton, with subregular triangular meshes formed by thin, filiform, solid rods (Fig. 1-10G).

Table 1-3. Orders and families of the Phaeodaria[a]

I. PHAEOCYSTINA: Skeleton absent or incomplete, composed of numerous single scattered pieces	
Skeleton completely absent	A. Phaeodinida
Skeleton composed of numerous scattered pieces, not of radial tubes	B. Cannorrhaphida
Skeleton composed of numerous hollow radial tubes, the proximal ends of which are in contact with the central capsule	C. Aulacanthida
II. PHAEOSPHAERIA: Skeleton a simple or double lattice shell, not bivalved, without a peculiar shell mouth	
Shell composed of a simple, nonarticulated lattice plate, without astral septa in the nodal points	
Network very robust and coarse, with irregular polygonal meshes; bars very thick, partly hollow (Fig. 1-10F)	D. Orosphaerida
Network very delicate and fragile, with subregular, triangular meshes; rods very thick filiform, always solid (Fig. 1-6E)	E. Sagosphaerida
Shell composed of numerous hollow, tangential cylindrical tubes, which are separated by astral septa in the nodal points	
Shell articulated, with astral septa, without a simple central shell (Fig. 1-10I)	F. Aulosphaerida
Shell double, composed of two concentric shells; the outer articulated, the inner simple	G. Cannosphaerida
III. PHAEOGROMIA: Skeleton a simple lattice shell, not bivalved, constantly provided with a peculiar large shell mouth placed on the oral pole of the main axis	
Structure of the shell not porcellanous (without needles imbedded in a punctulate cement substance)	
Structure of the shell diatomaceous, with very delicate and regular hexagonal pores (Fig. 1-9K)	H. Challengerida
Structure of the shell alveolar, with hollow alveoles between a double plate	I. Medusettida
Structure of the shell of simple lattice work, neither diatomaceous nor alveolar	J. Castanellida
Structure of the shell porcellanous, with peculiar fine needles imbedded in a punctulate cement substance (a circle of pores around the base of each radial tube)	
Surface of the shell panelled or dimpled; peristome flat	K. Circoporida
Surface of the shell smooth, even; peristome prominent	L. Tuscarorida

Table 1-3. (continued)

IV. PHAEOCONCHIA: Skeleton a bivalved lattice shell, composed of a dorsal and a ventral valve which are completely separated		
The two valves of the bivalved shell thick and firm, regularly latticed (Fig. 1-9L)	M.	Concharida
The two valves of the bivalved shell very thin and fragile		
Galea without rhinocanna or nasal tube, without frenula	N.	Coelodendrida
Galea with a rhinocanna or nasal tube, both connected by an odd or paired frenulum	O.	Coelographida

[a] After Haeckel (1887). Family names are designated by capital letters.

The Orosphaerida represent some of the largest radiolaria and possess robust skeletons. Some species have radial spines protruding from the shell, whereas others are spineless.

The second set of two families in the order Phaeosphaeria encompasses the Aulosphaerida and Cannosphaerida. They are characterized by skeletons with numerous hollow, tangential, cylindrical tubes, which are *separated by astral septa* in the nodal points. The Aulosphaerida possess a single shell (Figs. 1-10H, I); however, the Cannosphaerida have a double shell composed of two concentric shells, the outer articulated and the inner simple. The shell in the majority of Aulosphaerida is a regular sphere or an endospherical polyhedron. The double-shelled skeleton of the Cannosphaerida consists of a spherical or ovate inner shell, solid or latticed with a peculiar mouth. It is connected by radial beams to the spherical or polyhedral outer shell bearing articulated rods and the astral plates. From the latter arise hollow radial tubes projecting outward from the surface of the lattice shell.

The third order among the Phaeodaria is the Phaeogromia possessing simple, latticed skeletons, not bivalved, and provided with a large, peculiar shell mouth. The families are divided into two groups, one group possessing a nonporcellanous shell the other, comprised of those species whose shells are porcellanous, formed of fine needles embedded in a punctulate cement substance. The first group subsumes three families: (1) Challengerida (Fig. 1-9K) with shells of diatomaceous structure exhibiting delicate and regular hexagonal pores; (2) Medusettida possessing alveolated shells of simple, ovate, hemispherical, or cap-shaped form, and hollow articulate feet surrounding the widely open mouth; the capsule is excentric, placed in the aboral half of the shell cavity; and (3) Castanellida exhibiting simple lattice work shells with circular or roundish pores—the mouth of the shell is large, usually circular, and armed with teeth. The central capsule is excentric, placed in the aboral half of the shell cavity.

The Medusettida are among the most remarkable of this order. According to Haeckel's observations, the alveolate shell exhibits unique properties.

Its reticulate appearance seems to indicate at first sight the usual lattice-shell pierced by numerous very small pores. But as soon as we make the shell dry, air always enters into its thin walls, and each apparent pore is found to be a small alveole or a separate compartment, which contains a small bubble of air. The thin wall of the shell is therefore double, composed of two parallel, very thin lamellae of silica, which are little distant from one another, and are connected by a network of small rods or septa. (Haeckel, 1887, p. 1,664).

The outer surface of the shell is usually smooth, sometimes papillate, studded with small, conical, radial spines or infrequently ornamented by prolonged cylindrical spines or tubes scattered over the surface of the shell. Some of the cap-shaped shells, moreover, look remarkably like small, delicate medusae (jelly fish), which further contributes to their unique architectural diversity and no doubt is the source of the family name—Medusettida.

The second group in the order Phaeogromia possessed porcellanous shells and encompasses two families: (1) the Circoporida with a panelled or dimpled shell surface (spherical or polyhedral), and (2) the Tuscarorida possessing smooth-surfaced shells (ovate or subspherical) supplied in some species with long curved spines bearing bristles. Both the Circoporida and Tuscarorida have excentric central capsules, placed in the aboral half of the shell cavity. The Circoporida possess a simple mouth in the shell and exhibit constant character denoted by their name, i.e., circles of pores surrounding the base of the radial spines. Although occasional pores of this kind are also observed in the Tuscarorida, the shells of the latter family are ovate and with prolonged main axis whereas the Circoporida have spherical or polyhedral shells.

The fourth and final order of the Phaeodaria is the Phaeoconchia with three families. The skeleton in this order is a bivalved, lattice shell resembling a tiny clam shell (Fig. 1-9L) with the central capsule enclosed between the two valves. The Family Concharida possesses a thick and firm shell, regularly latticed, without a galea or cupola on their apex. The valves are free, not connected to one another. The remaining two families, Coelodendrida and Coelographida, encompass species with very thin and fragile, scarcely latticed shells. Each valve with a conical cupola or a helmet-shaped galea on its sagittal pole or apex, and with hollow tubes attached to the shell. The Coelodendrida lack the peculiar "nasal tube" or rhinocanna connecting the base of the cupola to the valve as occurs in the Coelographida. Moreover, three or more hollow radial tubes arise from each valve of the Coelodendrida.

In both families, the capsule is so enclosed between the two inner valves, that its three pore openings (Astropyle and two parapylae) lie in the frontal fissue between the valves. Haeckel reports that the members of the Family Coelographida exhibit

The highest degree of morphological development, not only in this group, but among all Radiolaria. They attain also the greatest size of all members of the class, since the diameter of their body is sometimes more than 20 mm., and in a few species even more than 30 mm. The complexity of their structure attains at the same time such a high degree, that they may be regarded as the most complicated, and (in a morphological sense) as the most highly developed of all protozoa. (Haeckel, 1887, p. 1,739)

The hollow branched tubes ornamenting the surface are complex. Some appear dichotomously branched from the base, not verticillate, and are called "brushes," whereas others called "styles" are verticillate, much longer than brushes and bear cruciate (cross-shaped) or alternately cruciate pairs of branches. The distribution and number of styles are a significant attribute in defining genera among the Coelographida.

Although much modern research on systematics of radiolaria is clearly needed to clarify some of Haeckel's observations on skeletal morphology and to more rigorously examine natural variability within populations of living radiolaria to establish boundaries for species, his work remains one of the most comprehensive treatments of radiolarian taxonomy. Moreover, Haeckel's classification is the starting point for some modern revisions of radiolarian taxonomy, and this provides a useful framework for a critical understanding of the recent work. As these modern works are set forth, some of the limitations as well as the strengths of Haeckel's system will become apparent.

Modern Classificatory Systems

Among the modern revisions of Haeckel's classificatory scheme, Campbell's system (1954) is perhaps one of the more comprehensive treatises. It, however, has not been widely accepted; owing in part to the artificial quality of the categories, and certain incongruencies with modern taxonomic definitions of some of the species (Strelkov and Reshetnyak, 1971). His excellent illustrations, however, are a useful supplement to modern treatises. More recent taxonomic schemes give promise of developing a natural and coherent basis for radiolarian systematics.

Petrushevskaya (1977) recommended a classificatory scheme for the radiolaria and their close relatives, based on the phylogenetic assumption that axopod-bearing protista arose from benthic ancestors through modification of the rhizopodia to yield increasingly stiffened pseudopodia culminating in the axopodia of modern forms. She hypothesized polyphyletic lineages arising from primitive bottom-dwelling Actinopoda. The Phaeodaria, Acantharia, and Sticholonchea are each represented by separate phyletic lines. The polycystine radiolaria are represented by a fourth branching line with the primitive Albaillellaria arising first from the main trunk followed later by the Collodaria, Sphaerellaria, and finally the Nassellaria. The Albaillellaria are characterized by a very marked bilateral symmetry of frame and lattice shell. In some cases, the shell is a spiral cone supported on an internal A- or H-shaped frame. These radiolaria are present in fossil deposits of the Carboniferous (e.g., Holdsworth, 1969) and were described in the genus *Albaillella* by Deflandre (1952). Based on the foregoing phylogenetic assumptions, Petrushevskaya proposes the following taxonomic scheme for the major groups of Actinopoda:

Superclass:	Heliozoa
	Radiolaria
Class:	Acantharia
	Euradiolaria
Subclass:	Phaeodaria
	Polycystina
Superorder:	Nassellaria
	Sphaerellaria
	Collodaria
	Albaillellaria

In this scheme, the category of Radiolaria is maintained but is further divided into the Acantharia and the "true radiolaria" placed in the Subclass: Euradiolaria. Some of Petrushevskaya's emendations to taxonomy (1975a,b, 1977) have been incorporated in modern classificatory systems described hereafter.

In a nicely illustrated treatment of extant radiolaria, Nigrini and Moore (1979) have catalogued major quaternary radiolaria using a classificatory scheme based primarily on Riedel (1967a–c) with some emendations by Goll (1968), Foreman (1973), and Petrushevskaya (1975a). Their descriptions are concise, but thorough, and include comments on recent distributions of the species, in addition to critical remarks on the taxonomic categories with reference to observations of other researchers. This is a particularly useful reference for anyone working with extant radiolaria and recent fossil species. Their illustrations are largely light micrographs of cleaned skeletons supplemented with hand-drawn figures. Consequently, their treatise is conveniently applied in laboratories with only light optics. Although, scanning and transmission electron microscopic examination of skeletons and sections through the living substance can help to elucidate fine points of similarities, it is often a limitation for practical applications where large numbers of specimens need to be sorted on a daily routine basis or under fieldwork conditions.

Recent progress in elucidating the systematics of polycystine fossil and modern radiolaria has been made by Riedel and co-workers (e.g.,

Riedel, 1967a,b, 1971c; Riedel and Sanfilippo, 1977). Their research is based on a rationale of constructing a natural classificatory system using the most thorough phylogenetic evidence available based on fossil skeletal evidence. Riedel (1971c) summarizes the threefold sources of information that inform their scholarly endeavor: (1) "Detailed investigations of intraspecific variation are revealing that the type species of some genus-group taxa are conspecific or very closely related. As a result, some generic and subgeneric names can be placed in synonymy and the taxonomic system simplified to that extent." (2) "Evolutionary lineages are being worked out on the basis of evidence from the fossil record, to reveal phylogenetic relationships. This results in a very significant and useful taxonomy." (3) "The entire polycystine radiolarian literature, and all available living and fossil assemblages, can be reviewed, and an attempt made to recast completely the broad lines of the classification. . . ."

As a consequence of this approach, many of Haeckel's classical definitions of taxa are no longer acceptable, and new family and genus definitions have emerged. The species of many Haeckelian genera have been redistributed among other genera in the new system, and the generic names of necessity followed the type species in this redistribution. The classification is only partially complete as new evidence is continually being gathered to permit expansion of the taxa based on sound empirical evidence. Consequently some genera, particularly among the Nassellaria, are not at present assigned to families and only the generic definitions are cited within these suborders. A summary of the classification by Riedel and Sanfilippo is presented and where appropriate compared to Haeckel's system and the system of Levine *et al.* (1980).

The Radiolaria (Subclass) are placed in the Class Actinopoda (Riedel, 1967c). This is in contrast to the general classificatory scheme of Levine *et al.* (1980), which no longer includes Radiolaria as a natural group. The Polycystina (Superorder) subsuming the Orders of Spumellaria and Nassellaria stand in correspondence to the Class Polycystinea of Levine *et al.* (1980) subsuming the Orders of Spumellarida and Nasselarida. The Suborders of Sphaerocollina and Sphaerellarina representing colonial and solitary spumellarian radiolaria in the classificatory scheme of Levine and co-authors are represented by family designations in the classification of Riedel and Sanfilippo (1977). Only the basic structure of the family and/or genus level classification is presented and, therefore, detailed references to species descriptions must be obtained from the original source which is nicely illustrated. This classification is potentially useful to biologists, as it permits incorporation of living species within this largely fossil-based system. An outline of the system of Riedel and Sanfilippo with figure citations for representative species is presented.

Class: Actinopoda
Subclass: Radiolaria Müller, 1858
Superorder: Polycystina Ehrenberg 1838, emend. Riedel 1967b

Radiolaria with skeleton of opaline silica without admixed organic compounds. The bars forming the skeleton are generally solid, rather than hollow.

Order: Spumellaria Ehrenberg, 1875

Skeleton generally in the form of a sphere, or derived from a sphere (e.g., ellipsoidal, discoidal, lenticular, spiral). Many species have several concentric shells, the robust outer one being termed "cortical" and the usually much smaller, more delicate inner one(s), termed "medullary."

Family: Orosphaeridae Haeckel, 1887 (e.g., Fig. 1-10F)

Large robust skeletons consisting of a lattice shell, or two approximately concentric shells, and radial spines generally circular in cross section. In some forms, a spicule of usually six to eight rays is present at or near the shell apex. Shell in some members subspherical, but actually bipolar (and allopolar) in all well-known forms. The family ranges from Eocene to Recent, having its greatest diversity in Oligocene and Early Miocene assemblages.

Family: Collosphaeridae Müller, 1858 (Fig. 1-15)

Colonial forms, the spherical or subspherical shells of which, though dissociated in sediments, are usually recognizable by having thin shell walls with relatively wide flat areas between pores, commonly conical or tubular protuberances, spines (if present) circular in section, and never possessing medullary shells. The family is known to range from the *Calocycletta virginis* Zone (Early Miocene) to the present, and may extend back as far as the Late Eocene.

Family: Actinommidae Haeckel, 1862; *sensu* Riedel, 1967b (e.g., Fig. 1-7D)

Solitary spumellarians with shells spherical or ellipsoidal (or modifications of those shapes), usually without internal spicule, and generally much smaller than orosphaerids. This polyphyletic family ranges through the entire Cainozoic and includes a great number of species, the relationships of most of which have not yet been worked out.

Subfamily: Saturnalinae Deflandre, 1953 (e.g., Fig. 1-7G)

Spherical shell with two opposite spines joined by a ring.

This subfamily had its principal development in the Mesozoic, but is represented throughout the Cainozoic by one or a few species. Forms with spines on the periphery of the ring are very rare in Palaeocene (and perhaps Eocene) assemblages, but are more common in Mesozoic faunas.

Subfamily: Artiscinae Haeckel, 1881; *sensu* Riedel, 1967b

Cortical shell ellipsoidal, usually equatorially constricted and enclosing a medullary shell; opposite poles of the shell generally bear spongy columns and/or single or multiple latticed caps.

Genus: *Cannartus* Haeckel, 1881; *sensu* Riedel, 1971a

Ellipsoidal cortical shell, enclosing a double medullary shell and commonly with an equatorial constriction, and with two opposite spongy columns in the main axis. No polar caps between cortical shell and spongy columns.

Genus: *Ommatartus* Haeckel, 1881, *sensu* Riedel, 1971a (Fig. 1-7H)

Equatorially constricted cortical shell, enclosing a double medullary shell, and bearing polar caps which in some species are surmounted by spongy columns.

In the course of the evolution of this lineage, the polar caps increased in size and the spongy columns decreased.

Family: Phacodiscidae Haeckel, 1881

A lenticular cortical shell, not surrounded by spongy or chambered structure, encloses a much smaller, single, or multiple medullary shell. The margin (and less commonly the surfaces) of the cortical shell may bear radial spines. This family apparently ranges through the entire Cainozoic. The present generic taxonomy is not satisfactory, and the two genera treated below include only a small proportion of Cainozoic phacodiscids.

Genus: *Heliostylus* Haeckel, 1881; *sensu* Sanfilippo and Riedel, 1973

Lenticular cortical shell joined to the medullary shell by two opposite bars in the equatorial plane (often continued as strong marginal spines) and by shorter bars in the central area. Phyletic relationships of the species have not been investigated, and the genus ranges from Late Palaeocene or earlier to the *Dorcadospyris ateuchus* Zone (Late Oligocene).

Genus: *Periphaena* Ehrenberg, 1873; *sensu* Sanfilippo and Riedel, 1973

Disc circular, triangular, or elliptical in outline, with smooth surface and numerous small pores. Margins with spines, or a poreless girdle, or both. Without two opposite bars as *Heliostylus*. This genus may be polyphyletic, and its total range is not yet known; however, some of the included species are useful stratigraphically.

Family: Coccodiscidae Haeckel, 1862

Discoidal forms consisting of a lenticular cortical shell enclosing a small single or double medullary shell, and surrounded by an equatorial zone of spongy or concentrically chambered structure. As it is defined below, the genus *Lithocyclia* includes the great majority of forms assignable to this family. This is probably not a satisfactory long-term solution for the taxonomy of the family at the generic level, but phyletic.

Genus: *Lithocyclia* Ehrenberg, 1847a; *sensu* Riedel and Sanfilippo, 1970.

Coccodiscids with the spongy or chambered equatorial girdle either entire or reduced to a number of separate arms (commonly three or four). Radial spines are present in some species. Members of this genus may be coarsely classified into the *Lithocyclia ocellus* group (*sensu* Riedel and Sanfilippo, 1970), the *L. aristotelis* group (*sensu* Riedel and Sanfilippo, 1970), *L. angusta* (Riedel, 1959b), and *L. crux* Moore (1971).

Family: Porodiscidae Haeckel, 1881; *sensu* Kozlova in Petrushevskaya and Kozlova, 1972

Discoidal forms comprising a small central chamber surrounded by concentric, interrupted or spiral latticed bands. In some, the disc is small, and the major part of the shell is represented by two to four arms in the equatorial plane. The family ranges through the entire Cainozoic, and includes a large number of species. The generic taxonomy is presently undergoing a profound change.

Family: Spongodiscidae Haeckel, 1862; *sensu* Kozlova in Petrushevskaya and Kozlova, 1972

Similar in general form to Porodiscidae, but the skeletal structure is predominantly irregularly spongy, concentric chambers (when present) being confined to the central region. As is the case with the Porodiscidae, this family ranges through the Cainozoic and is currently undergoing profound revision.

Family: Pyloniidae Haeckel, 1887 (e.g., Fig. 1-9C)

Shell consisting of an ellipsoidal central chamber of complicated construction surrounded by a series of successively larger elliptical latticed girdles in three mutually perpendicular planes, the major diameter of each girdle being the minor diameter of the next larger one. This family ranges from the Eocene to the present, but is well developed and common only in Miocene and later assemblages.

Family: Tholoniidae Haeckel, 1887

Shell completely latticed, without larger openings, and with two or more annular constrictions or furrows separating dome-shaped protuberances. Representatives of this family are rare, and range from the *Cannartus petterssoni* Zone (Middle Miocene) to the present.

Family: Litheliidae Haeckel, 1862 (e.g., Fig. 1-7L)

Shell usually ellipsoidal to spherical in outer form, and of spiral structure (at least internally). This probably polyphyletic family ranges throughout the Cainozoic.

Order: Nassellaria Ehrenberg, 1875

Skeleton heteropolar and generally bilaterally symmetrical, commonly in the form of a spicule, with rays arising from a median bar, a ring (often D shaped), or a cap-like construction often of several segments uniserially arranged.

Suborder: Spyrida Ehrenberg, 1847b; emend. Petrushevskaya, 1971a
  (e.g., Fig. 1-9G)

Skeleton generally possessing a complete sagittal ring, and commonly also latticed lateral chambers forming a bilobed cephalis. Families within this suborder, which ranges from Late Palaeocene or earlier to the present, are not yet satisfactorily defined.

Genus: *Dorcadospyris* Haeckel, 1881

Shell thickwalled, small, bearing two (or in some species more) larger cylindrical feet and usually a strong apical horn.

Genus: *Gorgospyris* Haeckel, 1881

Cephalis lacking horns, and thorax generally latticed, usually terminating in numerous lamellar feet.

Genus: *Liriospyris* Haeckel, 1881; *sensu* Goll, 1968

Species of this genus exhibit a variety of morphologies that cannot be summarized briefly.

Genus: *Psychospyris* Riedel and Sanfilippo, 1971

Large, discoidal forms with a degenerated cephalic wall and a peripheral spongy zone.

Genus: *Rhabdolithis* Ehrenberg, 1847b

Skeleton consisting mainly of a rod, circular in section, commonly thickened at one end and the other end in one species attached to the median bar. More delicate lateral branches commonly arise from the main rod. The genus ranges from the Late Palaeocene or earlier to some undetermined level probably in the Late Eocene.

Suborder: Cyrtida Haeckel, 1862; emend. Petrushevskaya, 1971. (Fig. 1-8)

Skeleton generally conical or cap shaped, consisting of segments uniserially and usually rectilinearly arranged.

Family: Plagoniidae Haeckel, 1881; emend. Riedel, 1967b. (Figs. 1-2G and 1-8G)

Cephalis consisting entirely of the fundamental spicule, or having a latticed skeleton including a large cephalis within which this spicule is well developed. This large, probably polyphyletic, family ranges through the entire Cainozoic.

Family: Theoperidae Haeckel, 1881; emend. Riedel, 1967b (Fig. 1-8H, I, K, M)

Cephalis relatively small, approximately spherical, often poreless or sparsely perforate. The internal spicule, homologous with that of the plagoniids, is reduced to a less conspicuous structural element than in that group. This diverse family ranges through the entire Cainozoic, and includes a considerable number of species of which the stratigraphic ranges and phyletic relationships are understood.

Genus: *Artophormis* Haeckel, 1881; *sensu* Riedel and Sanfilippo, 1970

A partly artificial grouping of three stratigraphically useful forms in which the abdomen bears numerous short teeth, or a postabdominal segment tending to be more coarsely latticed proximally than distally.

Genus: *Bekoma* Riedel and Sanfilippo, 1971

Cephalis bears apical and vertical horns, and is not separated from the upper part of the campanulate thorax by an external stricture. Three feet usually subcylindrical, with branches in the distal portion which may join to form a ring.

Genus: *Buryella* Foreman, 1973

Four or more segments, without appendages, not flared terminally, and with pores aligned transversely on at least one segment.

Genus: *Calocyclas* Ehrenberg, 1847b

Shell with strong apical horn, subspherical cephalis, hemispherical or inflated thorax, and numerous lamellar feet sometimes joined proximally to form an abdomen. The genus ranges from the *Buryella clinata* Zone to the *Thyrsocyrtis bromia* Zone (late Early Eocene to late Late Eocene).

Genus: *Cyrtocapsella* Haeckel, 1887; emend. Sanfilippo and Riedel, 1970

Shell generally drop shaped or pyriform, consisting of three or four segments of which the terminal one has a very constricted aperture. Some specimens have an additional, usually more delicate, generally closed conical segment below the constricted one.

Genus: *Eusyringium* Haeckel, 1881

Cephalis subspherical, with a robust cylindroconical horn. Thorax pyriform, with a constricted aperture, which may be prolonged as a subcylindrical tube.

Genus: *Lamptonium* Haeckel, 1887; *sensu* Riedel and Sanfilippo, 1970

Shell generally robust, pyriform with inflated thorax, and usually a strong apical horn. Three spines extend laterally from the thorax in one species, and some have three indistinct feet terminally, or the thoracic aperture closed by a lattice plate, or a short, narrow, subcylindrical abdomen.

Genus: *Lithochytris* Ehrenberg, 1847a

Three-segmented, closed distally, tending toward tetrahedral in general form. Short, robust apical horn and three robust feet terminally.

Genus: *Lithopera* Ehrenberg, 1847a

Terminally closed, two- or three-segmented forms in which the thorax forms the major part of the shell.

Genus: *Lychnocanoma* Haeckel, 1887; *sensu* Foreman, 1973

An apparently polyphyletic genus, ranging through the entire Cainozoic and comprising two-segmented forms with two or three feet and without corresponding ribs in the thorax. In some, abdominal lattice (usually delicate) joins the feet.

Genus: *Orbula* Foreman, 1973

Shell of two segments with rather large cephalis, and thorax constricted distally, with ribs terminating in short, subterminal feet. Only

three species have been described in this genus, which is restricted
to Late Palaeocene (and perhaps older) assemblages.

Genus: *Pterocanium* Ehrenberg, 1847a (Fig. 1-8M)

Cephalis bearing an apical horn; thorax including three ribs continued as well-developed, usually proximally latticed feet. Most species have delicate abdominal lattice work connecting the proximal
parts of the three feet. The genus is extant, and ranges back at least
as far as the Early Oligocene (*Theocyrtis tuberosa* Zone).

Genus: *Stichocorys* Haeckel, 1881

Many-segmented forms in which the first three or four segments
constitute a conical upper portion of the shell and the narrower subsequent segments constitute a cylindrical lower portion. The genus
ranges from the *Calocycletta virginis* Zone to the *Spongaster pentas*
Zone (Early Miocene to Pliocene).

Genus: *Theocorys* Haeckel, 1881

Three-segmented forms with an apical horn, distal margin undif-
ferentiated or with a simple peristome, and pores irregularly or quincuncially arranged. A polyphyletic genus including many species;
the evolutionary relationships of most of which are not understood.

Genus: *Theocotyle* Riedel and Sanfilippo, 1970

Three-segmented forms with an apical horn, pores quincuncially
arranged, and a distinctly differentiated peristome bearing short, broad
feet in early representatives. The genus ranges from the *Bekoma
bidartensis* Zone (early Early Eocene) to the lower part of the *Thyrsocyrtis bromia* Zone (Late Eocene).

Genus: *Thyrsocyrtis* Ehrenberg, 1847b; *sensu* Riedel and Sanfilippo,
1970

Phyletically related forms with three segments, robust apical horn,
pores quincuncially or irregularly arranged, and usually three feet
arising from a differentiated peristome—though in the terminal mem-
ber of the principal lineage there is no differentiated peristome and
the feet number four or more. The genus ranges from Late Palaeocene
into the *Thyrsocyrtis bromia* Zone (Late Eocene).

Family: Carpocaniidae Haeckel, 1881; emend. Riedel, 1967b

Shell consisting usually of cephalis and thorax, only a few species
having considerable abdominal structure. Cephalis small, not sharply
distinguished in contour from thorax, and tending to be reduced to
a few bars within top of thorax. The genus extends through most of
the Cainozoic, and is extant. Its evolutionary origin is not yet clear,
but the Eocene species may have originated from a form close to
Cryptoprora ornata (Ehrenberg, 1873).

Genus: *Carpocanistrum* Haeckel, 1887; *sensu* Riedel and Sanfilippo,
1971 (Fig. 1-9J)

Cephalis merged with ovate thorax, pores often longitudinally
aligned, and somewhat constricted peristome often bearing numerous

teeth. This genus is probably polyphyletic, and ranges from approximately the *Thyrsocyrtis bromia* Zone (Late Eocene) to Recent.

Genus: *Carpocanopsis* Riedel and Sanfilippo, 1971

Shell robust, with abdomen separated from thorax by a well-developed internal lumbar stricture. The phylogenetic relationships of the species included here have not been determined, and the group may be unnatural.

Family: Pterocorythidae Haeckel, 1881; emend. Riedel, 1967b (Figs. 1-8N and 4-7)

In the cephalis, two lateral lobes are separated from the main cephalic lobe by two obliquely downwardly directed lateral furrows originating from the apical bar. This family is still extant, and its earliest known representatives are forms evidently closely related to *Podocyrtis papalis*, in the earliest Eocene or latest Palaeocene.

Genus: *Anthocyrtidium* Haeckel, 1881; *sensu* Petrushevskaya *in* Petrushevskaya and Kozlova (1972) (Fig. 1-8N)

Delicate, two-segmented forms with subcylindrical cephalis bearing a bladed apical horn. A distinct peristome is usually present, and commonly also terminal and/or subterminal teeth. The origin of the genus is obscure, and it ranges from the *Lychnocanoma elongata* Zone (Late Oligocene or Early Miocene) or earlier, to the present.

Genus: *Calocycletta* Haeckel, 1887; *sensu* Moore, 1972

Three-segmented shell with robust apical horn, inflated hemispherical thorax, subcylindrical or distally tapering abdomen, and commonly lamellar or triangular feet.

Genus: *Lamprocyclas* Haeckel, 1881; *sensu* Petrushevskaya and Kozlova, 1972

Three-segmented forms with subcylindrical cephalis usually bearing a stout apical horn. Thorax campanulate, and abdomen commonly longer and wider than thorax, in many species with a peristome and with terminal and subterminal teeth. The origin of this genus has not been determined. It ranges from the *Theocyrtis tuberosa* Zone (Early Oligocene), or earlier, to the present.

Genus: *Podocyrtis* Ehrenberg, 1847a; *sensu* Riedel and Sanfilippo, 1970 (e.g. Fig. 4-7)

Most members of the genus have a robust apical horn and three broad lamellar feet, but the other skeletal features undergo marked evolutionary change—in the subgenus *Podocyrtis* the abdomen becomes broader and the pores lose their pronounced longitudinal alignment; in *Lampterium*, the thorax decreases in size and the abdomen becomes subcylindrical with few large pores, and in both subgenera, the feet are ultimately lost.

Genus: *Theocyrtis* Haeckel, 1887

Three-segmented, generally campanulate forms, in which the termination of the shell is ragged (without differentiated peristome).

Family: Amphipyndacidae Riedel, 1967a

Many-segmented forms in which the cephalis, generally with a poreless wall, is divided into two chambers by a transverse internal ledge.

Genus: *Amphiternis* Foreman, 1973

Amphipyndacids in which the collar structures extend downward into the thoracic cavity.

Family: Artostrobiidae Riedel, 1967a, *sensu* Foreman, 1973 (e.g. Fig. 1-9H)

Shell with six collar pores, a well developed vertical tube, no lateral or terminal appendages, the last segment not flared, and the pores of at least one segment arranged in transverse rows.

Family: Cannobotrythidae Haeckel, 1881; *sensu* Riedel, 1967b

Skeleton consisting of cephalis, thorax, and occasionally an abdomen, the cephalis consisting of several lobes of complex structure. The family evidently arose in the Late Cretaceous or early Paleogene, but the origins of most of the genera remain obscure.

Genus: *Acrobotrys* Haeckel, 1881; *sensu* Riedel and Sanfilippo, 1971

Two or more prominent tubes arise from the cephalis, their origins being above the level of the collar pores.

Genus: *Botryocella* Haeckel, 1881; *sensu* Petrushevskaya in Petrushevskaya and Kozlova, 1972

On one side of the eucephalic lobe (the darkly outlined spherical chamber) is a large antecephalic lobe, and on the other side, a postcephalic lobe. Tubes, if present, are at approximately the level of the collar pores.

Genus: *Botryocyrtis* Ehrenberg, 1860; *sensu* Riedel and Sanfilippo, 1971

Cephalis consisting of three or four obvious lobes not markedly differing in size, often covered by spongy meshwork.

Genus: *Centrobotrys* Petrushevskaya, 1965

The eucephalic lobe is surrounded by a large chamber not subdivided into ante- and postcephalic parts.

No revisions of the systematics of Phaeodaria are included in the scheme of Riedel and Sanfilippo as very few fossil remains are to be found. Additional sources of information on radiolarian systematics from a micropaleontological perspective can be found in Campbell (1954), Riedel (1971a,b), Petrushevskaya (1971a,b), Pessagno (1977), Kling (1978), Goll and Merinfeld (1979), and Nigrini and Moore (1979). Although the fossil remains of many radiolaria are rich in information about skeletal morphology, they are limited by the lack of organic remains, thus making it difficult to elucidate the relationship between the living matter and the skeleton. Recent cytological and fine structure research has contributed to our knowledge of the cytoplasmic correlates of skeletal morphology, thus clarifying the relationships of some taxa especially among extant species.

Cytological and Fine Structure Contributions to Radiolarian Systematics

Light Microscopic Studies of Solitary Radiolaria

In an elegant and comprehensive light microscopic study of collodarian and sphaerellarian radiolaria, Hollande and Enjumet nee M. Cachon (1960) examined the cytological details of 65 sphaerellarian species, 7 collodarian species, and presented critical comparative observations with 4 nassellarian species. They attended particularly to the organization of the cytoplasm within the central capsule and its relationship to the extracapsular ensemble of pseudopodia and axopodia. Among the significant topics studied were (1) the form and development of the nucleus (including detailed observations of chromosomal organization) during vegetative and reproductive growth, (2) the composition and pattern of distribution of reserve bodies and vacuoles within the intracapsulum, (3) the physical characteristics and form of the central capsular membrane (wall) including its porosity and surface distribution of fusules, and (4) the origin of the axopodia within the intracapsulum, their arrangement in relation to the nucleus, and their pattern of distribution in relation to the central cell body. They also carefully examined the relationship of the skeleton with the soft body parts of the radiolarian and determined the changes in spatial relationship between the skeletal elements and the vital cytoplasmic components (e.g., nucleus and capsular wall) as the radiolaria mature. This information was synthesized to elucidate the phylogeny and taxonomy of the species they examined. They worked with living specimens collected largely from the Bay of Algiers or at Ville Franche-sur-Mer and therefore their observations, though limited to the more abundant species, are particularly of interest to biologists who require systematic information about extant species. Their work, moreover, is also valuable to micropaleontologists, who need information about the soft parts of the living radiolaria as an aid in interpreting fossil evidence about phylogeny and taxonomy.

Their research in modern times is complementary to the seminal studies of Hertwig (1879), who first applied stain technology to thin sections of radiolaria as a means of studying their microanatomy. In many respects, Hollande and Enjumet have continued in the excellence of his tradition and clarified some of the enigmatic points raised by his early observations.

A synopsis of their major findings is presented here, particularly those results that are pertinent to radiolarian systematics. One of their most significant findings was the observation that the Sphaerellarians can be divided into three major groups, based on the origin of the axopodia in the intracapsulum, their spatial arrangement and location relative to the nucleus, and their cohesion in some species to form a flexible, whisker-like cytoplasmic process called an axoflagellum. The axoflagellum, when present, appears as a slender filament protruding radially from the central capsule and sometimes extending beyond the perimeter of the halo of individual axopodia.

According to the distributional pattern of the axopodia and the location and form of their intracapsular organizing center, the axoplast, Hollande and Enjumet identified the following taxa: (1) the anaxoplastidies lacking a well-defined axoplast—the radial axopodia arise in the central capsule cytoplasm from isolated points, sometimes near the papillae of the nuclear membrane; (2) the periaxoplastidies possessing a distinct axoplast occurring outside the nucleus either in a juxtanuclear position (Fig. 1-11B) or in a concavity formed by the slight invagination of the nuclear envelope (Figs. 1-11C, D), or variously embedded within the fold-like center of a capiform nucleus (Fig. 1-11A); and (3) the

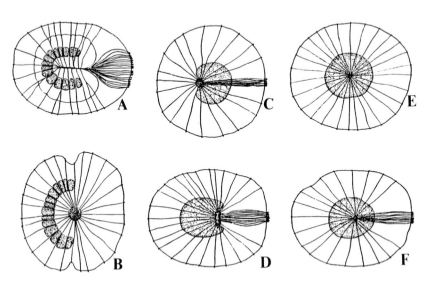

Fig. 1-11. Diagrammatic interpretation of the axoplast and associated microtubular rods radiating out through the cytoplasm of the periaxoplastidies (A–D) with the axoplast in a juxtanuclear position, and the centroaxoplastidies (E and F) with the axoplast enclosed within the nuclear membranous envelope (Hollande and Enjumet, 1960)

centroaxoplastidies, exhibiting a centronuclear axoplast (Figs. 1-11E, F). In the latter group, the axoplast is surrounded by the nucleus. The axopodia and, when it is present, the axoflagellum radiate outward through fine membrane-lined passages in the nucleus.

A detailed account of their many findings is not possible here, and the reader is encouraged to examine their monograph for specific information. A review of pertinent cytological findings, however, is presented as background for the outline of their classification scheme to be presented hereafter.

The endoplasm of the Sphaerellaria is relatively homogeneous and compact in the young individuals, but with a tendency to form radial lobes in the adult stages. Among the peri- and centroaxoplastidies, contrary to that observed in the anaxoplastidies, the endoplasm, whether compact or organized into lobes, is very uniform morphologically and physiologically. The cytoplasm is frequently rich with clear, spherical vacuoles arranged in radial arrays with the axopodial strands passing in between them. The lipid droplets, which are always present, and also some of the aforementioned clear vacuoles may contribute to the buoyancy of the organisms. The lipid droplets are not necessarily organized in a very precise pattern; however, among certain species they occupy a fairly predictable pattern. For example, in *Rhizoplegma* they may be found at the apex of each lobe of the central capsule. In other species (e.g., Rhizosphaera), much of the lipid is found intracapsullarly; however, some globules that are more strongly osmiophilic may occupy a cortical position. Differences in staining properties with osmic acid suggest a variability in chemical composition among the lipid inclusions. Albuminoid spherules, containing internal concretions, were observed solely among some centroaxoplastidies. These are delicately colored yellow in the living specimen; however, in optical sections, they appear colorless and possess a small eosinophilic granule or a crystal. Their chemical composition is not known. Among other inclusions noted in the centro- and periaxoplastidies are the very characteristic eosinophilic, cubical inclusions. These are either uniformly distributed as in *Hexacontium* or more generally grouped around the nucleus as with *Spongosphaera* or *Plegmosphaera*.

In the Anaxoplastidies, the endoplasm of the young stages appears much like that of young centroaxplastidies or Periaxoplastidies. It is much easier to observe the form of the endoplasm in adult stages. This gives rise to a concept of a morphological and physiological duality of the endoplasm. There is invariably one general plasmatic mass, very vacuolated, filling the region, homogeneous, finely granular, disposed in radial columns that terminate at the very large fusules of the capsule. The vacuolated plasm is homologous to the endoplasm of centroaxoplastidies (e.g., *Rhizosphaera*): clear vacuoles, albuminoid granules with internal concretions, and lipid globules are all intermixed. Within these inclusions, it appears the vegetative functions of anabolism and catabolism occur.

Based on the microanatomy of the cytoplasm and the structure and disposition of the siliceous shells and/or spicules of the various species, a classification of some sphaeroidian radiolaria was proposed by Hollande and Enjumet (1960). In general, this approach combining cytological and skeletal attributes has led to some reorganization of the classical categories constructed by Haeckel who largely attended to the geometry of the skeleton. Hollande and Enjumet place more emphasis on cytological structures as a major classificatory parameter and perceive skeletal details as a less significant, if not often confusing, basis for assigning species to higher order taxa. For example, they point out that the spongiose versus latticed character of the skeleton is not a reliable attribute to employ in erecting major taxonomic groups. The cytological data, based on nuclear structure and organization of the axoplast, yield more natural categories that often contain species with a mixture of shell types—some are spongiose, some latticed, and others possess both kinds. Moreover, they caution against placing too great of an emphasis on the number of concentric shells or particularly on counting spicules, as these attributes are ontogenetically variable and may change during maturation of the organism. Hence one must be clearly conscious of the stage of maturation in assessing species based on the number of concentric cortical shells. Their system is redacted here in a form that permits the cytological and skeletal distinguishing characteristics to be presented together. Only the major taxonomic features of each family are cited in the following summary, and descriptions of illustrative genera are presented following the exposition of the family characteristics.

For each family, a description of the skeleton is presented first, followed where appropriate by particular comments about the cytoplasm as summarized from the narrative in Hollande and Enjumet's monograph. A word about semantics may be helpful. In translating the French text, I have interpreted the term "Tissu spongieux" as spongiose matrix or material, since it is a nonliving skeletal matter rather than a living tissue, although it is often deposited on the skeleton as a siliceous layer not unlike a spongy tissue. The major group headings (e.g., centroaxoplastidies) have been retained in the French form out of deference to their origin; although it is possible to render them in an anglicized form such as centroaxoplastidiata. In the systematics, the term spicule pertains to a long rod-like projection of variable thickness and design, whereas a spine is more often a finer projection and shorter than spicules. Short, sharp "teeth" sometimes occur upon spicules. Triangular spicules refers to a spicule with a triangular cross section usually bearing a distinct ridge or keel along each of the three edges. A lattice plate ("plaque grillagee") usually refers to a thin, limiting surface layer deposited on the underlying skeletal elements. It is often a veneer-like layer. A *hexagonal frame* ("cadre hexagonal") usually means a network of hexagonal ridges deposited on the lattice, often adorning the bars delimiting the pores. A *verticil* is a shaft bearing lateral fine protuberances usually in the form of a rosette as with a verticillate influorescence in botany. Hollande and Enjumet (1960, pp. 56–57), moreover, use their own terminology for shell morphology. Of the innermost shells (called medullary shells by others), those that exceed, or not a diameter of 50 μm are termed correspondingly either a *macrosphere* or a *microsphere*. Those shells enclosing a macrosphere or a microsphere are termed *cortical shells*.

I. Periaxoplastidies.   The skeleton consists of either a solitary cortical shell or one cortical shell surrounding a microsphere of very fragile and primitive construction consisting in some cases of a simple assembly of siliceous fibrils; even reduced in some species to one or two juxta- or intranuclear fibrils. The solitary cortical shell bears neither spines nor spicules. It is generally irregular in shape composed of slender siliceous fibrils and limited by an irregular mesh often without definite geometric pattern. When it envelopes a microsphere, the cortical shell is spinose or supplied with spicules and possesses circular or polygonal pores in its mesh. The microsphere or its rudimentary fibrils give rise to divergent siliceous rods that attach to the cortical shell. With respect to the living part, the periaxoplastidies form a very homogeneous group. They possess an enormous axoplast which, depending on the individual case, either is independent of the nucleus or is juxtanuclear. The nucleus is sometimes hemispherical or capiform partially enclosing the axoplast. Numerous fine canals penetrate the nucleoplasm and contain the axopodia radiating from the axoplast. One axoflagellum or none is observed according to the genus. The central capsule is spherical or ovoid.

    A. **Family:** Cenosphaeridae

      Skeletal microsphere absent: peripheral shell without spicules or spines, composed of a simple, or double, tenuous network, peripheral mesh polygonal or irregular. Central capsule: spherical or ovoid, containing in the cortical region spherules of carotene-rich lipids or large refractile protein bodies. The capiform nucleus is not interposed between the axoplast and the axoflagellate cytoplasm (Fig. 1-11D)— the axoplast is drawn out into rods or in the form of a biconcave lens. Axoflagellar apparatus, well developed.

    B. **Family:** Stigmosphaeridae

      Medullary shell not present; however, in its place there are thin trabeculae occurring at the center of the capsule and providing an anchorage, in a diverse range of modalities, for the siliceous fibrils that emanate from it and produce the cortical shell. With respect to

the latter, some of the nodal points on the surface of the shell, where
the fibrils are united, produce one thin spine or spicule. The disposition of the fibrils is a distinguishing characteristic for genera. In
the endoplasm, the capiform nucleus is interposed between the axoplast and the axoflagellate cytoplasm (Fig. 1-11C).

C. **Family:** Heliasteridae

Microsphere clearly identifiable, intracapsular, and composed of a
loose network of delicate siliceous trabeculae, out of which arise a
certain number of fibrils affixed to the cortical shell which is provided
with spicules. It is a very general rule that four or five of the fibrils
converge at a common point in the cytoplasm near the periphery of
the axoplast where the microsphere originates. This important characteristic has not been observed anywhere else outside of the periaxoplastidies.

D. **Family:** Excentroconchidae

Single genus—*Excentroconcha* (Mast, 1910)

Skeleton peculiarly formed of two nonconcentric shells both ex-
tracapular; the smallest, adjacent to the capsule is cubical (15 µm);
the larger, 240 µm diameter, formed by ramification of the branching spicules arising from the smaller inner shell. Central capsule
originates as a subspherical form, but becomes lobate in the adult,
axoplast voluminous, spherical, centrally located, and slightly cra-
dled in the hollow of the broadly open capiform nucleus (Fig. 1-
11B), but separate from it. Axoflagella lacking.

II. Centroaxoplastidies. This group possesses very polymorphic skeletons, in some cases reduced to a mere spongiose web, very generally
composed of one or two microspheres and one or more cortical shells.
Microspheres are either lattices, or spongiose or spumose networks with
diameters of less than 30 µm. They are located intranuclear in the adult
stage. The cortical shells are likewise variable in number and organization sometimes spongiose, sometimes latticed, or even a combination
of lattice and spongiose shells. The spicules vary in number, always
arising from the central shell and originating at distinct nodal points
where the lattice elements converge. Axoplast surrounded by the nucleus with radially arranged axopodia traversing the nucleus and emanating from the endoplasm through the fusules within the capsular
wall. There are three families characterized by the organization of their
shells.

A. **Family:** Spongosphaeridae

Cortical shell spongiose (matrix spongiose or shell spongiose) microspheres always latticed. Cytological features permit identification
of three subfamilies, the first two, Spongosphaerinae and Spongodryminae, form a natural group possessing a lobed nucleus, with
irregular border, and internally the chromatin is generally arranged
in radial arrays. Often during maturation, the nucleus encloses the

two microspheres. The axoplast with few exceptions is spherical and gives rise to numerous, unbranched axopodia. Lipids are more or less abundant. Axoflagella present or absent, depending on the case. The Spongosphaerinae possess one to two latticed microspheres and radial spicules. The Spongodryminae possess one very small, latticed microsphere enclosed in a spongiose matrix independent of that of the cortex.

The third subfamily, Plegmosphaerinae, is characterized by a nucleus with a regular (smooth) border, containing very chromophilic chromosomes frequently with heterochromatin segments. The nucleus very generally is encompassed by numerous ergastoplasmic lamellae; axoplast poorly developed, cruciform giving rise to axopodia which ramify as with certain periaxoplastidies. Axoflagellate cytoplasm absent. Lipids are very abundant.

Distinguishing skeletal features are a single latticed microsphere, or *without microsphere*, and lacking radial spines which sets it apart from the Spongosphaerinae.

B. **Family:** Rhizosphaeridae

Microsphere spongiose or more exactly spumose, cortical shells seldom spongiose, but more generally latticed, and surrounded by an external, highly developed spongiose layer in mature specimens. Initially the alveoles of the spongiose matrix are bulky but the thick barrier that is produced ultimately is formed by the fusion of the branches, which thereby establishes the secondary partitions. During the course of growth, the other alveoles form on the outer side of the spumose shell. More peripherally, the radial spicules grow between them and become united with the barrier at their base, thus contributing to the enhancement of the microspheric skeleton. The microsphere encloses a very small cavity, but possesses a thickened wall made of a spongiose matrix, thus it may achieve diameters at times of 100 μm. Cortical shells latticed.

C. **Family:** Thecosphaeridae

Cortical and microspheres latticed spheres—one or two microspheres. The microsphere rarely exceeds a diameter of 30 μm. Further details are supplied in generic and species descriptions.

III. Anaxoplastidies.   The innermost shell is always extranuclear with hexagonal mesh, and often attains a diameter equal to 80 μm or greater (macrosphere). More rarely, the shell is small (microsphere), but in that case it is cubic, central, and extranuclear. The intracapsulum lacks an axoplast as implied by the name. Two families are recognized as very distinct, based on skeletal morphology, and among other cytological features, the shape of the nucleus.

A. **Family:** Macrosphaeridae

Lacking a microsphere, innermost shell always of a diameter greater than 50 μm, rigid, generally delicate, possessing a regular hexagonal mesh or more rarely thick-walled with circular pores; giving rise on the internal side, with few exceptions, to fine siliceous rods, variable in number, free of spicules, and each terminating separately in the endoplasm (intracapsulum). This shell bears the spines and spicules when present. Cytological features are very consistent within the group: nucleus nearly spherical during vegetative stages of growth, enclosed by a very thick membranous layer. Ectoplasmic pigment granules either uniformly distributed on the capsular wall or between its lobes or grouped at the two poles of the lobes.

B. **Family:** Centrocubidae

One, clearly cubical, microsphere, always intracapsular but situated external to the nucleus; the latter, surrounded by a thin papillate membrane compressed laterally by the microsphere and may have a large number of lobes in the adult form. At the edges of the microsphere, which may not be perfectly cubical, arise the robust, triangular spines, which by their branching give rise to an external spongiose shell. According to the species, this shell is simple or otherwise doubled by a very delicate, arachnoidal spongiose matrix. The primary spicules may give rise to the second order spicules. The latter may arise directly at the expense of the spongiose matrix as with *Spongodendron*.

Generic Descriptions.   Some illustrative genera and species within each family are presented as a further aid in clarifying the classificatory scheme.

Periaxoplastidies

Family: Cenosphaeridae
  Monogeneric: *Coenosphaera*
  No microsphere, cortical shell flexible made of a simple or double tenuous network, the thin trabeculae limited by an irregular or polygonal mesh.
  Coenosphaera reticulata Haeckel, 1862 (Fig. 1-12A)
    Cortical shell (diameter 180 μm) of irregular mesh, more often polygonal pattern, reinforced internally by relatively thick siliceous bars very irregularly ramified; central capsule spherical or largely ovoid (160 μm), slightly translucent. Endoplasm rich in lipoid deposits, more or less radially disposed. The cortex filled with opaque, nonbirefringent, hemispherical inclusions which may be lipid. Axoplast basophilic, thick, biconcave lens shaped, of large radius, and positioned at one pole of the nucleus (e.g., Fig. 1-11D) corresponding to the point where the axoflagellum originates. Numerous axopodia penetrate the nucleus and emanate from the fusules at the periphery of the central capsule.
  Coenosphaera tenerrima Haeckel, 1887
    Larger species than *C. reticulata* (600–800 μm), shell of irregular

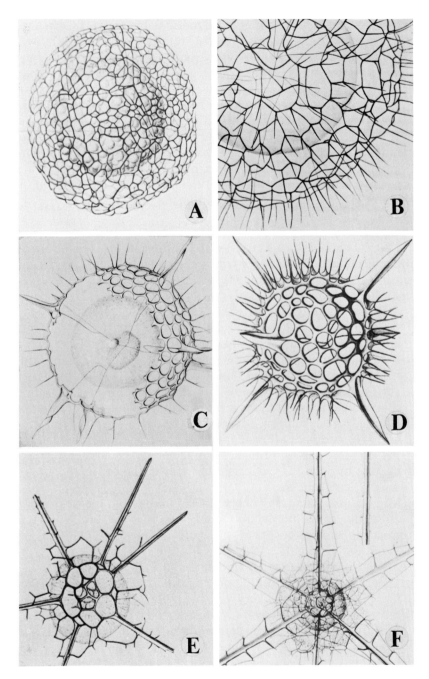

Fig. 1-12. Skeletal structure of **(A)** *Cenosphaera reticulata,* **(B)** *Stigmosphaera cruciata,* **(C)** *Stigmostylus ferrusi,* **(D)** *Hexastylus horridus,* **(E)** *Spongosphaera streptacantha,* and **(F)** *Hexadoras arachnoidale.* (Hollande and Enjumet, 1960)

form, composed of two interlaced trabeculated networks, mesh very irregular, without definite geometric form; central capsule ovoid and remarkably large (400 μm long axis), the cortical region dotted with yellow-orange lipid globules. Endoplasm generally homogeneous, more rarely with radially arranged lipid bodies. Axoplast and axoflagellate system particularly developed, enclosed within the massive hemispherical or capiform nucleus (Fig. 1-11D), which is surrounded by a system of vacuoles, and the whole mass penetrated by numerous fibrils that pass through the capsular wall and give rise to the axopodia; or, in the case of the more massive assembly, they are oriented out of the opening in the capiform nucleus, giving rise to the axoflagellum.

Family: Stigmosphaeridae

Genus: *Stigmosphaera*

Cortical shell delicate; attached to its inner surface are four or five rods, bifurcated at their tips where they attach to the cortical shell and are joined at the center to a thin rod (6 μm), representing a reduced medullary shell.

Stigmosphaera cruciata Hollande and Enjumet, 1960 (Fig. 1-12B)

Cortical shell (280 μm) latticed, mesh very irregularly polygonal and covered with spines straight or declined, sometimes curved, but their number less than the number of nodal points of the lattice from which they arise. Central capsule (175 μm) subspherical, surrounded by the zooxanthellae. Endoplasm, segregated into radial arrays of the numerous lipid bodies. Nucleus subspherical, adorned by a reniform axoplast, from which arise the numerous and large axopodia (Fig. 1-11C). The axoflagellum originating at the axoplast passes through the nucleus as a bundle of fibers and departs through the polar field of pores in the capsular wall.

Genus: *Stigmosphaerusa*

Microsphere composed of six fibrils issuing from the same central point. Four of the fibrils are forked; the other two are simple. All connect to the macrosphere at a point where they continue as a slender, radial spicule.

Stigmosphaerusa horrida Hollande and Enjumet, 1960

Cortical shell latticed, approximately 200 μm diameter, pores circular or ovoid with very variable diameters, and bounded by a bulky hexagonal framework; surface covered by long spines (ca. 70 μm) arising from the nodal points where the lattice bars meet; spines straight, or more or less curved, usually simple, but sometimes bifurcated and bearing small teeth; 11 radiate spicules are comparable in their arrangement to the spines. Structure of the central capsule as with *Stigmosphaera cruciata*, but axoplast spherical. Zooxanthellae present.

Genus: *Stigmostylus* Hollande and Enjumet, 1960

Microsphere reduced to four rods arising from a central point and crudely arranged in a cross. Each rod bifurcates to yield two siliceous fibrils which, contact the cortical shell and are prolonged at the surface of the shell into trigonal spicules resembling those of *Hexacontium*.

Stigmostylus ferrusi Hollande and Enjumet, 1960 (Fig. 1-12C)

Cortical shell (140 μm diameter) latticed, pores circular or ovoid, without hexagonal frame, bearing on the surface 20-μm spines straight or declining, often branched. Eight, more or less spiraled, radial spicules. A large axoplast adorns the nucleus; central capsule, 85 μm diameter.

Genus: *Hexastylus* Haeckel, 1881

Microsphere reduced to a rod, trifurcated at each extremity. Cortical shell bearing six stout, triangular spicules. With certain individuals, one of the extremities of the medullary rods bears four branches, the others, usually two. Each of the six fibrils emerges beyond the cortical shell and can be as long as the triangular spicules.

Hexastylus horridus Hollande and Enjumet, 1960 (Fig. 1-12D)

Cortical shell (85–95 μm) (pores circular or ovoid without a distinct hexgonal frame) bearing six stout, triangular spicules, twisted in a helix. Shell surface bristled with spines, straight or declined, and very strongly branched in the adult stages. Central capsule spherical (60 μm), fusules long, widely spaced, except in the region of the axoflagella, endoplasm with lipid (?) inclusions and rarely eosinophilic globules. Axoplast spherical, close to but not attached to the nucleus. Zooxanthellae present.

Family: Heliasteridae

Genus: *Heliaster* Hollande and Enjumet, 1960

The essential elements of the cortical shell in this genus encompass five siliceous radiating rods joined at a common point near the axoplast. These rods are arranged with four in one plane and the fifth in a plane perpendicular to the others as in a pentactine spicule of a sponge. The four primary rods, through ramifications, give rise on both sides of the nucleus to two siliceous arcs which elaborate solidly some trabeculae. The five branches of the pentactine spicules are the origin of five primary radial spicules. The very numerous secondary spicules arise from the rim of the siliceous arc. The radial spicules (slender and cylindical) vary in number from 6 to 12 according to the age of the radiolarian. In the neighborhood of the microsphere, they are twisted and often armed with spines. The cortical shell is latticed and covered with spines of the same shape and length as the spicules. Axoplast spherical to ovoid, in the neighborhood of the capiform nucleus.

Heliaster hexagonium

This species, the only one in the genus, resembles *Haliomma* hexagonium, a genus described by Haeckel (1887) as possessing two concentric and delicate shells connected by radial beams and bearing radial surface spines on the cortical (outer) shell. Microsphere (40–50 µm diameter); cortical shell (230 µm), with hexagonal pores in young stages and bounded by thin trabeculae, but becoming circular in mature stages as silica is deposited upon the internal edge of the primitive frame. Spicules, 5 to 12 depending on the age of the radiolarian, are slender and the length of those occurring on the cortical shell is approximately 75 µm. The longer of the spines that cover the cortical shell are of about the same length. Among the spines, some of them are more or less reflexed (decumbent), appear to be twisted, and are provided occasionally with some very small teeth. Central capsule (180 µm), speherical, limited by a very tenuous membrane (wall). Axoplast spherical or ovoid, very voluminous, immersed in the capiform nucleus. Axoflagellate cytoplasm passing through 1 of the 10 very large axopods. Pericapsular pigment, blue. Abundant zooxanthallae.

In addition, Hollande and Enjumet described two additional new genera: *Tetrapetalon* (sp. *T. elegans*) and *Arachnostylus* (sp. *A. tregouboffi*). The former genus possesses a microsphere resembling that of *Heliaster*; yet, four of the five branches of the pentactine spicules become divided into two before they join together, thus creating a form like a small rosette of four leaves borne atop the fourth spicule branch. The trabeculae arising from some of the spines form an irregular shell around the nucleus. Cortical shell latticed, bearing 5 to 10 large, straight, triangular spicules. Central capsule as in *Heliaster*.

The second genus, *Arachnostylus*, possesses an intracapsular microsphere, from whence arises a fixed number of simple, nontriangular spicules bearing in their region two or three verticils with branches that give origin to the cortical, latticed shell. The microsphere is of a primitive type. It comprises the origin of four siliceous rods, each one corresponding with a spicule radiating from a point near the axoplast. One is undivided, the others produce the trabeculae which give rise to some microspherical alveoli.

Family: Excentroconchidae
Solitary genus: *Excentroconcha* Mast, 1910
Family description, previously provided, summarizes the major characteristics of this genus; one species is presented.

Excentroconcha minor Mast, 1910
The microsphere is not cubic, but rather a complex space-enclosing structure of unique design. Its essential elements are two irregular pentagons, joined at one common edge and delimiting between them a dihedral angle of approximately 80°. The two pentagons are connected to one another by a raised arch spanning

across the corners opposite to the common edge. At the corners of the shell, six cylindrical, siliceous rods arise. They are covered at their base with small spines, and they very abruptly branch dichotomously or trichotomously. Two other rods, equally branched, arise from the convex border of the arch. The peripheral shell of the adult trophozoite (prereproductive stage) results from the anastomoses of the siliceous trabeculae of the various branches. It is relatively thick (30 μm), of the spongiose type, and not delimited on the external surface by a lattice plate or veneer. Central capsule (80 μm diameter) subspherical initially, but lobed in the adult stage. It is situated next to the microsphere and maintained in its position by the numerous siliceous bars associated with the system of spicules near the microsphere. Axoplast large, central, spherical. Capiform nucleus, slightly curved around the axoplast. Zooxanthellae absent.

Centroaxoplastidies

Family: Spongosphaeridae

Fourteen species in six genera were examined by Hollande and Enjumet. They were distributed in three subfamilies: (1) Spongosphaerinae with spongiose cortical shell and one or two latticed microspheres. There are three genera: *Spongosphaera*, *Spongodendron*, and *Hexadoras*. (2) Plegmosphaerinae with a single spongiose cortical shell and lacking a microsphere or when present not of a clearly defined form. The siliceous trabeculae, which are not dichotomous, ramify and dissipate in the spongiose net. There are two genera: *Plegmosphaera* and *Spongoplegma*. Subfamily (3) Spongodryminae possesses a spongiose siliceous, medullary matrix adjacent to the microsphere shell and independent of that on the cortical shell. Radial spicules are inserted upon the microsphere and exceed, or not, the periphery of the cortical shell.

Two genera are recognized: *Diplospongus* Mast, 1910 and *Spongodrymus* Haeckel, 1881

In total, this is a very important group of radiolaria due to their widespread abundance in surface water and their contribution to the sedimentary record.

The six genera are summarized and 10 representative species are described of the 14 species more fully analyzed by Hollande and Enjumet.

Genus: *Spongosphaera* Ehrenberg, 1847a,b

Two intranuclear microspheres in the adult are placed directly in contact with the peripheral spongiose matrix. The shells are united by the trabeculae with relatively regular contour but not a hexagonal mesh; spicules triangular and variable in number. Spongiose material not bounded at the periphery by a lattice plate. This genus is clearly separated from *Spongodendron* (not described here) by the long spines on the latter and, among other features, by its single microsphere separated from the spongiose matrix.

Spongosphaera streptacantha Haeckel, 1862 (Fig. 1-12E)

This species exhibits considerable polymorphism and is identified essentially by the morphology of the microspheres. The shells are always very distinct and of regular contour. Radial spicules variable in number (six or generally eight), robust (20 to 30 μm maximum thickness), straight or very rarely twisted in a helix, triangular, each ridge (keel) bearing very robust teeth directed toward the exterior. The mesh is always very irregular, compact, but the trabeculae which delimit it are of one consistency. Otherwise, the shell is perchance spherical or polygonal depending on whether or not the spongiose matrix is developed in a direction preferentially along the axis of the spicules. Central capsule may attain a very large diameter (40 μm) limited by one thin membrane bearing the fusules. Axoplast unidentifiable; however, the axoflagellum is very well developed. No zooxanthellae.

Genus: *Hexadoras* Haeckel, 1881

Medullary shell with a diameter as small as 20 μm or exceeding 40 μm and may attain 60 μm. The outer shell is spherical, spongiose, and encloses a simple lattice shell at its center. Originally placed by Haeckel in the Cubosphaeridae with six spicules disposed in three perpendicular planes—if indeed this is the same form as originally described by Haeckel.

Hexadoras arachnoidale Hollande and Enjumet, 1960 (Fig. 1-12F)

Microsphere spherical, pores circular or ovoid and bearing six robust spicules, long, large, and triangular with very thin spines. The spongiose matrix is extremely delicate, arachnoidal, immediately enclosing the medullary shell. The spongiose material is composed of very thin threads stretched between the tips of the spines and the extremities of the short branches associated with the triangular spicules. Maximum size of the spongiose shell, 60 μm diameter. Spicule length, 270 μm; breadth, 7 μm. Central capsule, 40 μm. One large centronuclear axoplast, lacking an axoflagellum.

Genus: *Plegmosphaera* Haeckel 1881

Among the Spongosphaeridae, these are of the easiest to recognize. Their skeleton is in effect reduced to a simple spongiose cortical shell. A lattice plate may delimit either the inner surface or the outer surface, or both surfaces of the spongiose cortical shell.

Plegmosphaera pachyplegma Haeckel, 1887 (Fig. 1-13B)

Haeckel characterized this species: Cortical shell (diameter, 200 μm) occurs free at the center of the radiolarian; the spherical shell with a radius equal to the thickness of the wall. Mesh of the spon-

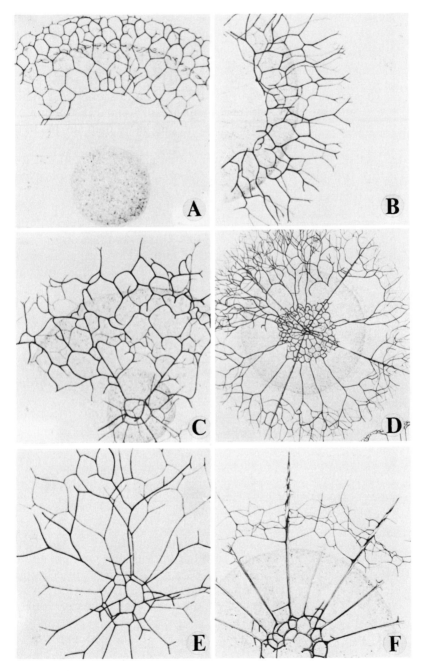

Fig. 1-13. Skeletal structure of **(A)** *Plegmosphaera coronata*, **(B)** *Plegmosphaera pachyplegma*, **(C)** *Spongoplegma rugosa*, **(D)** *Diplospongus dendrophorus*, **(E)** *Spongodrymus gracilis*, and **(F)** *Rhizosphaera helicoidalis*. (Hollande and Enjumet, 1960)

giose material 10 or 20 times as large as the delimiting bars. In the samples of Hollande and Enjumet (1960), the shells were of a diameter equal to or greater than 300 μm, which may reflect a more advanced age of the radiolarian.

Plegmosphaera leptoplegma Haeckel, 1887

This species is easily distinguished from the preceding one by the greater thickness of the spongiose layer (equal in thickness to the diameter of the central sphere), much more delicate due to the extreme tenuousity of the trabeculae by which it is bounded. Shell diameter, 430 μm, central capsule attains diameter, 190 μm with a thin capsular wall. Nucleus spherical or ovoid, axoplast well developed. Endoplasm with radially arranged vacuoles.

Plegmosphaera coronata Hollande and Enjumet, 1960 (Fig. 1-13A)

Spongiose shell (300 μm) limited externally by one latticed plate resembling that described by Haeckel in *P. exodictyon*, but here the shell is relatively little thickened and is deposited at the center of the protist as a sphere free of the spongiose matrix (diameter 170 μm). The lattice mesh is loosely woven and bounded by the thin trabeculae. Capsule, colored red; maximum size observed, 190 μm. Nucleus elliptical, axoplast small. A well developed ergastoplasm surrounds the nucleus forming 6 to 10 concentric lamellae. No zooxanthellae.

Genus: *Spongoplegma* Haeckel, 1881

This genus differs principally from that of *Plegmosphaera* through the possession of one microsphere which is immediately enveloped by the spongiose matrix produced by the ramification of the radial trabeculae. Pores of the central shell circular or more often very irregular.

Spongoplegma rugosa Hollande and Enjumet, 1960 (Fig. 1-13C)

Microsphere pores large, irregularly polygonal. The 8 or 10 radial trabeculae give rise to the spongiose matrix. The latter not limited at its periphery by a lattice plate.

Spongoplegma radians Hollande and Enjumet, 1960

Spongiose matrix, limited on the outer side by a series of "flying buttresses," thickened, and inserted on the long sinuous spines. Mesh regular and limited by delicate siliceous bars. Microsphere delicate with 8 or 10 radial trabeculae prolonged to form the sinuous spines. Inner shell diameter, 20 μm, outer shell diameter, 200 μm.

Genus: *Diplospongus* Mast, 1910

Microsphere very small (20 μm), irregularly latticed; pores ovoid, variable number of triangular spicules (7 to 12 in the samples reviewed by Hollande and Enjumet). Medullary spongiose matrix compact to the point of masking the microsphere. It is made of trabeculae produced at the ridges of the spicules. The mesh of the more peripheral spongiose matrix gives rise to a large number of radial shafts, cylindrical and thick, twisting as they ramify abundantly to their periphery. Cortical shell very clearly separated from the medullary spongiose tissue. Monotypic.

Diplospongus dendrophorus Mast, 1910 (Fig. 1-13D)

Microsphere, 20 μm; medullary spongiose matrix, 180 μm. Cortical shell, 650 μm. Eight to ten radial spicules, 400–500 μm long. They are narrow at the base, and thence abruptly widened, and clearly triangular on this side of the medullary spongiose matrix. Between the medullary spongiose matrix and the cortical shell, they are strongly armed with teeth and, at their apex, often twisted. The central capsule (ca. 400 μm) is irregularly covered with rounded protuberances and limited by one delicate membrane bearing numerous fusules. The endoplasm is not arranged in radial arrays, but is composed of four zones: (1) a thin perinuclear layer, with a loosely organized structure; (2) a clear alveolar zone, radially disposed of relatively significant dimension; (3) a dense finely granular, siderophilic layer; and (4) finally a cortical layer rich in lipid bodies. The centrally located nucleus is divided into unequal lobes and not clearly delimited. The large, spherical axoplast is partially segregated from the nucleus by the microsphere. A significant gelatinous layer encloses the central capsule and within it one observes solely some digestive products. It contains neither pigment granules nor zooxanthellae.

Genus: *Spongodrymus* Haeckel, 1881

Particularly separated from the preceding genus by the absence of radial, triangular spicules. The cortical shell is solely derived from the ramification of the thin siliceous fibers issuing from the spongiose matrix. The microsphere is always small. The spongiose matrix which encloses it is either compact and not very extensive or more loose and well developed. The central, small microsphere may be difficult to detect, owing to the large amount of spongiose material masking it.

Spongodrymus elaphococcus Haeckel, 1887

Microsphere, 20 μm, enclosed in a very compact spongiose matrix with small mesh, delimited by the fine trabeculae supported upon the radial fibrils which radiate from the central shell. The twisted, thick, cylindrical trunks arising from the spongiose medullary matrix produce, through their apical ramifications, a spongiose cortical shell (700 μm diameter). It may be delicate but very thick. Central capsule may attain a diameter of 450 μm.

Spongodrymus gracilis Hollande and Enjumet, 1960 (Fig. 1-13E)

Distinguishd from *S. elaphococcus* by the following: Microsphere, 20 μm, surrounded by a delicate spongiose matrix, well developed with a loose diamond-shaped mesh. Radial thin shafts very delicate. Cortical shell (500 µm diameter) spongiose, not very thickened (20 µm), and regular. Central capsule may attain a diameter of 400 µm. Capsular wall very fine with numerous fusules. Endoplasmic vacuoles radially disposed; a well-differentiated axoflagellum. Nucleus lobed and a large spherical axoplast.

Family: Rhizosphaeridae
Genus: *Rhizosphaera* Haeckel, 1860

Microsphere spumose (100 µm diameter), cortical shell either spongiose, or latticed, regularly spherical, or bossed according to the species, always perforated by numerous radial spicules, and plated or not on the external side by a spongiose matrix. The latter is the origin of the ramifications of spicules from whose bramble of branches arise the spines covering the shell. This is a polymorphic genus and the species are difficult to identify due to the variations in form. Two major groups are recognized depending on whether the external shell is spongiose or latticed.

Rhizosphaera heliocoidalis Hollande and Enjumet, 1960 (Fig. 1-13F)

Microsphere may reach a large dimension of 100 m. The alveoli peripheral to the microsphere are penetrated by numerous, radial, triangular spicules (bearing a keel or ridge on each edge) that emerge straight but are slightly twisted in a helix near their apex. The spongiose cortical shell (340 µm diameter), 40–45 µm thick, is composed of a very tenuous and irregular, loosely organized mesh. Mean length of the spicules, 180 µm; length of the exposed end, ca. 60 µm.

Rhizosphaera algerica Hollande and Enjumet, 1960 (Fig. 1-14A)

Microsphere less chaotically organized than that in the preceding species with numerous radial, triangular spicules straight to the apex. Cortical shell poorly developed, mature trabeculae thickened, connected to two verticils of branches arising from the edges (keels) of the spicules. Microsphere, 80-90 µm. Cortical shell diameter, 260 µm, 20 µm thick. Spicule length, 130 µm. Exposed portion, 20–30 µm.

Hollande and Enjumet also described three additional new species briefly summarized here:

1. *Rhizosphaera haeckeli*: Cortical shell regularly spherical, pores polygonal in young forms, and circular or ovoid in adult form. Numerous radial, triangular, straight spicules perforating the otherwise spineless shell. Cortical shell, 225–230 µm; microsphere, 100 µm; spicules, 130–135 µm; free end, 4–4.5 µm.
2. *Rhizosphaera spongiosa*: Cortical shell regularly latticed and lined with, but separate from, a spongiose matrix. Robust spines (60

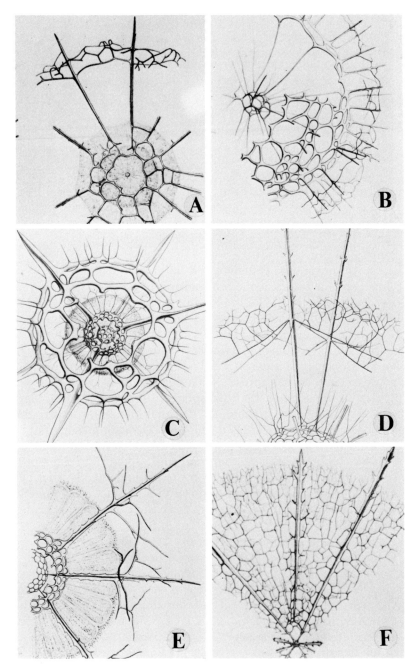

Fig. 1-14. Skeletal structure of **(A)** *Rhizosphaera algerica*, **(B)** *Actinosphaera capillaceum*, **(C)** *Hexacontium arachnoidale*, **(D)** *Diplosphaera cachoni*, **(E)** *Haeckeliella macrodoras*, and **(F)** *Centrocubus mediterranensis*. (Hollande and Enjumet, 1960)

μm) and spicules (100–120 μm) cover the cortical shell, the outermost 330 μm diameter.

3. *Rhizosphaera drymodes*: Skeleton relatively delicate with microsphere enclosing large central alveoli and surrounded by very
small peripheral ones. Cortical shell irregular, pores ovoid or
spherical. Microsphere bearing long, fine triangular spicules.
Spines and projecting portions of the spicules branched and anastomosing. Microsphere, 100 μm; outer shell, 240 μm; spicules,
140 μm; spines, 40–69 μm.

The description of *Rhizosphaera trigonacantha* Haeckel, 1862 is not
reported here in detail. It resembles *R. haeckeli* but differs in having
an irregular or bossed cortical shell. There are two forms differing
in size of the cortical shell and length of the spicules: (a) Microsphere, ca. 100 μm; cortical shell, 270 μm; spicule length, 100–110
um. (b) Microsphere, ca. 100 μm; cortical shell, 200 μm; spicule
length, 70 μm.

Genus: *Actinosphaera* Hollande and Enjumet, 1960

This genus is fundamentally a revision of a pre-existing one proposed by Haeckel; i.e., *Haliomma*, but included here in the Rhizos-
phaerides based on cytological studies. During development, the
primitively latticed microsphere is thickened little by little on the
external side and comes to possess a spongy matrix as in *Rhizosphaera*. However, *Actinosphaera* is distinguished from the latter by
the presence of radial spines on its cortical shell contrasted with
spicules on the shell of the *Rhizosphaera*. Likewise the cytology is
different. With *Actinosphaera*, the capsular membrane is very thick
and the fusules are remarkably developed. The endoplasm is radially
arranged, more or less equally distributed, with a large number of
hyaline vacuoles. Lipids are represented by some large osmiophilic
granules dispersed in the cytoplasm. There is no axoflagellar cytoplasm. The nucleus is subspherical with a distinct circumference.
Axoplast and axopods comparable to those in *Rhizosphaera*. Without
zooxanthellae.

Actinosphaera capillaceum Haeckel, 1862 (Fig. 1-14B)

Microsphere generally spumose. Cortical shell latticed with polygonal mesh in young stages or ovoid in older stages. Numerous
slender spicules in the form of cylindrical rods do not perforate
through the cortical shell. The latter uniquely bears the spines
which by way of its anastomosing lateral branches yields a spongy
matrix. Inner shell diameter, 27 μm. Outer cortical shell, 250 μm;
spine length, 35–65 μm.

Family: Thecosphaeridae

Genus: *Hexacontium* Haeckel, 1881

Three concentric latticed shells, the innermost (microsphere) is always intranuclear. Six radial, triangular spicules, straight or twisted, are arranged in opposite pairs along perpendicular axes. External cortical shell bears spines, simple or dichotomously branched. Cytological features are largely similar to other centroaxoplastidies.

Hexacontium arachnoidale Hollande and Enjumet, 1960 (Fig. 1-14C)

The second concentric shell bears slender dichotomous spines. The branches anastomose to form a loose, arachnoidal material within the space between the second and the third concentric shells. The diameters of the three shells are, respectively, 25, 50, and 175 μm. The second shell possesses circular pores, whereas the third shell has very irregular pores without a hexagonal frame on its surface. Spicules straight (free portion, 70 μm; thickness, 20 μm). Numerous accessory spines, short (30 μm).

The genera *Actinomma* Haeckel, 1862 and *Thecosphaera* Haeckel, 1881 are described only briefly in this summary.

Actinomma

Originally Haeckel distinguished among these genera based on the presence or absence of spines on the outermost of the three concentric shells. Hollande and Enjumet conclude that this is not a valid distinction and have erected a single genus to encompass those species formerly grouped in *Actinomma*, *Echinomma*, and *Pityomma* by Haeckel.

Actinomma mediterranensis Hollande and Enjumet, 1960

Form of the specimen relatively small, resembling at first, *Hexacontium asteraconthion*, but differing in the more numerous and more delicate spicules. Central capsule diameter, 98 μm; shells relatively thick possessing ovoid pores, all of similar design. Those of the external shell are prolonged on the internal side as a conical, very short tube. Outer shell with short spines. Number of spicules generally greater than 12.

Genus: *Thecosphaera* Haeckel, 1881

Distinguished from *Actinomma* principally by the absence of radial spicules. Peripheral shell surface smooth or provided with papillae or spines. Trabeculae thick, attached to the shell between them, but not connecting one shell to the other. Two new species were described: *Thecosphaera bulbosa* (microsphere, 20 μm), with large pores circular. Second shell, 40 μm, pores small and circular. Third shell, 95 μm, pores ovoid and irregular; the small bulbs each bear one very short spine. Connecting bars among concentric shells thick, often twisted or ramified. *T. radians*: External shell bristling with spines, inner shell 20 μm diameter, second shell, 41 μm, with ovoid pores; and third shell, 95 μm, pores circular and situated at the base of a funnel depression whose upper rim is hexagonal. One delicate spine arises from the nodal point where the edges of the rims meet.

Anaxoplastidies

Family: Macrosphaeridae
The genera of this family are conveniently summarized by a dichotomous key (Table 1-4), reproduced from Hollande and Enjumet (1960, p. 114).
Some representative species in the Family Macrosphaeridae are presented in the summary.
Haplosphaera spherica Hollande and Enjumet, 1960
Macrosphere, 220 μm, mesh large and hexagonal, delimited by thin trabeculae. Internal trabeculae short and not very abundant. Central capsule, 55 μm diameter.
Diplosphaera hexagonalis Haeckel, 1887
Inner shell (180 μm) with regular, hexagonal meshes and very thin thread-like bars bearing one bristle-like byspine at each nodal point. Outer shell (360 μm), twice as broad, with simple triangular

Table 1-4. Key to the genera in the Family Macrosphaeridae[a]

1. Macrosphere only present
    (a) Macrosphere without spines or spicules
        Genus: *Haplosphaera*
    (b) Macrosphere with spines, without spicules
        Genus: *Acanthosphaera* (?)
    (c) Macrosphere with triangular spicules or none
        (i) Spicules prickly or diversely branched
            Genus: *Cladococcus*
        (ii) Spicules regularly dichotomous
            Genus: *Elaphococcus*

2. One cortical shell
    (a) Cortical shell latticed
        (i) Cortical shell with loose triangular mesh
            Genus: *Diplosphaera*
        (ii) Cortical shell with hexagonal mesh
            Spicules inserted on the external shell by means of three pillars in the form of an arc
            Genus: *Haeckeliella*
            Spicules normal
            Genus: *Thalassoplegma*
    (b) Cortical shell spongiose
        (i) Formed by branching of the four primary verticils
            Genus: *Rhizoplegma*
        (ii) Formed of circular pores and a hexagonal plate—two concentric shells
            Genus: *Thalassoplegma*
        (iii) Formed as above, but only one shell
            Genus: *Lychnosphaera*

[a] Adapted from Hollande and Enjumet (1960).

mesh. Radial spines bear three smooth edges. Radially lobate central capsule, enclosed in inner shell, possessing a large nucleus, one-third its size.

Diplosphaera cachoni Hollande and Enjumet, 1960 (Fig. 1-14D)

Inner shell (170 μm) without regular hexagonal mesh, but a network of siliceous trabeculae forms two closely integrated, superposed layers, and bearing pores of a very variable design at the periphery. Surface hirsute with spiniform rodlets, simple or ramifying, always very irregularly curved or twisted, and often more or less obliquely directed. Spicules strong, triangular, 400 μm in length; verticillate.

Rhizoplegma verticillata (Spongopila verticillata) Haeckel, 1887

Macrosphere, 98 μm. The branches of the four major verticils produce the spongiose shell through their ramifications.

Haeckeliella macrodoras Haeckel, 1887 (Fig. 1-14E)

Extremely abundant. Relatively young specimens possess spicules, simply dentate, and the mesh of the external cortical shell is irregularly polygonal and delimited by the thin trabeculae. With age, the external cortical shell is thickened. The pores derived from the hexagonal pattern become "spherical" or ovoid. The spicules become branched through growth and dichotomization of certain spines. Various adaptations in shell spacing and spine length occur as a result of environmental variables including the time of year. Central capsule brown, spherical or lobed, in accordance with the age of the radiolarian. Capsular wall thin.

Family: Centrocubidae

Genus: *Centrocubus* Haeckel, 1887

Microsphere encompassed solely by the spongiose matrix with a peripheral arachnoidal layer. The closeness of the spongiose matrix to the microsphere varies. Spicules triangular, variable in number, and distinguishable according to their origin in three groups: (1) Primary spicules, eight in number, arise out of the angles of the microsphere; (2) secondary spicules are inserted at the base of the primary spicules; and (3) tertiary spicules originate directly from the spongiose matrix.

Two species were described by Hollande and Enjumet: *Centrocubus mediterranensis* with microsphere scarcely adjacent to the spongiose matrix; the latter forms a spherical shell with very thin siliceous trabeculae limited by a very compact mesh in the cortical region. Spongiose shell diameter 900–1000 μm, central capsule of large individuals exceeds 550 μm. Spicule total length, 725 μm (Fig. 1-14F).

Centrocubus ruber

Distinguished from *C. mediterranensis* by the very strong spongiose matrix that never forms a regular, spherical shell. The trabeculae are of equal thickness from the center to the periphery of the skeleton. The spicules of the pyramidal prolongations bear on their free ends six or seven verticils of spines. Microsphere, 20 μm; spongiose shell, 600 μm. Diameter of the mesh in the spongiose shell, ca. 30 μm. Central capsule of larger forms exceeds 500 μm diameter.

Genus:*Octodendron* Haeckel, 1887

Resembling *Centrocubus*, but essentially different by the presence of an arachnoidal cortical layer forming a spongiose matrix around the microsphere which, unlike that in *Centrocubus*, is clearly delimited on the inner surface.

Octodendron hamuliferum Hollande and Enjumet, 1960

Spongiose matrix slight, limited on the inner side by one distinct latticed plate (150 to 180 μm diameter sphere), formed by fused branches of the two or three major spiculate verticils. Delicate arachnoidal matrix supported by the branching of the 9 to 10 verticils. The branches (three per verticil) are expanded not far from their base to form a disc, which may be described as a spathule. Microsphere excentric, approaching cubic, small (diameter 15–20 μm). Maximum thickness of the spongiose matrix, 30 μm. Arachnoidal matrix, 400 μm. Spicules triangular, twisted in their free ends; total length, 1,000 μm. Peduncle of the spathule, straight. Central capsule diameter, 300 μm.

Octodendron arachnoidale Hollande and Enjumet, 1960

Microsphere clearly cubical; spongiose matrix more thick than that in the preceding species; not limited on the internal side by a latticed plate. Arachnoidal matrix as in *O. hamuliferum*. Microsphere, 20 μm diameter; spongiose shell, 250 μm diameter; spongiose matrix thickness, 60 μm; spicule total length, 700 μm; length of their free part, 150–300 μm. Number of verticils = 10. Central capsule, 600 μm diameter.

Hollande and Enjumet also present some interesting observations about the cytology of Discoides and Larcoides and describe three new collodarian species in two addenda to their report.

Light Microscopic Studies of Colonial Radiolaria

The artificial quality of Haeckel's systematics of the colonial radiolaria was recognized very early in the 20th century, owing largely to light microscopic research by Brandt (1885, 1902, 1905) and subsequently by Haecker (1908a,b) and Popofsky (1917), who employed Brandt's classification scheme and made small additional changes.

Brandt (1905) substantially revised Haeckel's systematics of colonial radiolaria. He proposed only two families: Collosphaeridae and Sphaerozoidae. The Family Collozoidae was eliminated, since Brandt had discovered the occurrence of skeletonless forms in the families Collosphaeridae and Sphaerozoidae. He reduced the number of genera from

Haeckel's 17 to 13 and eliminated nearly half of the species, from 84 to 44. The genera recognized by Brandt were Collozoum, Sphaerozoum, Rhaphidozoum, Belonozoum, Myxosphaera, Collosphaera, Buccinosphaera, Acrosphaera, Trypanosphaera, Clathrosphaera, Siphonosphaera, Mazosphaera, and Solenosphaera.

With the exception of some minor changes introduced by Brandt's students, Hilmers (1906) and Breckner (1906), who largely employed his classificatory scheme, little further research with colonial radiolaria was pursued until the latter half of the 20th century. This includes the work of Campbell (1954), which has already been mentioned, Riedel (1967a–c), Nigrini (1967, 1968), Petrushevskaya (1971b), and Nigrini and Moore (1979). Perhaps the most comprehensive and practical treatment of colonial radiolarian systematics has been presented by Strelkov and Reshetnyak (1971). Their work incorporates much of the insights from the research of Brandt and his contemporaries while adding significant new dimensions based on the substantial plankton samples that were collected in the western part of the Pacific Ocean, the South China Sea off the coast of Hainan, and in the Atlantic Ocean, during oceanic explorations by Dr. Strelkov. Additional samples from the Indian Ocean were obtained from the Institute of Oceanology, Academy of Sciences, USSR; and some samples from the plankton of the Pacific sector of the Antarctic were examined in their research.

The structure of Strelkov and Reshetnyak's systematics is as follows (Strelkov and Reshetnyak, 1971, p. 316):

Class Sarcodina
Subclass Radiolaria
Order Spumellaria
Suborder Sphaerocollidea
Superfamily Sphaerozoida
    I. Family Sphaerozoidae
      1. Genus Collozoum (Figs. 1-5A, B)
      2. Genus Sphaerozoum (Figs. 1-5C, D)
      3. Genus Raphidozoum
    II. Family Collosphaeridae
      1. Genus Collosphaera (Figs. 1-5E, F; 1-15A)
      2. Genus Acrosphaera (Figs. 1-15C, D)
      3. Genus Tribonosphaera
      4. Genus Siphonosphaera (Figs. 1-15G, H)
      5. Genus Solenosphaera
      6. Genus Buccinosphaera

In addition, the Family Collosphaeridae was divided into three tribes: tribe 1 (Collosphaerini) includes those genera possessing shells with a smooth inner and outer surface (type genus Collosphaera); tribe 2 (Acrosphaerini) possessing shells with a spine-covered surface (type genus Acrosphaera and the genus Tribonosphaera); and tribe 3 (Siphonos-

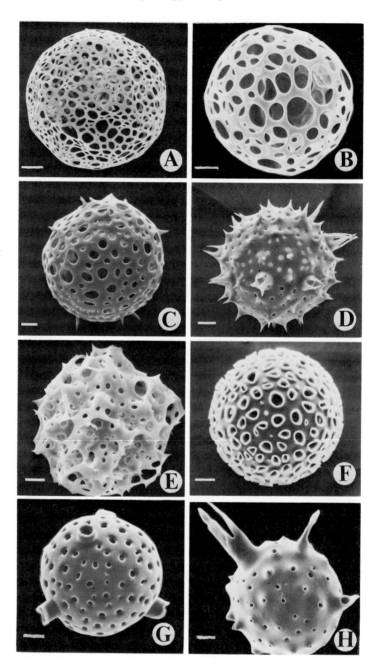

Fig. 1-15. Scanning electron micrographs of shells from colonial radiolaria. **(A)** *Collosphaera huxleyi*, **(B)** *Collosphaera polygona*, **(C)** *Acrosphaera cyrtodon*, **(D)** *Acrosphaera spinosa*, **(E)** *Acrosphaera circumtexta*, **(F)** *Siphonosphaera cyathina*, **(G)** *Siphonosphaera tubulosa*, **(H)** *Siphonosphaera socialis*. Scales = 15 μm.

phaerini), shells with tubules (type genus *Siphonosphaera* and the genera *Solenosphaera* and *Buccinosphaera*).

A brief synopsis of the families and some representative species is presented. Although their monograph is illustrated with line drawings, I include light and electron micrographs from my collection to illustrate species.

1. **Family:** Sphaerozoidae

The members of this family are devoid of shell-like skeletons, but possess spiny skeletons composed of simple or double symmetric spines surrounding the central capsules or scattered throughout the gelatinous sheath.

2. **Family:** Collosphaeridae

The skeleton consists of a latticed shell surrounding each central capsule in a mature colony. The ornamentation of the shell, including size and shape of the pores and surface decorations (spines and tubules on the pore rims), are used to determine genera and species.

Genera and some species within each family are briefly described.

Family: Sphaerozoidae

Genus: *Collozoum* Haeckel, 1862

Central capsules without skeletal elements; however, spicules may be scattered in the gelatinous sheath; usually, devoid of spicules.

Collozoum inerme (Müller, 1858a,b)

Colonies vary from spherical to filiform–cylindrical, the latter form often segmented with a series of constrictions demarking the node-like segments. The spherical central capsule has a thick, simple capsular membranous envelope. Diameter of the central capsule, 0.02–0.14 mm.

Collozoum ovatum Haeckel, 1887

Colonies vary from spherical to elongate. Central capsules are elongate–oval, the length exceeds its width two to three times and a single lipid droplet is at the center. Length of central capsule, 0.12–0.20 mm., width, 0.06–0.08 mm.

Collozoum amoeboides Haeckel, 1887

Colonies mostly spherical or rarely elongated. Amoeboid central capsules of iregular shape with blunt, pseudopod-like, or digitiform lobes. Diameter of central capsule, 0.04–0.15 mm.

Collozoum pelagicum Haeckel,1860a,b

Colonies elongated containing small, central, *irregularly polygonal* capsules, with a surrounding envelope of cytoplasm forming, according to Strelkov and Reshetnyak (1971), 8–10 digitiform lobes tapering at the distal ends, which are sometimes branching. Samples of living specimens observed by light microscopy at our laboratory at the Bermuda Biological Station possessed a simple entire disc-like envelope surrounding the central capsule. Diameter of central capsule, 0.025–0.08 mm.

Genus: *Sphaerozoum* Meyen, 1834

Skeletal elements are solely double (paired triradiate) spines composed of a central shank bearing two, three, four, or more lateral branches at the ends (Fig. 1-5D). The branches are either smooth and simple (*Spaherozoum punctatum*) or ramified and covered with spinules (*S. verticillatum*). Spines are frequently enclosing the central capsules or scattered in the surrounding gelatin.

Sphaerozoum punctatum (Meyen, 1834) (Figs. 1-5C, D)

Colonies are either spherical or ellipsoidal paired triradiate spines, each with three lateral branches, smooth or spiny, straight or slightly curved, length equivalent to the main axis. Length of main axis, 0.01–0.04 mm; length of lateral branches, 0.01–0.04 mm; length of colony, up to 5 mm; width, up to 7 mm.

Sphaerozoum verticillatum Haeckel, 1887

Colony oval shaped. Each spine has a short main axis with three straight, long branchlets, bearing lateral secondary branchlets. The latter may be arranged as four to six whorls. Length of main axis, 0.02–0.03 mm; length of lateral branches, 0.05–0.10 mm.

Genus: *Rhaphidozoum* Haeckel, 1862

The skeletal elements are represented by simple and radiate spines; sometimes both kinds are present depending on the species. Radiate spines bear long rays (most frequently four, more rarely three, five, or six) in a stellate arrangement.

Rhaphidozoum acuferum (Müller, 1855)

Colonies spherical or oval, with characteristically large individuals. Simple spines more abundant than the tetraradiate spines mixed among them and surrounding the central capsules. Surface of the spines is covered by small spinules. Diameter of central capsules, 0.15–0.25 mm; length of simple spines, 0.08–0.25 mm; length of tetraradiate spines, 0.10–0.16 mm.

Rhaphidozoum neapolitanum Brandt, 1881

Colonies are either spherical or oval. Simple spines only, either straight or more rarely curved, with fine surface dentation. Length of spines, 0.08–0.25 mm; diamater of central capsules, 0.15–0.25 mm.

Although Strelkov and Reshetnyak do not treat the genus *Belonozoum* in their systematics, it is included here, since it is commonly observed in the Sargasso Sea. The description is that largely of Haeckel (1887).

Genus: *Belonozoum* Haeckel, 1887

Spicula all of a simple needle-shape, varying from straight cylindrical rods with obtuse ends to slender curved spindles bearing sharp points on each end.

Belanozoum atlanticum Haeckel, 1887

Spicula simple rods, more or less curved or bent, pointed at both ends, and may bear numerous small spines placed vertically on the rods. Diameter of central capsule, 0.1–0.2 mm; spicule length, 0.07–0.15 mm.

Family: Collosphaeridae
Tribe: Collosphaerini
Genus: *Collosphaera* Müller, 1855

Large individuals with colonies of 1 cm or greater diameter, gelatinous envelope often opalescent in appearance by reflected light; shells smooth, latticed, lying at a considerable distance from one another. The form and size of the shells vary considerably among the species. Adult specimens with very large crystals and a light-blue pigment.

Collosphaera huxleyi (Müller, 1855)

Colonies usually spheroidal, shells have the form of regular or more rarely slightly crumpled spheres with large, irregularly sized or, more rarely, polygonal pores (Fig. 1-15A). There are 8 to 16 pores on half a diameter. Diameter of shells, 0.08–0.15 mm; diameter of pores, 0.009–0.01 mm; width of interporous septa, 0.005–0.009 mm.

Collosphaera tuberosa (Haeckel, 1862)

Colony spherical or slightly elongate. Shells are strongly crumpled, perforated spheres with deep depressions and rounded protrusions. The round, more rarely polygonal, pores vary in size. There are 10—16 pores per half a diameter of the shell. Diameter of shell, 0.05–0.31 mm; length of colony, 1.7–2 mm.

Collosphaera polygona Haeckel, 1862

Shells are characterized by large, open, polygonal meshes (Fig. 1-15B) of variable size separated by thin septa. Mesh hole width exceeds almost 4–12 times the width of the interporous septa. Diameter of shells, 0.1–0.3 mm; diameter of pores, 0.01 mm.

Tribe: Acrosphaerini
Genus: *Acrosphaera* Haeckel, 1881

Central capsules rather widely spaced from one another and each is surrounded by a latticed shell covered with spines on the outer surface. The latter are either short, solid spines as in *A. erinacea* and *A. setosa* or conical radial spines, straight or variously curved and bearing an arch-like pore at the base (*A. echinoidea*, *A. spinosa*, *A. collina*, and *A. inflata*). The pores can be simple, with entire, unornamented edges, or they may be provided with a spine or surrounded by a coronet of spines in all pores or only the larger ones.

Five species described by Strelkov and Reshetnyak are summarized here.

Acrosphera spinosa (Haeckel, 1862)

Shells spherical or more rarely oval with rounded pores of unequal sizes. The large pores less abundant than the small pores. The coronet of spines surrounding the large pores varies from a simple case of a single radial spine (Fig. 1-15C) or a large conical spine with an arch-like pore at the base (Fig. 1-15D). Large pores may be surrounded by 4–12 spines, dichotomously branched, rarely simple, and oriented in a parallel manner to form a coronet. If the coronet is absent, part of the large pores is elongated into short tubules not exceeding 0.0125 mm in length. Diameter of shell, 0.10–0.11 mm; diameter of pores, 0.005–0.01 mm; spine length, 0.01–0.02 mm.

Acrosphaera lappacea (Haeckel, 1887)

Colony spherical or more rarely sausage shaped. Shell uniquely bearing curved and obliquely oriented spines. The surface of the shell is further ornamented by massive laminated apophyses that are connected with one another by transverse spines producing a series of connected, arch-like ridges on the shell. Solitary, conical spines, bearing perforations at their base, project from the surface of the shell. Shell diameter, 0.06–0.16 mm; pore diameter, 0.002–0.0125 mm.

Acrosphaera cyrtodon (Haeckel, 1887)

Spherical shells with rounded or oval pores of unequal size. The large pores (six to nine) in contrast to the small pores, bear one single tooth or one single spine that is larger or thinner. The spine or tooth usually occurs tangential to the edge of the pore (Fig. 1-15C), and projects outward from the rim edge where it is attached. Occasionally, the solitary spine is slightly curved backward from the edge of the pore. Shell diameter, 0.08–0.09 mm; pore diameter, 0.01–0.02 mm; length of tooth, 0.03–0.06 mm; spine length, 0.0175–0.025 mm.

Acrosphaera circumtexta (Haeckel, 1887)

Shell spherical or distinctly prolate. The pores are surrounded by a ridge, and adjacent pores may be connected by a framework of thin bars resembling a set of flanges projecting from the surface. The conical spines are more short than long, and extend not only from the surface of the shell between the pores, but also from the pore ridges and their junctions. Diameter of shell, 0.06–0.15 mm; pore diameter, 0.002–0.02 mm.

Acrosphaera murrayana (Haeckel, 1887)

Spherical shells bear numerous large rounded pores and fewer small ones. All of the pores with the exception of the very small ones are surrounded by a coronet of one to three short spines broader at their base and tapered distally yielding a burr-like configuration to the shell. In occasional cases where the spines are lacking, the pores are surrounded by a collarette. Shell diameter, 0.07–0.09 mm; pore diameter, 0.007–0.02 mm.

Genus: *Tribonosphaera* Haeckel, 1881
Monotypic genus:
    Tribonosphaera centripetalis Haeckel, 1887
    Shells simple, latticed, and provided on the inside with radial, centripetal, spinose apophyses, which freely terminate near the center of the shell. Pores are rounded, rather large, and of unequal size. Inner surface of shell with 10–20 thin radial spines. Diameter of shell, 0.10–0.12 mm; pore diameter, 0.003–0.005 mm.
Tribe: Siphonosphaerini
Genus: *Siphonosphaera* Müller, 1858a,b
Latticed shells, mostly spherical; all or only part of the round, unequally sized pores are elongated into characteristic tubules projecting outwardly from the shell. Tubules with solid walls, not perforated.
    Siphonosphaera tubulosa (Müller, 1858a,b)
    Shell almost spherical, more rarely irregular with a small number (5–10) of short cylindrical tubules, irregularly distributed over the surface and widely separated from each other. The tubule length is one-fifth to one-sixth the diameter of the shell, and the length of the tubule is almost equivalent to its diameter. Pores without tubules are larger and closely distributed. There are two to three tubules per half-diameter of the shell. Diameter of shells, 0.12–0.15 mm; length and width of tubules, 0.02–0.03 mm.
    Siphonosphaera cyathina Haeckel, 1887
    Shells thick-walled, rounded, or more rarely oval in form; the rounded or less commonly oval pores are all elongated into cylindrical short tubes, which are slightly tapered toward the distal end with a flared, collar-like edge (Fig. 1-15F). There are 9–10 tubules on the half diameter of the shell. Diameter of shells, 0.10–0.14 mm; tubule diameter, 0.01–0.05 mm; length, 0.011–0.045 mm.
    Siphonosphaera socialis Haeckel, 1887
    The very small shells are usually spherical or more rarely ellipsoidal, and most rarely there may be double shells! The rounded pores are either large elongated into tubules (main pores), or small (secondary pores) seldom with tubules. The shape of the tubules is usually cylindrical (Fig. 1-15H); however, occasionally some have a flared distal end, or bear widened edges with tapered lobes. The number of tubules varies from 2 to 16. Secondary pores more numerous than main pores. Diameter of shells, 0.05–0.08 mm; pore diameter, 0.005 mm; tubule length, 0.015–0.025 mm; width, 0.0075–0.15 mm; diameter of small pores, 0.002 mm.
    Siphonosphaera compacta (Brandt, 1905)
    The thick-walled shell is almost regularly spherical bearing a few large pores elongated into thick-walled short tubules, and numerous regularly distributed small pores with thin-walled short tub-

ules. The latter bear a short spine on their distal end or exhibit a smooth edge. Occasionally, the pores or spines of the adjacent tubules merge. Diameter of shell, 0.06–0.125 mm; pore diameter, 0.0025–0.015 mm; tubule length, 0.0025–0.0075 mm.

Siphonosphaera martensi Brandt, 1905

The rather thick-walled mostly spherical shell bears large and small pores, *all* elongated into poorly developed, short tubules. Diameter of shell, 0.09–0.10 mm; pore diameter, 0.002–0.0175 mm; tubule length, 0.002 mm.

Siphonosphaera macropora Strelkov and Reshetnyak, 1971 (p. 357)

Diagnosis. Shells spherical or with slightly concave walls with large, rounded, equally sized pores, regularly distributed over the entire surface. The width of the interporous septa almost equals the diameter of the pores. A small part of the pores (7–8) are elongated into short and straight cylindrical tubules. The length of the latter equals their width.

Discriminated from *S. socialis* by the large pores (almost twice the diameter of those in *S. socialis*) and the short tubules. Shell diameter, 0.05–0.07 mm; pore diameter, 0.01 mm; length and width of tubules, 0.01–.015 mm.

Genus: *Solenosphaera* Haeckel, 1887

The central capsules are closely adjacent to one another and surrounded by shells bearing tubules perforated by pores or by tapered, truncated, conical processes bearing a large aperture at the distal end. The edge of the aperture can be either smooth or, more rarely, ornamented with several spines (teeth).

Solenosphaera zanguebarica (Ehrenberg, 1872)

The representatives of this species are typical for the genus. Shell, rather large, polygonal with three to four (more rarely two to six) tapered or short tubules, which are clearly delimited from the surface. The tubule aperture is smooth or more rarely, with one to two teeth. Shell diameter, 0.11–0.15 mm; large irregularly polygonal pores vary in diameter from 0.005 to 0.01 mm; width of taperings or tubules, 0.03–0.04 mm.

Solenosphaera pandora Haeckel, 1887

This species is characterized by the distinct demarcation of the tubes from the surface of the almost regularly spherical shell, more rarely polygonal. Long cylindrical tubes, varying in number from one to six, more frequently three to four, bear a smooth outer edge, but perforated by pores of varying diameters on their walls. Shell diameter, 0.07–0.10 mm; tubule length, 0.03–0.05 mm.

Solenosphaera collina (Haeckel, 1887)

Shell spherical with irregular polygonal openings, large conical taperings (8–20) of almost equal height and width. The aperture of the latter does not exceed the diameter of the largest pores of the shell; aperture provided with a transversely located single tooth (more rarely three to four). Shell diameter, 0.15–0.25 mm; pore diameter, 0.005–0.02 mm; height of taperings, 0.03–0.04 mm.

Solenosphaera inflata (Haeckel, 1887)

Shells with polygonal pores (width of pores two to six times that of septa) and large pyramidal taperings (6 to 12). Each tapering is almost of equal height and width and terminated on the distal end by a large oval pore. The edge of the pore provided with a transversely oriented thick, single conical spine, perforated at the base by three to six pores. Diameter of shell, 0.07–0.14 mm; large pore diameter, 0.02–0.05 mm; small pore diameter, 0.005 mm; spine length, 0.02–0.04 mm. Endemic to the Atlantic Ocean.

Solenosphaera chierchiae Brandt, 1905

The rather thick-walled shell is drawn out into three to four or rarely one to six taperings or short tubules that confer a decidedly angular shape to the shell. The free edge of the aperture bears one to three (or two to six) spines, more rarely teeth, and in very rare instances it is completely smooth. Pores are rounded, more rarely polygonal. Shell diameter, 0.065–0.10 mm.

Solenosphaera tenuissima Hilmers, 1906

Shells elongated–oval, bearing large polygonal pores varying in size, without taperings or tubules; one to two (more rarely three) large apertures on the slightly protruding parts of the shell. If only two large apertures are present; they occur on opposite poles of the shell giving it a distinctively barrel shape. Shell diameter, 0.055–0.07 mm; large aperture, 0.005–0.0125 mm; pore diameter, 0.005–0.0275 mm; spine length, 0.07–0.1126 mm.

Solenosphaera polysolenia Strelkov and Reshetnyak, 1971 (p. 364)

Diagnosis. Large, almost regularly spherical shell with clearly outlined, short conical tubules. Number of tubules from 10–14. The edge of the outer aperture of the tubules is smooth. The surface of the shell is represented in the form of a transparent net, consisting of polygonal, unequally sized meshes.

Shell diameter, 0.19–0.29 mm; diameter of large mesh openings, 0.005–0.007 mm.

Genus: *Buccinosphaera* Haeckel, 1887

The radial, perforated tubules are *directed toward the inside* of the porous shell.

Buccinosphaera invaginata Haeckel, 1887

Colony mostly oval. Shells appear as crumpled spheres with shallow depressions bearing short, conical tubules (oriented to the inside of the shell) with narrower inner apertures (one-half the diameter of the outer aperture). Pores of the shell and tubules are small, rounded, and unequal in size, with diameter almost equal to the septa separating the pores. Diameter of shell, 0.10–0.13 mm; pore diameter, 0.002–0.005 mm; tubule length, 0.02–0.03 mm; diameter, 0.01 mm.

Fine Structure Research and Taxonomy

Fine structure evidence has been applied increasingly to define taxa particularly among the protozoa, where light microscopy often fails to elucidate significant microanatomical features. However, care must be exercised in selecting appropriate fine structure features that clearly reflect stable species differences and not transitory states of the organism. Likewise, the practical appropriateness of fine structural features for taxonomic definition must be taken into consideration, even though they may be sufficiently stable to qualify as a defining attribute. There is little significance to erecting new species, based on a miniscule structure that may have little physiological discriminatory value or practical value within the limits of accessibility to electron microscopic data. A structure, for example, that occurs within a very limited region of a cell may require hours of detailed work to section thoroughly the specimen and locate the structure. Unless there are compelling reasons to employ fine structure evidence as a primary source of taxonomic criteria, it is wise to employ more accessible sources of discriminating characteristics such as light microscopic data. Under the best of conditions, fine structure evidence is particularly useful when it correlates with some macroscopic feature that can be used to identify species in field situations. Tyler (1979) has summarized some of the criticisms of applying fine structure evidence in taxonomy, including the likely phylogenetic insignificance of some ultrafine structures. Of particular concern is the possible variable or transitory nature of some physiologically influenced cytoplasmic inclusions, including composition and/or morpholoy of reserve substances (e.g. Grimstone, 1959; Kerkut, 1961; Pitelka, 1963; and Cheissin, 1965). These metabolically variable, cytoplasmic inclusions are of doubtful significance for taxonomic discriminations.

Nonetheless, fine structure evidence may be particularly useful when conventional optics provide insufficient data to elucidate broad lines of phylogenetic affinities and to inform us about the most natural grouping of species based on conservative microanatomical features. Corliss (1979), for example, has critically discussed the merits of fine structure evidence in ciliate taxonomy, including the contributions of electron microscopy to clarifying species designations based on the microtubule geometry in the locomotory apparatus of ciliates. Individual microtubules (ca. 300 Å diameter) are composed of molecular protein subunits and can be rapidly assembled or disassembled and thus in some parts of the cell represent highly variable structures without necessarily a predictable pattern of organization. Those associated with locomotory organelles, however, are often organized into superordinate structures that are more stable or at least are apparently reproducibly reassembled after transitory dispersed states. For example, the microtubule array within the axoneme of axopodia, although reversibly assembled and disassembled depending on whether the axopodium is extended or contracted, appears to have a species-specific pattern when fully organized. The intracellular "root" of the axoneme also may bear a characteristic organization, thus providing an alternate source of data on axopodial microtubule geometry. Some of the variations in form of cytoskeletal locomotory microtubles in a variety of protozoa are presented in standard texts (e.g., Grell, 1973). Much additional research is needed, however, to determine the specificity and stability of microtubule assemblies within radiolarian species.

Fine structure research by Jean and Monique Cachon at Villefranche sur Mer has contributed substantially to our knowledge of capsular wall and axopodial organization in some major radiolarian groups, particularly among solitary radiolaria.

Hollande et al. (1970) examined the origin and fine structure of the central capsular wall. They concluded that the segmented wall, formed of closely intercalated, complementary plates, was formed by secretion within a cortical layer of membranes surrounding the central capsule. They elucidated the relationship of the fusules to the capsular wall and its surrounding membranes and suggested that the plates of the capsular wall were homologous to the thecal plates secreted at the surface of some armored dinoflagellates. This evidence and other similarities to dinoflagellates during swarmer production in radiolaria have led to the suggestion that the radiolaria may have evolved from dinoflagellate-like ancestors. The cortical structures of Acantharia have been examined by Febvre (1972–1974), who has described the contractile myonemes, putatively mediating expansion and contraction of the extracapsulum, and other unique features of the capsular wall and associated cytoplasm. In their classical work on radiolaria, Hollande et al. (1970) have nicely illustrated the organization of the fusules and the presence of thin strands of cytoplasm passing through slits between the plates in the capsular wall. In a series of subsequent papers, Cachon and Cachon (e.g., 1971a,b, 1972a,b, 1973a,b) carefully examined the organization of the axoneme in the axopodia of polycystine and phaeodarian radiolaria and elucidated the relationship of the axoplast to the nuclear fine structure in the Sphaeroides, thus completing the earlier cytological research by Hollande and Enjumet (1960) that set forth the basic groundwork for the cellular organization in this group. The organization of the axoplast in the centroaxoplastidies (Cachon and Cachon, 1972a) consists of a fine fibrillar central mass that gives rise to the microtubule bundles passing outward, radially into the cytoplasm, and eventually departing through the fusules as the axonemes of the axopodia. The axoplast and proximal rays of microtubule bundles are enclosed within a cytoplasmic sheath which is embedded within the lobate nucleus. The latter is surrounded by a nuclear membranous envelope that segregates the nucleoplasm from the axoplast and microtubule shafts. Hence, the fine structure evidence clarifies the unresolved issue raised by the light microscopic analysis; e.g., how the axoplast and axopodial shafts can be seemingly embedded within the nucleus of the centroaxoplastidies.

The axoneme in the axopodia exhibits a very predictable pattern (Fig. 1-16). There are two centrally-located microtubules that give rise to a spiral-like arrangement of dextrally oriented arms composed of interconnected microtubules. Each curved arm bears a set of secondary and tertiary branchlets pointing counterclockwise from the surface of the arms. Cachon and Cachon (1972a) have carefully analyzed the angles of the interconnecting bridges between the arms and in relation to the central binary tubules thus further characterizing the unique pattern of the spiral. The organization of the axoplast and associated microtubullar bundles in the periaxoplastidies (Cachon and Cachon, 1972b) exhibits a similar general plan as the Centroaxoplastidies except that the axoplastic cytoplasm lies within a concavity formed by the infolding of the nucleus. The axoplast, nestled within this "pocket" in the nuclear envelope, sends out rays of microtubules passing through the cytoplasmic channels among the lobes of the nucleus. The axoflagellum originating from the opening of the nucleus in-pocketing, where the axoplast is lodged, extends directly outward radially and is surrounded by a more densely granular cytoplasmic sheath containing mitochondria and other intracellular organelles. By contrast, the anaxoplastidies have no axoplast; however, each microtubule shaft supplying the axopodia originates at an attachment site on the nuclear envelope.

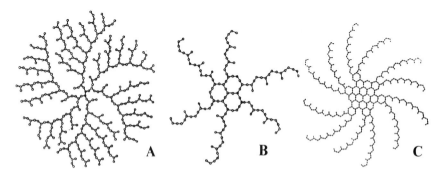

Fig. 1-16. Microtubular arrangement in the axonemes of species belonging to the Spumellaria **(A)** and Nassellaria **(B** and **C).** (Cachon and Cachon, 1971)

The microtubule shaft is composed of a regular array of microtubule bundles arranged in a hexagonal and prismatic mesh as also occurs in the Nassellaria (Cachon and Cachon, 1971a). This is illustrated in Fig. 1-16C. In the anaxoplastidies, the microtubule shaft contains two central tubules that each give rise to two arms as in the centroaxoplastidies. Both of the arms are equally disposed around the central tubule making an angle of 120°. Further high resolution electron microscopy of the microtubule cross sections has shown that each tubule is formed of 13 subunits (Cachon and Cachon, 1977a,b) as is typical for microtubules in a wide variety of organisms. The geometric organization of the microtubular arrays in the axonemes remains to be explained. It is clear, however, that much of the cytological bases for taxonomic discriminations proposed by Hollande and Enjumet have been clarified and confirmed by these fine structure studies as is nicely summarized by Cachon and Cachon (with Febvre-Chevalier and Febvre, 1973, 1974).

Subsequent research with the solitary collodarian species of *Thalassicolla* (Cachon and Cachon, 1976) showed that, unlike the previously studied Spumellaria, there is no intracapsular shaft of microtubules. The fusule consists of a strand of cytoplasm containing vesicles, but devoid of microtubules in the region of the capsular wall pore. The shaft of microtubules originates at a point somewhat distal to the pore in the capsular wall. Hence, Cachon and Cachon have classified these species as "Exo-axoplastidies."

Further contributions to systematics of the Nassellaria haves resulted from light and electron miscroscopic examination of the cytoplasmic and skeletal structure. Based largely on skeletal morphology, the Plectoidea were revised by Cachon and Cachon (1969). The family Plagonidiidae was amended to include four type-genera:

1. *Plagonidium*: spine with four branches (an apical one, a dorsal branch and two lateral ones) two arising as a pair from each pole of the central shaft or rod. Each pair of branches is oriented at 90' with respect to each other.
2. *Tetraplagia*: a spine with four branches disposed as in the preceding species but here arising from a central point or node.
3. *Plagiocarpa*: with four branches arising in pairs from the poles of a common basal rod and forming a tripodal arrangement enclosing the central capsule.
4. *Pentaplagia*: spine with five branches arising from a basal rod, two pairs at opposite ends of the rod oriented at 90' to each other (their planes oriented 90' to each other), and the fifth spine projecting laterally from the central rod.

Subsequent research (Cachon and Cachon, 1971a) on the fine structure of the microtubule organization in the axoplast and peripheral axopodia showed that the microtubules at the center of the axoplast are joined in a hexagonal array (resembling an hexagonal mesh), which gives rise at its periphery to a set of spiral-like arms, curved in a clockwise direction and bearing short secondary branchlets, also directed in a clockwise direction.

The organization of the complex and massive bundles of microtubules in the astropyle and parapylae of the Phaeodaria is the subject of a descriptive analysis in Chapter 2 under the general topic of phaeodarian fine structure and cellular specialization. The unique morphology of the specialized fusules is more logically treated as a general fine structure topic than as a part of the systematics included in this chapter.

A Perspective on the Current Status of Radiolarian Systematics

The systematics of radiolaria are far from complete, and may be viewed as preparadigmatic. There is no single integrated body of information drawing upon the cognate disciplines of micropaleontology, reproductive biology, genetics, ecology, and molecular and cellular biology that is of sufficient magnitude to serve as a paradigm for development of a clearly natural classificatory scheme. The work of Haeckel (1887) though impressive in its scope and detail is, as widely recognized, not a very satisfactory scheme when applied to species identification in natural populations where biological variability often makes assignment to taxa difficult. The elegance of his scheme based on neatly arranged geometric patterns of skeletal form, although aesthetically pleasing and seemingly justified by its orderliness and harmony in design, is clearly not reflective of natural biological affinities as repeatedly pointed out by Hollande and Enjumet (1960) and Riedel (1967) and others. His work nonetheless stands as a very useful foundation for constructing modern schemes as exemplified by that of Campbell (1954). To varying extents more recent, clearly divergent systems such as that of Riedel and Sanfilippo (1977) make use of some of the classical species descriptions although revised in newly constructed superordinate taxa.

Several sources of knowledge including micropaleontology, ecology, cytology, fine structure and biochemical physiology are contributing to our present partial understanding of a more natural classificatory scheme. The situation is analogous to a three-dimensional puzzle whose various complex parts are only partially assembled toward the requisite complementary form permitting their close assembly into the final structure.

One part of the puzzle has been assembled by researchers of fossil material (e.g., Kling, 1978; Riedel and Sanfilippo, 1977; Petrushevskaya, 1981; Boltovskoy, 1981; Goll, 1979; Foreman, 1973; DeWever et al., 1979) and their relationship to extant species. This contribution, though clearly advanced beyond classical views, can be strengthened by additional knowledge about radiolarian reproductive biology, comparative physiology, fine structure, and perhaps genetics.

Some of this essential biological knowledge is being assembled from the work of Hollande and Enjumet (1960) and subsequently Cachon and Cachon (e.g., 1971a,b, 1972a,b, 1973a,b) on cytology and fine structure, by Strelkov and Reshetnyak (1971) on naturally occurring populations and their ecology, and perhaps from research at our laboratory on physiology, biochemistry and fine structure (e.g. Anderson, 1976a–c, 1978a,b, 1980, 1981; Anderson and Swanberg, 1981; Swanberg and Anderson, 1981; Anderson and Botfield, 1983).

It is clear, however, that much additional information is needed to provide the necessary interlocking knowledge that will permit the synthesis of a satisfactory systematic classification. Among these are: (1) a more thorough understanding of radiolarian morphogenesis to permit clear delineation of changes in form during ontogeny and the effects of environmental variables on morphology; (2) population dynamics, including factors that influence their distribution in space and time, their longevity and abundance in relation to environmental variables; (3) their mode of reproduction and the relative contribution of asexual and perhaps sexual reproduction to their abundance and adaptiveness; (4) their genetics and cytogenetics coupled with knowledge of phylogeny; and (5) cellular and molecular biology to eludcidate their fundamental physiology and to identify molecular aspects that may complement fine structure differences used to separate species. The use of fine structure evidence in establishing species definitions is a worthwhile approach, provided adequate care is taken to identify stable, discriminating structures from among the variable features that may reflect transitory states (Tyler, 1979).

Until such interdisciplinary information is more abundant, we will undoubtedly have to be satisfied with partially complete taxonomic systems. The open system of classification based largely on fossil specimens presented by Riedel and Sanfilippo (1977) and the contributions by Goll (1979) and Nigrini and Moore (1979) on fossil and extant species are welcome additions, as they may eventually permit synthesis with the biological-based systematics of Hollande and Enjumet (1960) and Strelkov and Reshetnyak (1971).

Much additional work is needed on the Phaeodaria, which have been largely neglected by micropaleontologists, owing in part to the delicate quality of their skeletons, which are often poorly preserved or nonexistent in the fossil record. Some are deep dwelling species, and a complete account of their biology and distribution will require a careful sampling program throughout the water column. Haeckel (1887), for example, reports samples of some Phaeodaria from depths well over 2,000 fathoms.

It is becoming increasingly clear that radiolaria are a diverse and complex group of organisms, notwithstanding recent trends to question the authenticity of Haeckel's thousands of species. There are certainly sufficient numbers of species to present ample opportunity for research by those who have the talent, motivation, and resources to join in. This is obviously true with respect to the fundamental biology essential to development of a comprehensive, natural-based system of classification.

Chapter Two
Radiolarian Fine Structure and Cellular Specialization

Protozoan Cellular Specialization

Among the pseudopod-bearing protozoa, radiolaria exhibit a remarkable degree of cellular specialization. The clear compartmentalization of their cytoplasm into well-defined, structurally stable regions (intracapsulum and extracapsulum) separated by a specialized membranous barrier is unique among the protozoa. This significant structural and physiological specialization becomes all the more apparent when analyzed by electron microscopy. A conceptual scheme relating radiolarian fine structure to that of other pseudopod-bearing organisms is presented as a general introduction to the more specific descriptions of fine structure among various groups of radiolaria. The pseudopod-bearing protozoa can be classified into three major groups based on the degree of stable compartmentalization of their cytoplasm (Anderson, 1983) as outlined in Table 2-1. The "diffuse organized" protozoa exemplified by some amoebae (Fig. 2-1) and their close relatives exhibit a cytoplasmic fine structure without major compartmentalized regions bounded by stable membranes or nonliving barriers. Their cytoplasm with the exception of subcellular organelles (nucleus (N), mitochondria (M), Golgi bodies (G), and digestive vacuoles (DV), etc.) exhibits few major compartmentalized zones of specialized function. The endoplasm and ectoplasm, when present, represent diffuse regions of cytoplasm intergrading with one another and largely differentiated by regions of varying macromolecular organization. These macromolecular differentiating structures include the presence of a microfilament network contributing to the ectoplasm (MF, Fig. 2-1), immediately adjacent to the cytoplasmic surface of the plasma membrane, and aggregates of intracellular organelles (vacuoles, Golgi bodies, and vesicles) more densely organized within the endoplasm. These organelles may

Table 2-1. Categories of cellular specialization in protozoa[a]

Diffuse specialized	Transitional specialized	Zonal specialized
The cytoplasmic organelles are dispersed widely throughout the cell or form transitory ensembles. There are no specialized large regions enclosed by stable compartmentalizing barriers such as living membranes or nonliving boundaries secreted by the cell. Regions of differentiation, when they occur, are often transitory depending on the physiological state of the organism. They intergrade with the surrounding cytoplasmic regions rather than being distinctly separate (Fig. 2-1).	Specialized regions of cytoplasm occur to varying degrees with stable, clearly identifiable organellar structures; however, these regions merge with one another without permanent compartmentalizing barriers between the zones of specialization. Cellular secretory products may form perforated walls or boundaries partially segregating the zones from one another; however, specialized connecting structures across the boundaries are rudimentary or lacking (Fig. 2-2).	There are clearly identifiable differentiated zones of cytoplasm, often separated by stable boundaries such as compartmentalizing living membranes containing specialized structures mediating connections between the zones (Figs. 2-3, 2-4). Structural differentiation is complemented by physiological specialization within the zones.

[a] Adapted from Anderson (1983).

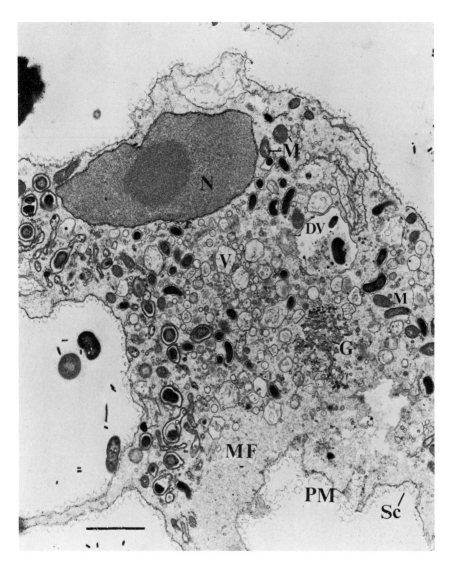

Fig. 2-1. Fine structure of a marine amoeba (*Hartmanella* sp.) exhibiting a diffuse cytoplasmic organization with a peripheral layer of microfilaments (MF) near the plasma membrane (PM) coated with scales (Sc). Scattered vacuoles (V), subcellular organelles including a prominent nucleus (N), mitochondria (M), Golgi bodies (G), and digestive vacuoles (DV) occur in the central cytoplasm. Scale = 2 μm. (Anderson, 1977b)

be stabilized within this general region by cytoskeletal elements; however, their mobility within the cytoplasm and general fluidity of endoplasm accompanying locomotion are well-known characteristics of amoeboid organisms. Given the fluidity of cytoplasmic organization, the lack of major compartmentalized zones and the gradual intergradation of cytoplasmic regions within the cell, these organisms are classified as diffuse organized protozoa.

Protozoa with more clearly demarked cytoplasmic regions, exemplified by the foraminifera and some testate rhizopoda that possess shells or theca partially segregating the more dense central cell cytoplasm from the peripheral cytoplasmic network, are classified as "transitional organized" protozoa. Even though they possess a substantial, usually nonliving barrier separating intrashell cytoplasm from the more frothy or web-like external cytoplasm, they do not have specialized cytoplasmic structures connecting these regions. The intrashell cytoplasm intergrades with the external cytoplasm by a continuous thread of cytoplasm passing through pores and apertures in the shell (e.g., Fig. 2-2). The fluidity of the cytoplasm in these two regions (reminiscent of the amoebae) contributes in part to the transitional quality of the cytoplasmic organization. In planktonic foraminifera, for example (Anderson and Bé, 1979), the intrashell cytoplasm, which is usually dense and compact in healthy, well-fed specimens, may extend out of the aperture and form a bulge or a cytoplasmic layer surrounding the shell. The network of rhizopodia arise from this layer, and no membranous barrier with specialized connections exists between these two regions. Specimens that are poorly nourished or lack substantial intrashell cytoplasm exhibit a frothy or web-like organization of cytoplasm within parts of the shell, and this intergrades imperceptibly with the extrashell cytoplasmic network. The fluidity, moreover, of the intrashell and extrashell cytoplasm contributes to the intermingling of the two regions. The shell bears pores of highly specialized organization (Bé et al., 1980); however, these do not effectively delimit intrashell from extrashell cytoplasm as only fine strands of cytoplasm pass through the micropores in the pore plate; however, the bulk of the intrashell cytoplasm passes outward through the open aperture. Algal symbionts, when present, occur in the intrashell and extrashell cytoplasm. They are sequestered within perialgal vacuoles within the dense regions of cytoplasm, but may be sequestered within a thin cytoplasmic envelope outside the shell in the rhizopodial network. On the whole, however, these modes of symbiont sequestration intergrade depending on the physiological state of the host–symbiont association. Given the presence of a graded cytoplasmic continuity between intrashell and extrashell cytoplasm, these protozoa are grouped in the transitional organized category.

Protozoa with definite differentiated zones bounded by stable complex barriers with specialized connections between the zones are clas-

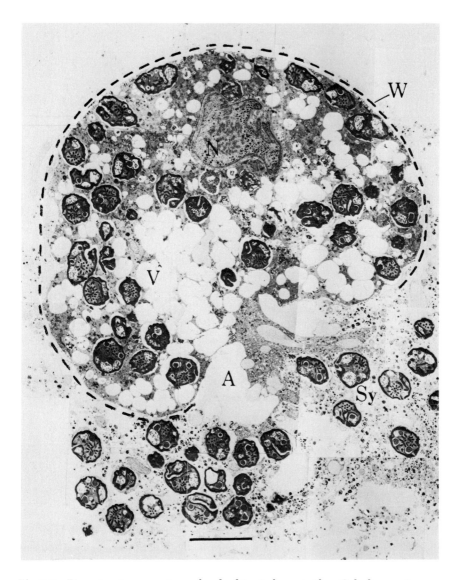

Fig. 2-2. Fine structure montage of a planktonic foraminifera *Orbulina universa* showing the chamber aperture (A) with an emergent mass of cytoplasm containing numerous dinoflagellate symbionts (Sy) that are scattered throughout the intrashell and extrashell cytoplasm. The gradual intergradation between intrashell and extrashell cytoplasm is characteristic of transitional specialized protozoa. The calcareous wall (W) was dissolved prior to sectioning and is indicated by a dashed line. Numerous vacuoles (V), symbionts (Sy), and a nucleus (N) occur within the intrashell cytoplasm. There is a clear gradation between intrashell and extrashell cytoplasm passing through the aperture. Scale = 20 μm. (Adapted from Anderson, 1983)

sified as "zonal specialized" organisms. The Acantharia and radiolaria are included here. The distinct differentiation of intracapsular and extracapsular cytoplasm exhibited by radiolaria, for example (Fig. 2-3), stands in clear contrast to the foregoing diffuse and transitional organized protozoa. The intracapsular cytoplasm may change in organization during ontogeny and exhibit moderate changes over time, depending on the physiological state of the radiolarian; however, it is clear that the intracapsulum is more differentiated than in many amoeboid organisms. Moreover, the capsular wall clearly segregates the intracapsulum from the extracapsulum, but permits continuity by way of the elaborate fusules connected with the capsular wall membranes. Algal symbionts and large digestive vacuoles are exclusively in the extracapsulum, and the nucleus or nuclei, Golgi bodies, and much of the reserve substance are in the intracapsulum. The Acantharia may also exhibit additional examples of differentiation in the extracapsulum where putative contractile elements (myophriscs) in the cytoplasmic sheath as observed in *Acanthometron* may mediate changes in shape, and regulate buoyancy; however, this has not been confirmed experimentally (Febvre, 1971) Elaborate variations in capsular wall organization, and in the arrangement of the intracapsular cytoplasm, are observed among the radiolaria. Some of these variations are presented to illustrate the diversity of specialized forms that have arisen.

Fine Structure of Solitary Radiolaria

The fine structure of some representative, skeletonless species is presented followed by comparative data from skeleton-bearing Spumellaria.

Skeletonless Spumellaria

Thalassicolla nucleata is a large surface-dwelling solitary radiolarian achieving diameters up to 3 mm (Fig. 1-1A). It is fairly typical of the larger skeletonless Spumellaria. Due to its large size, it is difficult to obtain a sufficiently low power electron microscopic view of its cross section; therefore, a light micrograph of a thin section (Fig. 2-4A) is presented as an overview of general cytoplasmic organization. An electron micrograph of an ultrathin section obtained near the periphery of the central capsule (Fig. 2-4B) exhibits corresponding fine structure features (Anderson, 1976a). In Fig. 2-4A, a centrally located nucleus (N) is surrounded by very compactly arranged radial lobes of cytoplasm bearing densely stained inclusions (D) proximal to the nucleus and hyaline vacuoles distributed radially from the perinuclear region through the zone of densely stained bodies and distally to within about 25 μm of the perimeter of the central capsule. This distal zone of cytoplasm

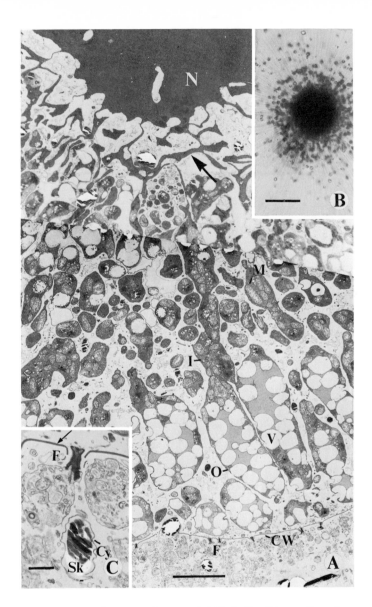

Fig. 2-3. A composite electron micrograph **(A)** of ultrathin sections from a large spumellarian radiolarian. A living specimen is shown in inset **(B)** (marker = 100 μm). The fine structure, characteristic of zonal specialized protozoa, exhibits a thin capsular wall (CW) enclosing the intracapsulum, and connected to the peripheral extracapsulum by thin strands of cytoplasm (fusules = F) that pass from the intracapsulum to the extracapsulum. A large nucleus (N) with lacy peripheral lamellae (arrow) is surrounded by radially arranged lobes of cytoplasm bearing an inner segment (I) rich in cytoplasmic organelles and an outer segment (O) containing electron-lucent vacuoles surrounded by a finely granular substance. Scale = 10 μm. A higher-magnification view **(C)** of the capsular wall exhibits the thin organic wall (arrow), fusule (F) and a skeletal element (Sk) surrounded by the cytokalymma (Cy). Scale = 1 μ. (Based on Anderson, 1976a)

(DZ) is composed of closely spaced lobes separated by fine radial canals. A fairly robust, nonstained capsular wall (CW), about 10–12 μm thick, separates the intracapsulum from the granular extracapsulum (EC), which appears as a densely stained cytoplasmic envelope proximal to the capsular wall and emits strands of rhizopodia (R) radiating toward the periphery. The fine structure view (Fig. 2-4B) displays the details of the cytoplasmic organization near the capsular wall. The narrow cisternae (arrows) separating the lobes of cytoplasm extend outward to the perimeter of the central capsule. Each lobe possesses a densely granular cytoplasm (ground substance) with a meandering network of endoplasmic reticulum (ER) and Golgi bodies (G) occurring generally near the central axis of the lobes. The mitochondria (M), with typical protozoan tubular internal cristae, occur largely at the edges of the lobe in close proximity to the membranes lining the cisternae. This may permit efficient gas exchange between the surrounding cisternal milieu and the mitochondria within the cytoplasm. The mitochondria are known to be respiratory centers requiring oxygen and producing carbon dioxide as a respiratory product. The extensive radial cisternae may serve as conduits connected with the external environment to mediate more efficient exchange processes deep within the cytoplasm of these large cells. The capsular wall (CW) bears numerous canals through which the thin strands of cytoplasm (fusules, F) pass connecting intracapsulum with extracapsulum. There are also fine slits within the capsular wall which segregate the surface of the capsule into closely articulated plates joined in the region of the slits, but apparently not cemented together. Hence, it is possible for fine strands of cytoplasm to pass through and perhaps to permit exchange processes to take place

----

Fig. 2-4. *Thalassicolla nucleata.* **(A)** Thin section light micrograph showing a large centrally located nucleus (N), surrounded by vacuolated cytoplasm arranged as radial lobes (Lb) containing translucent and dense vacuoles (D) which are less abundant in the distal zone (DZ) of the intracapsulum. An unstained capsular wall (CW) is surrounded by the granular extracapsulum (EC) containing rhizopodia (R). Dinoflagellate symbionts (Sy) enclosed within perialgal vacuoles occur in a thin layer of cytoplasm near the capsular wall. Marker = 100 μm. **(B)** An ultrathin section through the central capsule exhibits narrow lobes of cytoplasm separated by cisternae (arrows). Mitochondria (M), Golgi bodies (G), and occasional segments of endoplasmic reticulum (ER) occur in the cytoplasm. A thick capsular wall (CW) bearing fusules (F) and fine slits (S) surrounds the central capsule. Marker = 1 μm. **(C)** An enlarged view of the vacuolated extracapsulum shows a digestive vacuole (V) with partially digested algal prey (Pr) and a nearby benign vacuole, shown in partial view, containing an intact algal symbiont (Sy). Scale = 4 μm. (Anderson, 1976a and 1978b)

between the free space between the lobes and the surrounding seawater. In some cases, it appears that the slits are situated opposite the distal segment of the interlobular cisternae, thus enhancing exchange of dissolved material between the central capsulum and the surrounding environment. The envelope of vacuolated cytoplasm immediately surrounding the capsular wall contains an assortment of large vacuoles (V

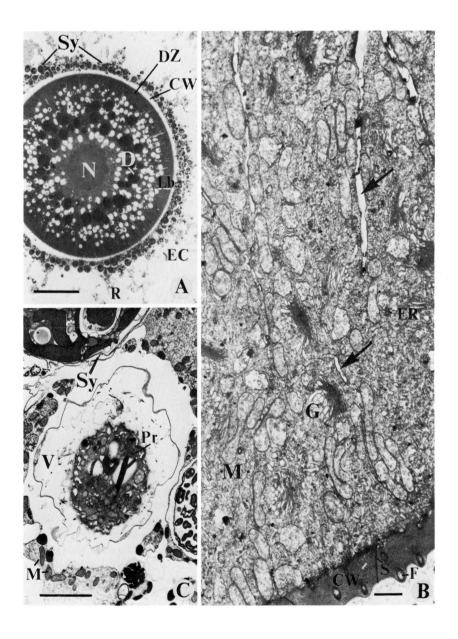

in Fig. 2-4C). Some are apparently digestive vacuoles as they contain partially degraded prey, and others may be temporary food storage vacuoles based on their densely stained lipoidal material. Algal symbionts (Sy in Fig. 2-4C) occur in nearby perialgal vacuoles; however, these apparently are not digestive vacuoles, as no sign of symbiont lysis was observed in these specimens of *T. nucleata*. The thin envelope of cytoplasm between the vacuoles possesses mitochondria, small vesicles, and densely stained pigment granules of unknown physiological function. At a more distal region of the extracapsulum, the strands of rhizopodia and attached thin cytoplasmic envelopes become elaborated to form the bubble-like alveoli (A in Fig. 1-1A). These cytoplasmic structures are surrounded by a very tenuous envelope of cytoplasm about 0.8 to 1.0 μm in thickness. They are probably active metabolic regions, however, as mitochondria are frequently observed within the thick layer of cytoplasm surrounding the alveolar space. The gelatinous envelope encompassing the ectoplasm, as observed by light microscopy, does not accept the electron microscopic stains; therefore, it appears transparent in the electron micrographs. The fine structure evidence clearly substantiates the highly differentiated cytoplasmic structure of the intracapsulum compared to the extracapsulum in this large species.

Skeleton-Bearing Spumellaria

The skeleton-bearing species pose a particular problem in fine structure studies. Species with bulky skeletons are difficult to section, unless the skeleton is dissolved at some point prior to sectioning. However, this makes it difficult to examine the relationship of the skeleton to the cytoplasm. An alternative method is to examine species with fine, spongiose skeletons or those with thin shells.

A light micrograph of a thin section through the central capsule of *Spongodrymus* sp., a spongiose skeletal species, exhibits the general plan of the cytoplasm (Fig. 2-5A). A large nucleus, ca. 50 μm diameter occupies the center of the cell and produces a complex assembly of lace-like protrusions from its surface which connect to the surrounding

---

Fig. 2-5. *Spongodrymus* sp. **(A)** Light microscopic section showing the radially arranged lobes of cytoplasm (Lb) within the central capsule surrounding a prominent nucleus (N) with lace-like extensions projecting into the cytoplasm of the lobes. Each lobe (Lb) is composed of an inner segment (I) and an outer segment (O). The granular extracapsulum (Ec) surrounds the very thin capsular wall. Scale = 20 μm. **(B)** Fine structure of the intracapsular lobes exhibits the vacuolated outer segment (O) and the thinner inner segment (I) rich in mitochondria and other subcellular organelles. The thin capsular wall (CW) is clearly visible. Scale = 6 μm. (Anderson, 1976a)

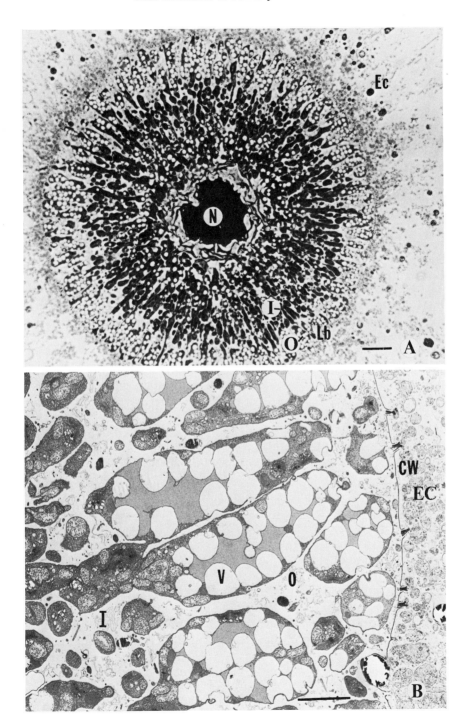

cytoplasm. These membrane-bound lacy lamellae are more clearly presented in Fig. 2-3A, which is a low-magnification electron micrograph. The surrounding membrane exhibits a typical double membrane configuration characteristic of the nuclear envelope. It interdigitates with a thin layer of cytoplasm surrounding the nucleus that gives rise to the radial intracapsular lobes. The intracapsular cytoplasm is organized into radial lobes that interconnect laterally at places (Fig. 2-5A). There are two segments to the lobes: an inner less vacuolated densely stained portion forming the inner segment (I) ca. 50 μm in length, and a more expanded, vacuolated portion near the capsular wall, forming an outer segment (O) ca. 25 μm long. The inner segments are more compactly arranged and exhibit more lateral connections than the outer segments. The capsular wall is very thin in this species and is difficult to detect in light micrographs. The extracapsulum (Ec) forms a halo of granular cytoplasm enclosing the capsular wall. An electron micrograph of the lobes (Fig. 2-5B), corresponding to an enlarged segment of Fig. 2-5A, clearly exhibits the vacuolar organization of the outer segment (O) and the fine structure of the inner segment (I). The vacuoles (V) are surrounded by a finely granular substance of unknown chemical composition. The inner segment is richly supplied with mitochondria and other subcellular organelles. These organelles are presented more clearly in Fig. 2-6. The large mitochondria are surrounded by granular hyaloplasm containing Golgi bodies, endoplasmic reticulum, microtubules (Mt) running parallel to the long axis of the lobe, small vesicles, and occasional single-membrane-bound organelles that appear to be microbodies. The microtubules situated along the main axis of the lobes may provide cytoskeletal support for the lobes. The microtubules also continue outward through the fusules and into the axopodia as shown by light microscopy (Hollande and Enjumet, 1960). There is no specialized structure where the inner segment of the lobe connects to the outer segment; although, there appears to be frequently a slight neck or constriction where the two segments join. However, otherwise, the cytoplasm forms a continuous gradation between the two regions. The outer segment of the radial lobes connects to the cytoplasmic strands of the fusules in the capsular wall membranes.

The capsular wall, ca. 0.06 μm thick, consists of a fine osmiophilic substance secreted within the cisterna between the membranes of the capsular wall (arrow, Fig. 2-3C). The fusules are complex consisting of a perforated collar (Po, Fig. 2-14B) surrounding the fusule strand connecting intracapsulum to extracapsulum. The pores in the fusule collar are round to ovoid with a diameter of approximately 0.04 μm. Microtubules pass through the fusule. In the extracapsulum, the fusules connect to the vacuolated cytoplasm and give rise directly to the radially disposed axopodia (Ax, Fig. 2-7). Digestive vacuoles containing prey in varying stages of digestion occupy the pericapsular region. Segments

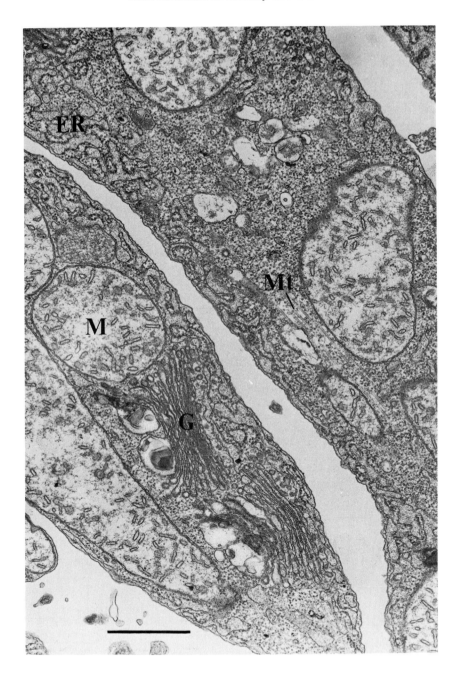

Fig. 2-6. High magnification fine structure image of the inner segment of the intracapsular lobes containing mitochondria (M), Golgi bodies (G), endoplasmic reticulum (ER), and microtubules (Mt) lying parallel to the axis of the lobes. Scale = 1 μm. (Anderson, 1976a)

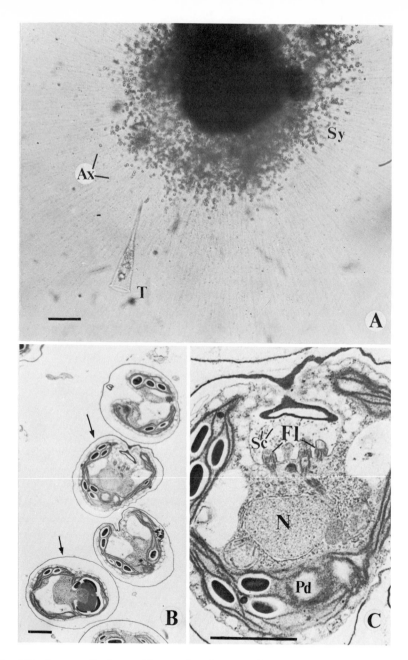

Fig. 2-7. A spumellarian and its prasinophyte symbionts. **(A)** Light microscopic view of a living solitary radiolarian (*Spongodrymus* sp.) showing the corona of radially arranged axopodia (Ax), numerous symbionts (Sy), and a tintinnid prey (T). Scale = 0.2 mm. **(B** and **C)** Fine structure images of sections through the prasinophyte symbionts exhibit their characteristic features including multiple flagella (Fl), coated with scales (Sc), nucleus (N), and peripheral plastids (Pd). Each algal cell is surrounded by a thin organic theca (arrow). Scales = 2 μm. (Anderson, 1976c)

of the skeleton (Sk, Fig. 2-3C) are enclosed by a cytoplasmic sheath, which is generally thickened at the periphery of the skeleton, but is somewhat thin or incomplete on the innermost part of the skeleton embedded within the - intrecapsulum. Segments of the skeleton occur in a central location within the lacunae between the nuclear lamellae, thus confirming the light microscopic observations of Hollande and Enjumet (1960) that the skeleton may be embedded "within the nucleus." The algal symbionts are abundant in the peripheral regions of the rhizopodia (e.g., Fig. 2-7) and are described more fully in the section on radiolarian–symbiont associations.

Nassellaria

The Nassellaria are frequently small species and are best viewed by electron microscopy of ultrathin sections to determine their cytoplasmic microanatomy. A composite electron micrograph of the fine structure of a nassellarian specimen with a simple basal tripod is shown in Fig. 2-8 (Anderson, 1977). The large nucleus (N), 15 μm diameter, enclosing a denser nucleolus lies in proximity to the apex (Ap) of the podoconus, which is shown here to be an assembly of microtubules arranged as a cone. The apex and central zone contain what appears to be microtubule monomers — protein subunits that are assembled in a tightly wound helix to produce each fine tubule. The podoconus rays (Ry) consist of bundles of microtubules that diverge toward the porochora (pore plate bearing the fusules) at the base of the central capsule. These microtubule bundles continue through the fusule strand (F) and into the axopodia at the periphery of the central capsule. They may be homologous to the radially arranged microtubles in the lobes and fusules of the Spumellaria.

The intracapsulum (Fig. 2-8A) contains a densely granular cytoplasm with numerous mitochondria, Golgi bodies, microbodies containing a fine granular matrix and surrounded by a single membrane, endoplasmic reticulum appearing as a fine network of reticulated cisternae and small vesicles (Ve) near the periphery of the intracapsulum next to the capsular wall. The microbodies (1.5 × 1.0 μm) are substantial and are as large as or bigger than the mitochondria. The capsular wall (70 nm thick) contains fine slits that sometimes contain thin strands of cytoplasm and may permit exchange of dissolved substances including discharge of waste products between the intracapsulum and the surrounding environment.

Depending on the size of the specimen and perhaps its age, the central capsule contains more or fewer vacuoles and lipid droplets, which undoubtedly also vary according to the nutritional state of the organism; the more mature and well-nourished specimens would be expected to possess larger amounts of lipid inclusions. The extracapsular cytoplasm, proximal to the capsular wall (Fig. 2-8A), contains lipid droplets

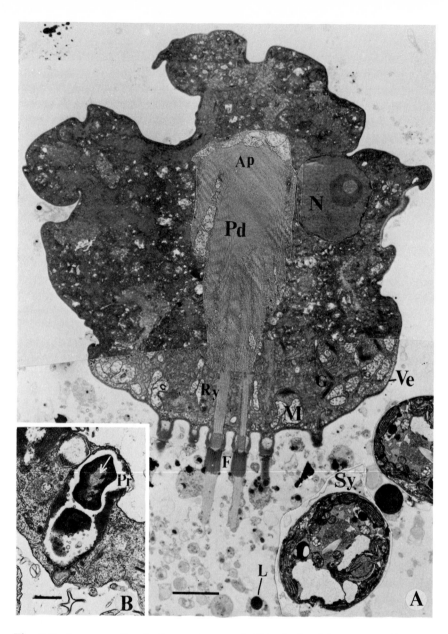

Fig. 2-8. A composite electron micrograph **(A)** of ultrathin sections through the central capsule of a nassellarian exhibits the cone-shaped podoconus (Pd) with ray-like rods of microtubules (Ry) projecting through the fusules (F). The nucleus (N) with a prominent nucleolus lies near the apex (Ap) of the podoconus. Golgi bodies (G), mitochondria (M), and peripheral vesicles (Ve) occur within the intracapsulum. Symbionts (Sy), and lipid reserves (L) occur in the extracapsulum. Marker = 4 μm. Inset **(B)** shows an enlarged view of a rhizopodial digestive vacuole containing a partially digested bacterial prey (Pr) which still contains a characteristic whorl of centrally located DNA (arrow). Scale = 0.4 μm. (Anderson, 1977a)

(L), osmiophilic granules that may be rich in unsaturated lipids, and digestive vacuoles (Fig. 2-8B) containing prey including bacteria. The fusules give rise to the axopodia with an internal shaft of microtubles. There are 12 radially arranged arms, composed of microtubules, attached at the center to a reticulate core of bundles. The roots of the axonemes in the intracapsulum also consist in cross section of 12 radially arranged arms (Fig. 1-16). There are apparently eight divisions of microtubule segments extending clockwise from the arms. This pattern appears to be continuous from the intracapsulum, through the fusule to the axopodia. The fusule fine structure is discussed more thoroughly in the section on "comparative fusule structure."

Phaeodaria

The fine structure of phaeodaria particularly with reference to the microanatomy of the openings in the capsular wall has been studied by Cachon and Cachon (1973b). Neither the main aperture (astropyle) nor the accessory openings (parapylae) exhibit structures resembling axopodia or fusules. Hence, the Phaeodaria are rightly classified as very different from the polycystine radiolaria. The parapylae which classically have been compared to large fusules are quite different. They consist of a stout bundle of microtubules enclosed in a thin cytoplasmic sheath through which they exit the specialized opening in the capsular wall. The flared base of the microtubule bundle rests within a specialized concave depression in the intracapsulum and arises from a massive axoplast cushioned within the depression. The astropyle is in some respects much more complex than the parapylae. It has been considered to be a cytopharynx. At the level of the capsular wall where the astropyle emerges from the intracapsulum, it is shaped as a funnel-like structure with the broad open end directed toward the exterior (Fig. 2-9A). The cytoplasm within the astropyle forms a complex set of folds or pleats lining the inside of the funnel. Microtubules are arranged in parallel bands along the edges of the folds on the intracytoplasmic side of the membrane. The number of microtubules increases considerably in the more distal part of the astropyle where it is more or less prolonged into a proboscis. At this level, the microtubules are very abundant in the central region of the cytoplasm where they occur as a closely spaced bundle in the form of a cylinder. Mitochondria and vesicles are abundant in the cytoplasm surrounding the microtubules and especially in the arms of the folds of cytoplasm (Fig. 2-9D). The main function of this elaborately folded cytoplasmic structure, richly endowed with microtubules, is apparently phagocytosis. Such a specialized cytopharynx has not been observed in either solitary or colonial polycystine radiolaria, which yields further evidence of the highly specialized and possibly advanced phylogenetic status achieved by these radiolaria.

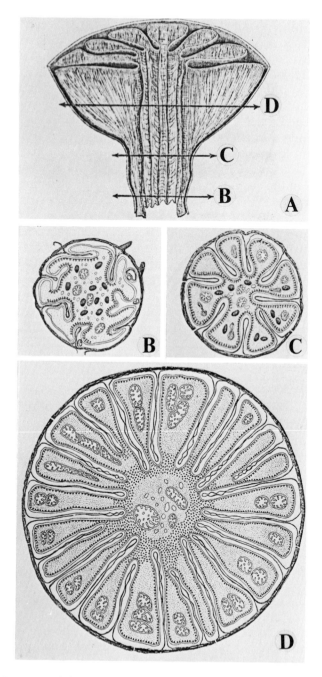

Fig. 2-9. Diagram of the organization of the astropyle in Phaeodaria showing the arrangement of the cytoplasmic "pleated lobes" in a longitudinal view **(A)** and in cross-sectional organization at successive levels **(B–D)** progressing distally from the central capsule outward. (From Cachon and Cachon, 1973b)

Fine Structure of Colonial Radiolaria

The colonial radiolaria are of interest from a fine structure viewpoint, as they are considerably more complex than most solitary spumellarian species. The union of many central capsules within a common rhizopodial network suggests an advanced level of structural development. It is of interest, therefore, to determine what fine structure differences, if any, exist among colonial species compared to solitary species, and to survey the variations in fine structure among colonial species.

Skeletonless Species

Collozoum inerme is an abundant near-surface-dwelling colonial radiolarian consisting of numerous spheroidal to ellipsoidal central capsules interconnected by cytoplasmic strands and bound within a common translucent gelatinous envelope (Fig. 1-5A). Although the colony may be several centimeters in length, the individual central capsules are small (ca. 30 × 60 μm). During some stages of development, they may contain a centrally located oil droplet; however, their cytoplasm shows little detailed differentiation with light optics when viewed in the living state. The fine structure of a central capsule (Fig. 2-10A) exhibits the general plan of cytoplasmic organization (Anderson, 1976b). Mature cells are multinucleate, and most of the nuclei (N) occur in the central part of the cytoplasm. The nuclei contain nucleoli which are often observed at the periphery of the nucleoplasm. The number of nuclei vary, but may be as many as four or more undoubtedly depending on the stage of reproductive maturation. Nuclei divide and proliferate in many colonial radiolaria as they approach reproductive maturity. The cytoplasm is segregated into closely spaced lobes (Figs. 2-10B and 2-11A). Each of the cytoplasmic processes (ca. 9 μm in length) is oriented radially, thus in a tangential section (Fig. 2-10B), the cross sections of the lobes are visible, whereas a more radial section (Fig. 2-11A) exibits the orientation of the lobes around the nuclear region. A network of cisternae (0.1–0.2 μm wide) separates the lobes. The composition of the intracisternal milieu is not known. Three kinds of vacuoles are detected within the intracapsular lobes (Fig. 2-10A). In the near vicinity of the nuclei, there is a layer of electron-lucent vacuoles (T) scattered among the subcellular organelles. Somewhat more distal from the nuclei, slightly larger vacuoles filled with electron-dense granular material (D) are predominant. At the very perimeter of the central capsule, elongate vacuoles (V) form a regular pattern at the tips of the lobes and confer a lace-like pattern to the edge of the central capsule. The chemical composition of these vacuoles has not been determined. Mitochondria, Golgi bodies, and endoplasmic reticulum occur throughout the cytoplasm within the lobes.

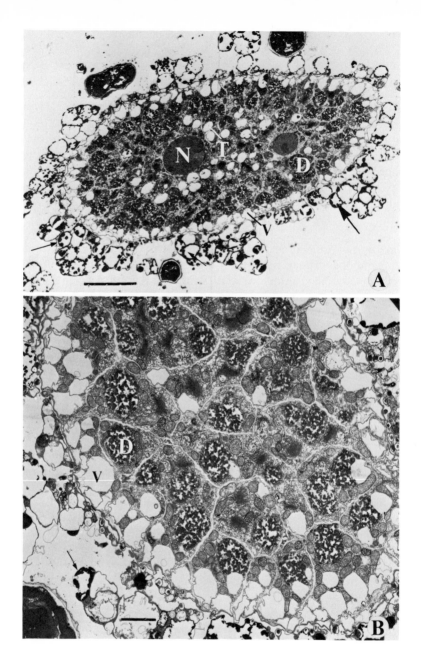

Fig. 2-10. Fine structure of *Collozoum inerme*. **(A)** Central capsule section containing nuclei (N), vacuolated lobate cytoplasm containing translucent vacuoles (T), granular-containing vacuoles (D), and peripheral vacuoles (V). A lacy layer of vacuolated cytoplasm (extracapsulum) surrounds the thin capsular wall envelope and contains electron opaque deposits within the vacuoles (arrow). Scale = 10 μm. **(B)** A tangential section through the central capsule exhibits the organization of the lobes viewed in cross section containing numerous mitochondria and vacuoles with granular deposits (D). Scale = 2 μm. (Anderson, 1976b)

The capsular membranes surrounding the cell enclose a very thin space which is electron lucent. Thus, there is no evidence of a substantial organic capsular wall in this species. The ensemble of membranes surrounding the central capsule are complicated due to the juxtaposition of several membranous elements as exhibited in Fig. 2-14D. The outermost membrane, peripheral unit membrane, corresponds to the plasma membrane (PM) surrounding the cell. The next innermost membrane (IM) represents the inner unit membrane of the capsular wall envelope. Thus the plasma membrane and its internal membrane (PM + IM) constitute the capsular membrane envelope, which is joined to the fusules (F) and the peripheral rhizopodia of the extracapsulum. Immediately adjacent to the double membranes of the capsular wall, there is a cisternal space which is bounded on one side by the innermost membrane (IM) of the capsular wall envelope and on the other side by a plasma membrane delimiting the intracapsular lobe. The innermost membrane (VM) is the vacuolar membrane that delimits the large vacuole (V) at the distal part of the lobe. Consequently there are four cytoplasmic membranes in close juxtaposition delimiting the central capsule.

Immediately adjacent to the central capsular membranes, there is a vacuolated layer of extracapsular cytoplasm (Figs. 2-10 and 2-11) that connects to the rhizopodial network. No Golgi bodies are observed in the extracapulsum; however, the rhizopodia are supplied with mitochondria and a variety of vacuoles. The mitochondria in the rhizopodia (0.5–0.9 μm) appear on the average to be smaller than the intracapsular mitochondria (1–1.5 μm). The rhizopodial vacuoles, however, are larger than the intracapsular vacuoles. Some of the larger rhizopodial vacuoles (ca. 3–4 μm diameter) contain dense osmiophilic deposits on the inner surface of the vacuolar membrane (arrow, Fig. 2-10A), whereas smaller vacuoles contain fragments of digestion products presumably carried from more peripheral larger digestive vacuoles. Small vesicles containing whorls of granular material (W) are also distributed throughout the extracapsulum and may be a source of the gelatinous substance secreted around the colony (Anderson, 1976b). Dinoflagellate symbionts, enclosed within perialgal vacuoles, are also present in the extracapsulum (Fig. 2-11B).

The fine structure of *Collozoum caudatum* (Swanberg and Anderson, 1981) exhibits features characteristic of other *collozoum* sp. The large spheroidal central capsules (100–300 μm diameter) possess multiple nuclei with a finely fibrillar chromatin, surrounded by a thick layer of cytoplasm delimited from nearby, radially arranged lobes by a wide cisterna (Fig. 2-12A). The perinuclear cisternae are interconnected with the narrower cisternae delimiting the radially arranged lobes of cytoplasm. The cytoplasmic lobes containing mitochondria, Golgi, microbodies (Mb) and a network of endoplasmic reticulum (ER) possess distinct

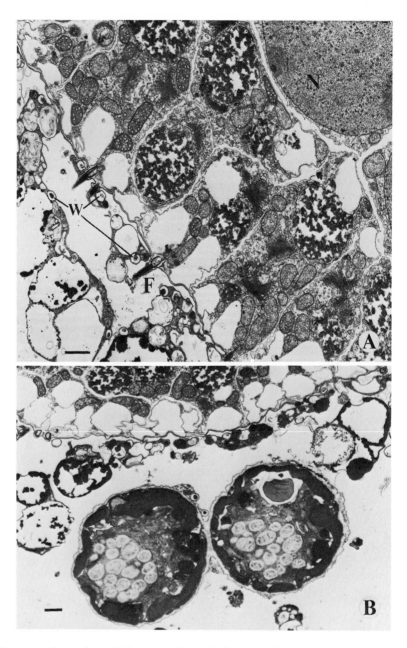

Fig. 2-11. A nearly radial section through the central capsule of *C. inerme* **(A)** shows the arrangement of the lobes near the capsular wall bearing fusules (F) and the vacuolated extracapsular cytoplasm. Small vesicles contain whorls (W) of osmiophilic substance that may be precursors of the gelatinous coat secreted by the extracapsulum. Symbionts in the extracapsular cytoplasm proliferate by fission as exhibited in **(B)**, showing two recently separated daughter cells still enclosed in a continuous host cytoplasmic sheath. Scales = 1 μm. (Anderson, 1976b)

vacuoles at their distal ends near the capsular membranous envelope. These vacuoles enclose osmiophilic granules that may be the purple-pigmented inclusions observed near the capsular wall with light optics. These vacuoles are not as regularly arranged as those in C. inerme. Electron-lucent vacuoles of varying dimension (ca. 3–5 μm diameter) are dispersed through the cytoplasm in the lobes.

The capsular wall membranes appear very much as those in C. inerme, and the surrounding layer of extracapsular cytoplasm is vacuolated and supplied with osmiophilic granules. Some of the granular dense bodies in the extracapsulum, however, do not exhibit the bead-like whorls observed with C. inerme. The general cytoplasmic ground plan of radial lobes and peripheral vacuoles may be characteristic of many *Collozoum* sp., and should be investigated more thoroughly among a wide variety of species in this genus.

Spicule-Bearing Species

Sphaerozoum punctatum (Figs. 1-5C, D) is a commonly observed spiculate colonial radiolarian bearing symmetrical, triradiate spines. The distinctly spherical central capsules (120–150 μm diameter) are clearly larger than those of C. inerme. A light micrograph of a central capsule (Fig. 1-5C) displays the large lipid inclusion at the center of the cell surrounded by the vacuolated cytoplasm, containing lipid droplets and cytoplasmic organelles too fine to be resolved by light optics. Zooxanthellae (Sy) are clustered close to the capsular wall. The fine structure of the cytoplasm (Anderson, 1976c) exhibits the major features of a mature, nonreproductive cell (Figs. 2-12B, C). The cytoplasm is not organized into distinct lobes as was observed with C. inerme. The peripheral cytoplasm (ca. 30 μm thick), surrounding the central lipid inclusion (ca. 60 μm diameter), contains numerous vacuoles (largely electron lucent) interspersed with the nuclei. Some of the vacuoles contain dense, osmiophilic granules, others contain dispersed organic matter. A layer of irregular vacuoles (Vc, Figs. 2-12B, C) occurs at the periphery of the central capsule reminiscent of those observed in C. inerme, but not so regularly arranged. Mitochondria, Golgi bodies, and microbodies are abundant in the peripheral region of the central capsule near the layer of vacuoles (Vc) and interspersed among the other cytoplasmic organelles more centrally located in the cytoplasm. A meandering network of narrow cisternae ramifies throughout the intracapsulum and segregates it into closely interrelated masses of cytoplasm. At some places, however, the cisternae are dilated (asterisk, Fig. 2-12B) and are clearly discernible.

The nuclei possess prominent cord-like masses of chromatin (Fig. 2-12B), giving them a distinctly mottled appearance. There are multiple nuclei within the cells and some are surrounded by a thin layer of cytoplasm delimited by the canal-like cisternae. Similar nuclei with

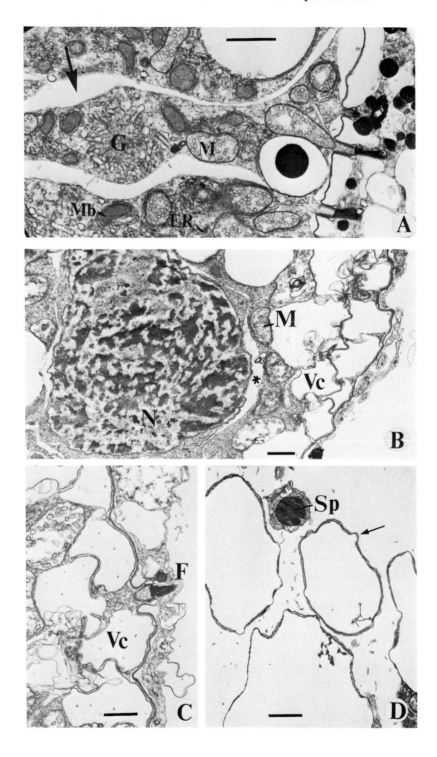

cord-like chromatin have been reported in *Sphaerozoum neapolitanum* (Hollande, 1974). In *S. punctatum*, the capsular wall membranes form an envelope ca. 60 nm thick (Fig. 2-12C). The total ensemble of peripheral membranes is organized very much as in *C. inerme* with four layers of membrane. The fusules (F, Fig. 2-12C) within the capsular membranes connect with a thin layer of vacuolated cytoplasm surrounding the central capsule and give rise to the rhizopodia that ramify throughout the gelatin. The alveoli (Fig. 2-12D) delimited by a thin layer of cytoplasm (29 to 80 nm thick) are attached to the rhizopodial network. The thin cytoplasmic envelopes surrounding the alveoli often contain mitochondria which protrude prominently on the surface of the cytoplasmic partition surrounding the alveoli. The siliceous spicules (Sp, Fig. 2-12D) are enclosed within a cytoplasmic sheath (cytokalymma, Anderson, 1981) that secretes the silica and may serve as a template or mold to determine the shape of the spicules.

Shell-Bearing Species

Collosphaera globularis possesses a spherical, porous shell surrounding the central capsule (Fig. 1-5F). It is abundant in the Sargasso Sea especially during midsummer (June–August), and is a convenient representative of the collosphaerid colonial radiolaria illustrating the fine structure features of shell-bearing species. The central capsule (ca. 300 μm diameter) contains a large central lipid inclusion (L, Fig. 2-13A) approximately 30 μm diameter, surrounded by a multinucleated cytoplasmic layer (ca. 18 μm thick). The nuclei (N), 6 μm. diameter, possess cord-like masses of chromatin similar to that in *S. punctatum*. The strands of chromatin appear more densely aggregated near one side

---

Fig. 2-12. Fine structure of the peripheral cytoplasm in the central capsules and extracapsulum of *Collozoum caudatum* **(A)** and *Sphaerozoum punctatum* **(B–D)**. **(A)** The intracapsulum of *C. caudatum* contains radially arranged lobes bearing peripheral vacuoles with an osmiophilic granule (possibly a pigment granule). Mitochondria (M), Golgi bodies (G), microbodies (Mb), and endoplasmic reticulum (ER) occur within the cytoplasm. Radially arranged cisternae separate the lobes (arrow). Scale = 2 μm. (Swanberg and Anderson, 1981) **(B)** A nucleus (N) with cord-like masses of chromatin enclosed within a thin cytoplasmic layer bounded by a cisterna (asterisk). Nearby peripheral vacuoles (Vc) occur next to the capsular wall membranes in *S. punctatum*. Scale = 1 μm. **(C)** A more detailed view of the peripheral vacuoles in *S. punctatum*. Scale = 1.3 μm. **(D)** Alveoli of the extracapsulum are surrounded by an envelope of cytoplasm that is extremely thin at points (arrows). Spicules (Sp) surrounded by a cytokalymma occur commonly in the interstices between the alveoli. Scale = 1 μm. (Anderson, 1976c)

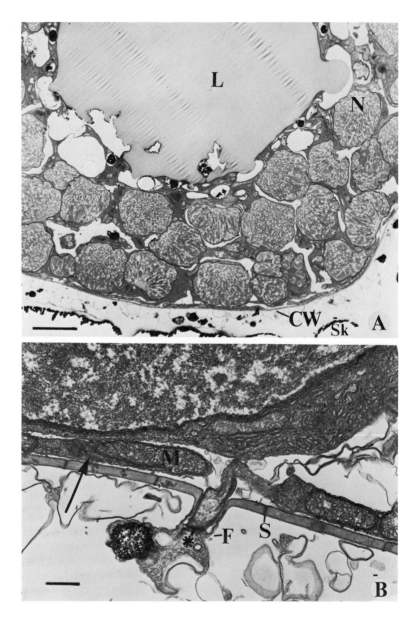

Fig. 2-13. *Collosphaera globularis.* **(A)** Fine structure of a portion of a central capsule with a large lipid droplet (L), surrounded by a layer of multinucleated cytoplasm (N) bounded by a thin capsular wall (CW). Fragments of the siliceous shell (Sk) occur at the periphery. Scale = 10 μm. **(B)** A detailed view of the capsular wall with a fusule (F) emanating from a thin layer of intracapsular cytoplasm (arrow) bearing mitochondria (M). Fine slits (S) occur in the organic wall. A thin osmiophilic layer (asterisk) is present at the juncture of the intra-capsulum with the extracapsulum within the fusule (F). Scale = 0.5 μm. (Anderson, 1978a)

of the nucleus, where a thickened "plaque" on the inner membrane serves as an attachment point (Fig. 2-13A). Each nucleus is surrounded by a thin layer of cytoplasm delimited by the broad cisternae that ramify throughout the cytoplasm and segregate it into irregularly shaped lobes interconncted by cytoplasmic bridges. At the perimeter of the central capsule, the cisternae delimit a thin layer of cytoplasm (3 μm thick) devoid of nuclei, but richly supplied with mitochondria (arrow, Fig. 2-13B). No large vacuoles are present at the periphery of the central capsule as observed in C. inerme and S. punctatum, and on the whole, the cytoplasm is remarkably compact. It appears granular and electron dense, owing in part to the abundant endoplasmic reticulum and densely granular hyaloplasm surrounding the nuclei. At the periphery of the central capsule, the thin cytoplasmic layer connects to the fusules (strands of cytoplasm) passing through pores (Fig. 2-13B) in the distinct capsular wall (ca. 140 nm thick). The fusule consists of an elevated cylindrical rim formed by the outgrowth of the capsular wall (F, Fig. 2-13B) surrounding the cytoplasmic strand. A dense, osmiophilic partition (asterisk) is situated at the same level as the attachment site of the capsular wall membranes to the cytoplasmic strand of the fusule. The capsular wall is segmented by fine slits (S, Fig. 2-13B) reminiscent of those observed in Thalassicolla nucleata.

A highly vacuolated layer of cytoplasm bearing opaque osmiophilic granules arises from the fusules and loosely encloses the central capsule. The siliceous skeleton (Sk, Fig. 2-13A), which is partially fractured during sectioning, is enclosed within a thin cytoplasmic sheath produced by the extracapsular strands of cytoplasm.

Fusule and Capsular Wall Fine Structure

The fusules, and capsular wall to which they are attached, are unique structures characteristic of the radiolaria and Acantharia. Their fine structures, therefore, merits special attention. The diversity of fusule structures among solitary and colonial radiolarian species is of interest for a number of reasons including possible taxonomic value in providing additional evidence to delineate taxa, morphological significance in elucidating the microanatomical relationships between cytoplasmic elements and capsular wall, and physiological interest in providing data about the functional relationship between intracapsulum and extracapsulum. The fusules are a major organic link between extracapsulum and intracapaulsum. In comparison to the fine capsular wall slits, the fusules may be the single most significant structures providing physiological and structural continuity between the two major compartments of the radiolarian cell. It is difficult to estimate the total volume of cytoplasm passing through the slits based on fine structure evidence.

However, if all of these channels were filled with cytoplasmic connections, they could contribute significantly to the total communication between intracapsulum and extracapsulum. In the Spumellaria, however, the sheer abundance of fusules in the sieve-like surface of the central capsular wall must account for a significant if not predominate part of the cytoplasmic exchange between the two regions. In the Nassellaria and Phaeodaria, the "fusule pore fields" are more localized, but nonetheless are large and are undoubtedly the major cytoplasmic conduit across the capsular wall.

Some comparative fine structure data are presented as a means of highlighting the significant structural features characteristic of major radiolarian groups. Among the Spumellaria, as exemplified by the solitary species in the genus *Spongodrymus*, the central capsule is relatively large with a diameter of ca. 200 μm; however, each of the numerous fusules, which must be of the order of hundreds or thousands, has a diameter of ca. 750 nm. In *Thalassicolla nucleata* wih a central capsule diameter of about 400 μm, the fusules are likewise very numerous and have a diameter of about 350 nm.

In a typical small cyrtoidean, nassellarian the central capsule diameter may be as large as 80 μm (Fig. 2-8A). The fusules limited to the basal pore field can have diameters as large as 1.3–1.5 μm at their broadest part. This is about two to three times the diameter of fusules in large spumellarian species. Among the shell-bearing colonial spumellarian species, the central capsules with a diameter of ca. 150–300 μm possess fusules with a diameter of ca. 500 nm as in *Collosphaera globularis*. The fusules of the spicule-bearing colonial radiolarian, *Sphaerozoum punctatum* have a diameter of approximately 400 to 420 nm and the central capsule diameter is 150 μm. Similarly the skeletonless colonial species *Collozoum inerme*, whose central capsules are small (30 × 60 μm), have fusules with a diameter of 380–400 nm. *Collozoum caudatum* (central capsule diameter: 100–300 μm) possesses fusules with a diameter of 450–500 nm. In comparison to some of the larger solitary Spumellaria whose central capsules are densely punctuated with fusules, the colonial species appear to have more widely spaced fusules.

Some comparative high magnification electron micrographs of fusules in some solitary and colonial radiolaria are presented in Fig. 2-14. In all cases observed thus far, the nonliving organic wall when present is enclosed within a membranous envelope in close contiguity with the wall. It is sometimes difficult to detect the membrane on the surface of the wall particularly in species with thick capsular walls (e.g., *Thalassicolla nucleata*); however, the membranes are clearly evident in the region of the fusule where they fuse with the cytoplasmic membrane surrounding the fusule strand. The architecture of the fusule in *Thalassicolla nucleata* is dominated by the thick capsular wall through which it passes. The wall (ca. 1–1.5 μm thick) is perforated by cylindrical pores (Fig. 2-14A) with an elevated conical rim on the outer side. The fine strands of cytoplasm penetrate the pores and emerge on the distal side where they are continuous with the extracapsular cytoplasm. The enclosing membranes of the capsular wall fuse with the plasma membrane of the fusule immediately above the rim of the conical pore (arrow). The considerable thickness of the capsular wall imposes a rather slender and elongate configuration to the fusule strand. In contrast, the delicate capsular wall of *Spongodrymus* sp. composed of a very thin deposit within the cisternae of the living membranes may be described best as a capsular membranous envelope. It possesses very remarkable fusules of elaborate design. The capsular wall forms an outward directed, flared, collar whose walls are perforated by micropores (Po, Fig. 2-14B). The fusule strand passes through the collar-like pore and may have fine lateral cytoplasmic threads connected to the walls of the pore, particularly near its proximal rim. In other cases, the distal rim of the fusule pore (collar) may be partially fused with the membrane surrounding the fusule strand. On the whole, however, it is clear that there is a continuous free space linking the extracapsular milieu with the space surrounding the lobes in the intracapsulum by way of the channel extending through the collar-like pore. This system may permit rapid exchange of substances between the intracapsulum and the surrounding seawater in contact with the extracapsulum.

Given the large size of the cell and the presence of numerous mitochondria in the inner segment of the lobes, there could be a severe deficiency in oxygen availability and waste product release, unless special adaptive structures existed to mediate efficient exchange processes. The widely spaced, radial lobes provide substantial free space for exchange processes between the lobes and the surrounding fluid. The open collar-like pores of the fusules, moreover, permit continuity of the interlobular space with the external milieu surrounding the extracapsulum, thus further enhancing exchange with the external environment. In contrast, *Thalassicolla* sp., with a thick capsular wall and narrow cisternae between the densely spaced intracapsular lobes, distributes its mitochondria most densely near the perimeter (outer layer) of the intracapsular cytoplasm and most particularly in close contact to the plasma membrane next to the cisternae. This position of the mitochondria can enhance gas exchange in what otherwise would be a rather closed cytoplasmic environment. Thus, this large cell has evolved an architecture that permits maximum development of intracapsular cytoplasmic volume while also maximizing exchange processes in the vicinity of the thick capsular wall. The *Spongodrymus* sp. in contrast possess widely spaced, radial lobes with substantial free space for diffusion between the lobes. This may be an adaptive mechanism to permit efficient exchange between the mitochondria-rich inner

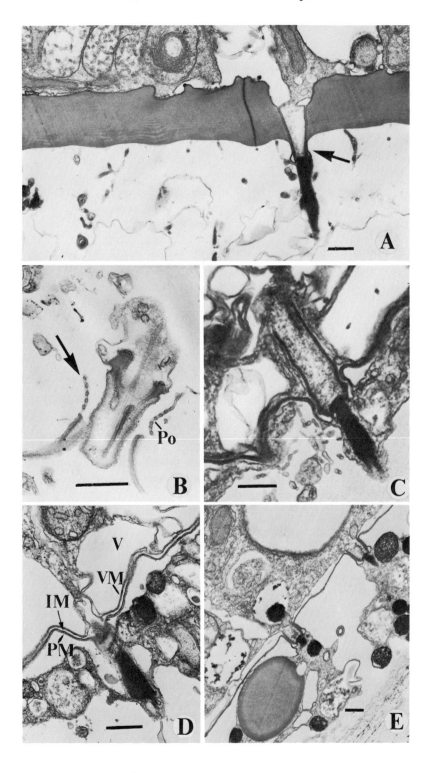

segment of the lobes and the surrounding milieu. The fusules complement this structural differentiation by providing a large diffusion space in the collar-like opening of the fusule, thus facilitating exchange between intracapsular and extracapsular free space. It is not known why the *Spongodrymus* sp. has evolved bipartite lobes possessing a large storage space in the outer segment (evidenced by numerous vacuoles and granular deposits) and a respiratory center in the inner segment, rich in mitochondria. This is in contrast to *Thalassicolla*, which has large vacuoles near the center of the intracapsulum and a mitochondrial-rich cytoplasm at the periphery. Nonetheless, the capsular wall and fusule structure in *Spongodrymus* sp., providing enhanced free space for diffusion, complement the structural and functional specialization of the intracapsulum. Additional research is needed to determine the generality of this fusule structure among a wider variety of large spongiose skeletal species, and particularly among species of *Spongodrymus*.

Some representative fusules of colonial spumellarian radiolaria (Fig. 2-14) exhibit certain features in common. The fusule projects outward from the surface of the capsular wall to varying degrees among the illustrated species. In *Collozoum caudatum* and *Collosphaera huxleyi*, the outward prolongation is more marked than that in *Collozoum inerme* or in *Sphaerozoum punctatum*. In *Collozoum* sp. and in *Sphaerozoum punctatum*, the distal portion of the fusule, lying immediately outside of the thin osmiophilic disc at the juncture of the capsular wall, possesses an osmiophilic conical zone. It is considerably longer in *Collozoum* sp. than in *Sphaerozoum punctatum*. In all of the cases, it projects into the peripheral envelope of extracapsular cytoplasm. There is some evidence in *Collozoum caudatum* (Swanberg and Anderson, 1981) that this densely staining cone may be permeable to small vesicles passing through the fusule. The reason for its osmiophilic properties, however, is not immediately clear. In some preparations, it presents a finely granular or fibrillar composition and occasionally exhibits clear evidence of microtubules oriented parallel to the long axis of the cone

Fig. 2-14. Comparative fine structure of fusules in *Thalassicolla nucleata* **(A)**, *Spongodrymus* sp. **(B)** with a perforated outwardly directed collar (arrow) containing pores (Po), *Sphaerozoum punctatum* **(C)**, Collozoum inerme **(D)**, and Collozoum caudatum **(E)**. The complex membranous system delimiting the capsular wall in C. inerme consists of an outer membrane (PM) equivalent to the cell plasma membrane, a cisternal space delimited by membranes (IM) and the peripheral intracapsular vacuoles (V), surrounded by a vacuolar membrane (VM). Scales = 0.5 μm. (**D** from Anderson, 1976a; **E** from Swanberg and Anderson, 1981)

(Anderson, 1976b). These dense cones were not observed so prom-
imnently in the fusules of *Collosphaera globularis*. Hypothetically,
these osmiophilic zones, occurring at the juncture of the fusule with
the capsular membranes, may serve as cytoplasmic gates controlling
the permeability of the fusules and thus regulating communication
between the intracapsulum and extracapsulum. A similar thin osmiophilic partition is observed in solitary radiolaria (Hollande, *et al.*, 1970;
Anderson, 1976a, 1977a). In the Nassellaria, it becomes particularly
expanded as an osmiophilic dense rod projecting toward the distal side
(Fig. 2-15).

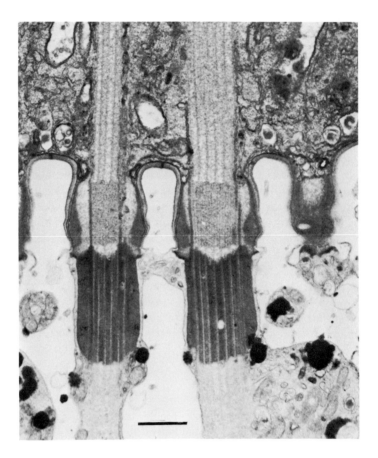

Fig. 2-15. Nassellarian fusule fine structure exhibits the outwardly directed,
collar-like pores of the capsular wall containing cytoplasmic strands with shafts
of microtubules extending from the intracapsulum through the fusules into the
extracapsulum. Scale = 1 μm. (Anderson, 1977a)

The general organizational plan of some nassellarian fusules (Figs. 2-8 and 2-15) is remarkable for its complexity. The capsular wall surrounding the fusule projects outward as a slightly expanded collar-like sheath. The fine membranes enclosing the capsular wall fuse with the fusule membrane at the rim of the collar-like opening. A thin envelope of extracapsular cytoplasm may also bridge between closely spaced fusules at the level of the rim of the capsular wall collar. The proximal part of the fusule enclosed within the capsular wall collar exhibits several zones of organization. The most peripheral zone next to the capsular wall collar is an osmiophilic, finely granular or fibrillar layer of cytoplasm encircling the fusule. It is thin at the proximal end near the base of the capsular wall collar and somewhat thickened at the distal end near the outer rim of the collar. A dispersed layer of fine fibrillar matter lies immediately within the dense layer. The chemical composition of these fibrils is unknown. They exhibit an arachnoidal or spongy fine structure. At the center of the fusule and occupying most of its volume, a shaft of microtubules, extending from the rays of the podoconus, passes through the fusule. A finely granular matrix is interspersed among the microtubules within the collar-like pore of the capsular wall. On the distal side of the fusule, the microtubules continue into the extracapsular cytoplasm and pass through a densely osmiophilic plug resembling the conical projections observed in spumellarian species; however, here it is much more cylindrical and blunt at its distal end. The microtubules within this part of the axopodium appear more stable than those extending outward in the distal axoneme. They are not disassembled when the axopodium is withdrawn as sometimes occurs during fixation. No mitochondria are observed within these large fusules, although occasional electron-lucent vesicles are observed in between the parallel microtubules. Mitochondria are frequently observed, however, in the thin layer of cytoplasm immediately surrounding the axopodium at the tip of the dense osmiophilic plug within the fusule.

The fine structure of the fusules raises some interesting questions about communication between the intracapsulum and the extracapsulum. In general, the diameter of the fusules, particularly in many Spumellaria, is too small to permit passage of normal-sized mitochondria, most vacuoles, and other major membrane-bound organelles. To what extent, therefore, is there exchange of organelles between the intracapsulum and extracapsulum? With the exception of some small vesicles, no organelles have been observed within fusules. Thus, we know little about their passage between the two zones. It is possible that promitochondria (very small mitochondrial precursors) may pass through the fusules. Certainly, small primary lysosomes (vesicles containing digestive enzymes) may be able to pass the fusule. This possibility is discussed more fully in Chapter 3 when digestive processes are described. The general complexity of the fusule structure is additional evidence of the high level of cellular specialization achieved by these protista. The physiology of fusule function is an intriguing area of research that is rarely explored.

Host-Symbiont Fine Structure

Light microscopy clearly exhibits numerous algal symbionts associated with the extracapsulum of many radiolarian species, particularly among the Spumellaria and Nassellaria (Figs. 1-2, 2-7, 2-8, and 2-16). It is not possible to resolve the fine details of the host–algal association or to observe the cytoplasmic structure of the symbionts using light microscopy. Fine structure evidence (Taylor, 1974; Anderson, 1976a–c, 1977, 1978a,b) shows that many of the radiolaria possess dinoflagellate symbionts as observed, for example, in *Thalassicolla nucleata* (Fig. 1-1A), *Collosphaera globularis*, and Nassellaria (Fig. 2-8A).

Based on light microscopic and fine structure characteristics, Taylor (1974) identified the symbionts isolated from *Collozoum inerme* collected from the bay of Naples and those from *Sphaerozoum punctatum* as *Amphidinium* sp. The genus *Amphidinium* is characterized by the presence of thylakoids within the pyrenoid(s), and plastid lamellae with only two rather than three thylakoids as observed in other dinoflagellates (Dodge, 1973). *Amphidinium* sp. resembles *Amphidinium chattonii*, but differs only in the possession of a single large pyrenoid. The symbiont associated with *Collozoum inerme* collected in the Sargasso Sea (Anderson, 1976b) exhibits similar fine structure as *Amphidinium* sp. Symbionts observed in *Collozoum caudatum* (Swanberg and Anderson, 1981) possess at least two pyrenoids with included thylakoids, which suggests among other characteristics that they are different from those in *Collozoum inerme* and may be more closely related to *Amphidinium chattonii*. The large pelagic, solitary radiolarian, *Thalassicolla nucleata*, bears symbionts with fine structure characteristics of the genus, *Amphidinium*, at least among specimens collected in the Sargasso Sea (Anderson, 1976a, 1978).

At present, it is wise to be cautious in generalizing about the species of algal symbionts associated with particular genera of radiolaria, as we do not know how variable their associations are nor how significant varying geographical locations with different water masses, and possibly differences in abundance of putative symbiotic algae may be in determining the species of alga associated with the host. We know very little about how radiolaria become infected with algal symbionts nor how specific the host–algal association is for varying radiolarian species. In general, the individual specimens examined thus far possess only

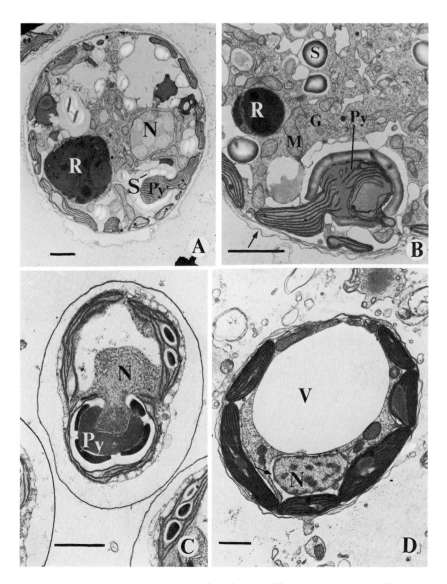

Fig. 2-16. Algal symbionts associated with spumellaria. **(A** and **B)** Dinoflagellate symbiont fine structure exhibiting the nucleus (N), reserve substance (R), starch (S), pyrenoid (Py), mitochondria (M), and Golgi bodies (G). An assembly of peripheral membranes produced by the host (arrow) encloses the symbiont plasma membrane and underlying thecal vesicles that contain the thecal wall in free-living forms. Scales = 2 μm. **(C)** Prasinophyte symbiont with centrally located nucleus (N) containing an invagination in the pyrenoid (Py). Scale = 2 μm. **(D)** A brown-pigmented symbiont associated with *Spongodrymus* sp. showing the large vacuole (V), nucleus (N), and nuclear membrane cisterna (arrow) enclosing the peripheral chloroplasts. Scale = 1 μm. **(D,** Anderson and Swanberg, work in progress)

one kind of algal symbiont. Whether this is due to host–algal specificity or superior competitiveness of one algal species over another during early stages of infection remains to be determined.

In both solitary and colonial species, the dinoflagellate symbionts are enclosed in a thin perialgal envelope of cytoplasm produced by the host rhizopodial system. It is generally very thin (ca. 600 Å at the thinnest part), particularly in the peripheral rhizopodia, and may be considerably thicker (600 nm or more) in regions near the central capsule. The thinnest envelopes are comparable to those observed in other rhizopod-bearing protista (eg., planktonic foraminifera) (Anderson and Bé, 1976). However, the symbionts of radiolaria are rarely enclosed within massive cytoplasmic sheaths as occurs near the shell aperture of some planktonic foraminifera.

The symbionts exhibit typical dinoflagellate fine structure including a large mesocaryotic nucleus (N, Fig. 2-16A) with characteristic puffy, coiled chromosomes (Dodge, 1973). The surrounding cytoplasm contains mitochondria with tubular cristae, Golgi bodies, endoplasmic reticulum, vacuoles of varying size and internal composition, and large, osmiophilic reserve bodies. The plastids (Pd) occur abundantly near the periphery of the cell and may have lobes extending inward toward the center of the cell. The structure of the thylakoids (pigment-containing membranes within the plastid) is shown in Fig. 2-16B.

Prominent pyrenoids (Py) are connected to the plastids and contain internal thylakoids continuous with those in the lobes of the plastid. A starch sheath of varying thickness surrounds the pyrenoid. It is in part an indicator of the vigor and productivity of the symbiont. The starch is usually abundant when the symbiont is highly productive.

The periphery of the symbiont is enclosed by a fine layer of thecal vesicles (arrow, Fig. 2-16B). These membranous, flattened saccules are the site where thecal plates are secreted in the free-living form. In symbiotic stages, however, the wall of thecal plates is absent and only a fine thecal, organic membrane remains. This very thin lamella appears as a fine line in high-magnification views of the thecal vesicles. Dinoflagellate cells lacking thecal plates and appearing somewhat rounded are typical of symbiotic stages observed in a variety of invertebrate hosts including protozoa and coelenterates (Taylor, 1974). They are called coccoid forms. The host perialgal cytoplasmic envelope may be closely appressed to the periphery of the symbiont, thus making it difficult to distinguish host membranes from those of the symbiont. At other places, the host cytoplasmic envelope is more loosely associated with the symbiont, forming an undulating sheath, and it can be clearly distinguished. Mitochondria and small vesicles are occasionally observed in the thicker portions of the perialgal envelope. This very thin cytoplasmic envelope is undoubtedly quite translucent, thus permitting penetration of light to sustain symbiont photosynthesis. It also provides host control over the symbiont position and perhaps the availability of nutrients to the alga. Light microscopic observations show that the symbionts are moved about by rhizopodial streaming. A regularly observed diel movement of the symbionts occurs in the host extracapsulum. At the onset of daylight, the symbionts are distributed in the peripheral rhizopodia; however, at sunset, the symbionts are withdrawn near the capsular wall (Fig. 2-17), thus further substantiating the close structural and functional relationship between symbiont and host.

In the spongiose skeletal species (e.g., *Spongodrymus* sp.), there are at least two types of symbionts: dinoflagellates resembling those described in the foregoing species and prasinophytes. The prasinophyte symbionts (Fig. 2-16C) are only loosely attached to the host axopdia as observed in electron microscopic ultrathin sections. Nonetheless, they are under close control by the host, as they also are dispersed outward at daybreak and collected inward at night as observed with dinoflagellate symbionts. During the day at least, they are enmeshed among the axopodia and make surface contact with fine rhizopodial strands among the axopodia. There is little evidence of their enclosure within cytoplasmic sheaths. The fine structure of the symbionts exhibits typical prasinophyte features (Manton, et al., 1963; Manton and Parke, 1965; Dodge, 1973). The nucleus (N, Fig. 2-16C) with finely fibrillar chromatin invaginates into the pyrenoid (Py), which is prominent and electron dense. It bears a distinct starch cap. The plastids at the periphery of the cell contain dense osmiophilic inclusions typical of many prasinophytes. Multiple flagella (Fl, Fig. 2-7C), reduced in symbionts to mere stubs bearing scales on their outer membrane, emerge from the cytoplasm near the nucleus. The nuclear invagination of the pyrenoid, multiple, scale-bearing flagella, and the general cytoplasmic features of these symbionts are characteristic of prasinophytes. These symbionts are similar to prasinophytes observed in the turbellarian (flatworm) *Convoluta roscoffensis* by Parke and Manton (1967) and those observed to be associated with the extracapsular cytoplasm of the large, skeletonless radiolarian *Thalassolampe margarodes* (Cachon and Caram, 1979).

A distinct osmiophilic wall surrounds the radiolarian prasinophyte symbionts (Fig. 2-7B,C). It invaginates in the cavity where the flagella emerge; however, since the flagella have been reduced to mere stubs (flagellar bases), it is not possible to determine the spatial relationship between the thecal cavity and the emergent flagella. The function of the very thin theca in this symbiotic association is not known. It is, however, very similar to the theca of *Platymonas convolutae* (Parke and Manton, 1967), but the latter bears distinctive scales on its surface. The pyrenoid structure, moreover, is less osmiophilic in *Platymonas convolutae*, and the plastids in Parke and Manton's preparation appear to have more closely spaced thylakoids. The symbiont enclosed within the tissue of the turbellarian lacks a theca which is in contrast to the

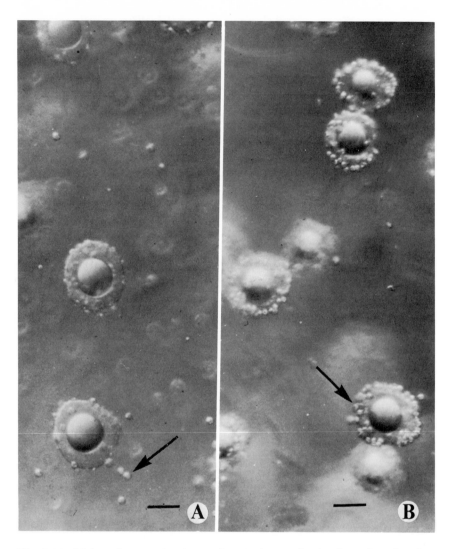

Fig. 2-17. Light microscopic views of central capsules in the colonial radiolarian *Collozoum* sp. During the day **(A)** the symbionts (arrows) are dispersed in the rhizopodial network and at night **(B)** many of the symbionts are gathered close to the central capsules. Scale = 100 μm.

radiolarian prasinophyte. The pyrenoid structure in the radiolarian prasinophyte is also different from that in *Prasinocladus marinus* (Parke and Manton, 1965), which is somewhat less osmiophilic and is invaginated by a slender and lobate intrusion of the nucleus.

The plastid fine structure is also different exhibiting more loosely arranged, undulating thylakoids than in the radiolarian prasinophyte.

On the whole, its features are closer to those of *Platymonas* sp. (Manton and Parke, 1965); however, additional research on isolated and cultured cells of the symbiont will be required to make a more definitive identification. The sheer number of these decidedly yellow-green cells (Fig. 2-7A) among the axopodia as observed by light microscopy (Anderson, 1976a) is impressive and suggests a very stable and productive association with the host. More recently, a brown-pigmented alga of unknown taxonomic position (Fig. 2-16D) has been found associated with a species of *Spongodrymus* (Anderson and Swanberg, work in progress). This large alga (ca. 10–20 μm diameter) possesses a massive central vacuole surrounded by a thin layer of cytoplasm containing the nucleus, whose double membranous envelope (arrow) encloses the plastids. This arrangement of the plastids is reminiscent of the nuclear–plastid organization in some Haptophyceae (Dodge, 1973). The symbionts vary considerably in diameter and with the light microscope give the impression of a small, brown-tinted glassy bead. The larger cells have only a thin layer of cytoplasm srrounding a massive central vacuole that occupies the major part of the cell volume. This may contribute to the glassy appearance of the brownish algae when observed with light optics. With the inclusion of these yellow-brown algae, the count of different algal symbionts in the astrosphaeridian radiolaria (spongiose skeletal form) comes to three (i.e., dinoflagellate, prasinophyte, and the brown alga). More may be discovered, as this group of extant, large radiolaria has not been investigated thoroughly by biologists.

As an aid in establishing a context for the range of symbionts associated with species of radiolaria and other invertebrates, some commonly observed host–algal associations are cited in Table 2-2. For purposes of comparison with radiolaria, the fine structure of symbionts associated with Acantharia has been described by Febvre and Febvre-Chevalier (1979). The fine structure of the yellow-brown symbionts, located within the endoplasm of Acantharia, is remarkably similar to that of the brown symbionts in the extracapsulum of some radiolaria. Among the spumellarian radiolaria, dinoflagellates appear to be the most common type of algal symbiont. Prasinophyte algae are apparently associated largely with some spongiose skeletal species. The yellow-brown algal symbionts have been observed only infrequently in SCUBA collected species, and these were obtained in locations near Barbados. Nassellaria frequently possess dinoflagellate symbionts, and no other algal symbiont has been observed in our collections.

Parasitism

Dinoflagellate parasites infect some species among the Polycyttaria especially *Thalassicolla* sp. and the Collodarian colonial radiolaria (e.g., *Collozoum inerme*). The parasites in *Collozoum inerme* are usually

Table 2-2. Some marine symbiotic algae and their hosts[a]

Alga	Host

Dinophyceae (dinoflagellates)
  Symbiodinium (= *Gymnodinium*)
  microadriaticum and *Endodinium*
  (= *Amphidinium*) *chattonii*
	Coelenterates (corals, medusae,
	  Vellela), molluscs (giant clams),
	  planktonic foraminifera, colonial
	  and solitary radiolaria, and
	  acantharia

Bacillariophyceae (diatoms)
  Licmophora sp.
  Nitzschia panduriformis,
  Amphora tenerrima, and others
	Turbellaria (*Convoluta convoluta*)
	  and benthic foraminifera[b]

Cryptophyceae *Erythromonas haltericola* ?	Ciliates (*Mesodinium rubrum*)
Haptophyceae or Prymnesiophyceae Yellow-brown spherical cells of uncertain generic position	Acantharia and solitary radiolaria (?)
Rhodophyceae *Porphyridium* sp.	Benthic foraminifera

Prasinophyceae
Platymonas convolutae
	Turbellaria, flat worms, (*Convoluta*
	  roscoffensis) and solitary
	  radiolaria (*Spongodrymus* sp.)

Chlorophyceae
Chlamydomonas hedleyi, *C.*
  provasolii, *Chlorella* sp.
	Benthic foraminifera and planktonic
	  foraminifera

[a] Based on data from Anderson (1981), Anderson and Bé (1976), Dodge (1973), Febvre and Febvre-Chevalier (1979), Hibberd (1977), Hollande and Carre (1974), Lee (1980), Taylor (1974), and Trench (1980).
[b] For more detailed information on symbionts of foraminifera see Lee (1980).

species of *Merodinium*. *Merodinium brandti* typically invades the nucleus of *C. inerme* where it forms a plasmodial stage. Eventually, the *Merodinium* nuclei divide profusely leading to necrosis of the radiolarian nucleus which becomes filled with parasite daughter nuclei and some black residual products of host nuclear decay. The parasite nuclei become segregated from the plasmodium to form a flagellated, motile stage, exhibiting typical dinoflagellate features including an epicone and hypocone, obliquely joined by a pronounced girdle bearing one of the two flagella within its groove. In *Merodinium vernale*, the nuclear disintegration of the host, *C. inerme*, is accompanied by release of nucleated plasmodial fragments that fill the central capsule and eventually are released through the capsular wall, forming a stage with intracapsular and extracapsular plasmodia. These eventually give rise to motile dinoflagellate swarmers that flee the host and invade another colony to reinitiate the infective cycle. Some early researchers (e.g., Brandt, 1885) erroneously interpreted the dinoflagellate swarmers as radiolarian gametes, thus producing considerable confusion about their identity, as is explained more fully in the section on radiolarian reproduction.

The colonial species *Myxosphaera coerulea* is reported to be infected with *Merodinium mendax*, whose motile stage exhibits a distinctively oblique groove separating the epicone from the hypocone and is somewhat more conical in shape than *M. brandti*. *Thalassicolla* sp. are infected by a decidedly different dinoflagellate parasite that has been assigned to the genus, *Solenodinium* (Chatton, 1920). *Solenodinium fallax*, for example, invades the centrally located nucleus of *Thalassicolla* where it forms plasmodia that develop into tubular inclusions within the host nucleus. These plasmodial tubules eventually erupt from the disintegrating host nucleus, and invade the surrounding intracapsular cytoplasm displacing the host cytoplasm. The numerous radially arranged parasite tubules eventually give rise to motile stages bearing a typical dinoflagellate groove and unequal flagella.

Skeletal Structure and Morphogenesis

The siliceous skeleton of many radiolaria is the most prominent feature immediately recognized by light microscopists and has become an object of significant scientific interest as a major feature used in radiolarian taxonomy and as a useful tool in geological investigations of stratigraphy and paleoecology. The diverse and remarkably complex geometric forms of the radiolarian skeleton composed of amorphous silica raise intriguing questions as to how these single-celled organisms secrete their skeletons and how the unusually complex organizational pattern is dictated by the cell. As with most biological phenomena, there is variability in skeletal form within a species; however, we do not know the range of this variation nor how much is attributable to variability in cellular control of silica deposition versus environmental effects on biological activity. The larger biological question of the ontogenetic and phylogenetic origins of form in biological systems becomes particularly acute when we consider radiolarian skeletons. If we can solve the riddle of how solitary cells create such remarkably complex secretory structures and what selection factors in phylogenetic history have influenced the long course of its development, we undoubtedly will have achieved considerable progress in understanding fundamental cell processes pertinent to a wide variety of organisms. From the perspective of fundamental cell biology, the radiolarian skeleton is a convenient physical record of cell pattern production; and alterations in its composition and/or geometry, as a result of experimental interventions, may elucidate basic questions of morphogenesis and cellular control mechanisms in a unique way not possible with soft structures alone. Investigations of variation in form and ennvironmental influences on skeletal morphology of living species are clearly significant in interpreting the fossil record where skeletal structures are the single-most important source of information. Regrettably, our knowledge is meager in contrast to the magnitude of the opportunity this problem presents. Some recent developments in the quest for a fuller understanding of these phenomena are presented and related to historical perspectives and achievements.

Skeletal Secretion

The mechanism of silica secretion in radiolaria has long been of interest since the late 19th century, when Haeckel (1887) and contemporary scholars speculated on the origin of the delicate siliceous network of the skeleton. Owing to the limited resolution of conventional light microscopy, it was not possible to detect the fine cytoplasmic processes involved in skeletal formation. Nonetheless, Haeckel offered two hypotheses as to the possible origin of the skeletal elements published in the Challenger Expedition Report (p. cxxiv):

It may indeed be assumed that these skeletons arise directly by a chemical metamorphosis (silicification, acanthinosis, etc.) of the pseudopodia and protoplasmic network; and this view seems especially justified in the case of ther Astroid skeleton of the Acantharia, the Spongoid skeleton of the Spumellaria, the Plectoid skeleton of the Nassellaria, the Cannoid skeleton of the Phaeodaria, and several other types. On closer investigation, however, it appears yet more probable that the skeleton does not arise by direct chemical metamorphosis of the protoplasm, but by secretion from it; for when the dissolved skeletal material (silica, acanthin) passes from the fluid into the solid state, it does not appear as embedded in the plasma, but as deposited from it. However, it must be borne in mind that a hard line of demarcation can scarcely, if at all, be drawn between these two processes.

Although Haeckel appears to have favored the second hypothesis, it is not at all clear what specific mechanisms of silicification were envisioned by the process of being deposited from the protoplasm. This becomes all the more uncertain when he states that a hard line of demarcation cannot be drawn between these two processes. The reader is left wondering from what surface the silica is secreted and into what space within the plasma.

Further theoretical explanations were offered by Thompson (1942), who appealed to physical–chemical processes at cytoplasmic interfaces to explain the reticulate pattern of skeletal formation observed in many species. He began with the assumption that the close packing of the alveoli produced changes in the interfacial surface energy that favored deposition of amorphous silica at the interstices among the bubble-like alveoli. According to his reckoning, the polyhedral domains delimited by these interstices create the lattice-like canals where silica is induced to polymerize. His theory presumes that the form of the skeleton can be explained on purely mathematical principles of the solid geometry of closely packed membranous surfaces and the physicochemical properties of interfaces. Although his theory is appealing for its interdisciplinary elegance and seeming simplicity in drawing upon fundamental mathematical and physical chemical principles, there are certain issues that remain unresolved. Not all species possess closely packed alveoli yet exhibit rather complex skeletons. Some skeletons, moreover, possess long spicules, sometimes elegantly twisted or ornamented with teeth or verticillate barbels. These are complex structures which bear little resemblance to interfaces among close-packed spherical or spheroidal symplastic surfaces. Although they did not refute Thompson's theory, Hollande and Enjumet (1960) reported that the skeletal elements, including the long spicules of radiolaria, were covered with a layer of cytoplasm which upon disturbance, was rapidly withdrawn exposing the bare skeleton. They suggested that the skeleton was formed inside this living protoplasmic coat. Recent light microscopic and fine structure research (Anderson, 1976a, 1980, 1981) has confirmed the presence of a cytoplasmic coat on the skeletal elements and provided a refined analysis of the role of cytoplasmic membranous structures in skeletal deposition. It appears increasingly that the major assertions of Thompson's model are not correct. Light microscopic evidence of spine elongation in sphaerellarian radiolaria (Figs. 2-18A and 2-19), for example, consistently shows the presence of a cytoplasmic sheath surrounding the new growth. It is not alveolate and exhibits constant cytoplasmic streaming along the direction of the axis of the spine. As new silica is deposited increasing the length and girth of the spine, the cytoplasmic sheath expands and elongates concomitant with skeletal growth. The flowing and expanding cytoplasmic sheath appears to act as a dynamic mold, establishing the architectual ground plan for the skeletal structure as it is gradually built up and controlling the fine ornamentation deposited on the skeletal surface. Small vesicles stream through the cytoplasm within this sheath; however, their function has not been determined.

Transmission electron microscopic examination of ultrathin sections through skeletal containing cytoplasm further confirms the presence of a delicate cytoplasmic sheath enclosing the skeleton (Figs. 2-18B, D). It has been observed in a wide variety of solitary and colonial spumellarian radiolaria with skeletons composed of spicules, spongiose lattices, and perforated spheres. The term *cytokalymma* was coined (Anderson, 1980, 1981) to designate this cytoplasmic sheath that encloses, molds, and deposits the skeleton. The cytokalymma, illustrated by a line drawing reconstructed from light and electron microscopic evidence (Fig. 2-18C), forms a closely adhering sheath around the skeletal elements and is attached at places to the axopodia and/or rhizopodia. It is thicker on more distal skeletal elements of some solitary species where presumably new growth is occurring. However, it may be thin and incomplete on more internal, possibly more mature, skeletal components. Based on transmission electron microscopic images of ultrathin sections, the thickness of the cytokalymma may vary from a thin envelope (ca. 0.2 μm thick) to a rather substantial sheath (several micrometers thick). Mitochondria and vesicles are observed within the cytoplasm in the thicker regions of the cytokalymma. Considerable variability in thickness of the cytokalymma occurs across species. Those with large spicules and spines may have a thicker cytokalymma than species with finer skeletal elements. Variability in thickness may also occur according to the physiological state of the organism; however, additional research is needed to amplify present light microscopic data before definitive conclusions can be made.

Scanning eletron microscopic observations of developmental stages of the cytokalymma have begun to elucidate the events associated with its formation and relationship to the underlying skeleton (Anderson, 1981). When a large spine is broken during laboratory culture of some sphaerellian radiolaria, the cytoplasm at the base of the spine generates a cytokalymma which spreads over the surface of the spine and eventually encloses it within the silica-secreting sheath. During the earliest stages of cytokalymma development, fine threads of branching rhizopodia at the base of the spine begin to elaborate and cover the spine surface. These expanding rhizopodial filaments coalesce, forming a thin, but at first incomplete, cytoplasmic shroud around the spine. Expansion and complete fusion of the initially patch-like segments of the incipient cytokalymma produce an entire cytoplasmic sheath which flows forward and coats the distal part of the spine (Fig. 2-19B). At the very tip, the cytokalymma elongates to form fine filopodia that radiate distally or may be recurved and joined with the cytokalymma sheath

---

Fig. 2-18. Skeletogenesis. **(A)** Light micrograph of spine regeneration exhibits a thin regenerated segment (arrow) emerging from the broken spine stump (Sp) and surrounded by a very diaphenous living sheath of cytoplasm, cytokalymma (dashed lines). Scale = 20 μm. **(B)** An electron micrograph of a portion of skeleton surrounded by the cytokalymma attached to a fusule (F). Scale = 1 μm. **(C)** Diagram showing the spatial relationship between the cytokalymma (Cy) and enclosed skeleton (Sk) secreted within it. **(D)** A high-magnification view of an ultrathin section through a cytokalymma (Cy) and a region of new skeletal growth showing the granular like matrix (arrow) at the growing edges of the skeleton (Sk). Scale = 0.25 μm. (Anderson, 1981)

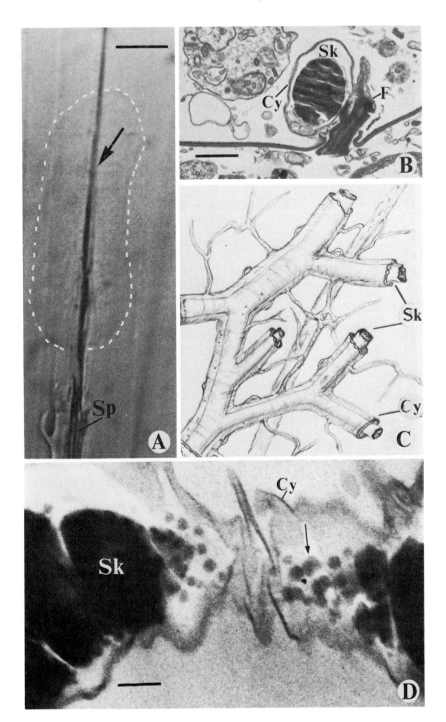

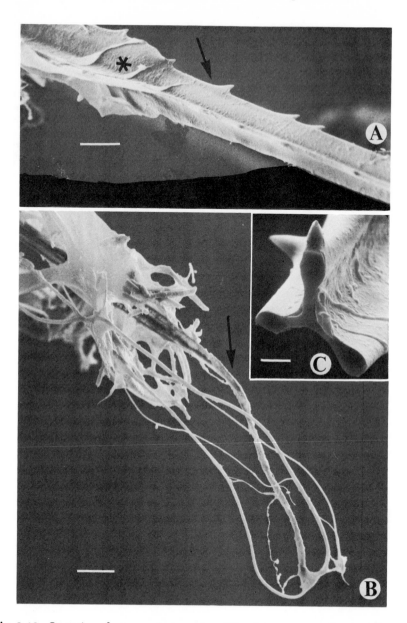

Fig. 2-19. Scanning electron micrographs of the thin segment of a regenerated spine shaft (arrow, **A**) attached to a thicker segment (asterisk) from which it emerged, and of a growing spine **(B)** enclosed by the cytokalymma (cytoplasmic sheath) that extends from the tip of the spine (arrow) and forms fine strands of cytoplasm that are folded back and attach to the thickened sheath around the spine. Scales = 10 μm. **(C)** A cleaved segment of a spine reveals its triangular cross section and the thin layer of cytoplasm enclosing the spine. Scale = 2 μm. (Anderson, 1981)

covering the spine. The new growth of the spine is clearly visible in both light microscopic views and scanning electron microscopic images as a decidedly thinner extension of the spine. Although it bears a similar morphology as the spine stub from which it elongates, the new growth is usually of smaller diameter (Fig. 2-19A). Thus, it is possible to monitor the course of new growth over time by light optics and make estimates of the rate of elongation. In Fig. 2-19C, a segment of new growth has been cleaved to illustrate the thin cytokalymma surrounding the spine, and the continuous covering of the cytokalymma extending from the unbroken, broader part of the spine to the newly developing thinner segment that has been deposited within the cytokalymma. A tooth on the edge of the spine is visible near the site of cleavage. It also is enclosed by the thin envelope of the cytokalymma. In these preparations, the specimens were freeze-dried prior to heavy metal coating for scanning microscopy, and therefore the cytokalymma may be somewhat shrunken compared to the living state. Nonetheless, the preparations appear to be preserved in a fairly natural state as the fine rhizopodial filaments look very much as they appear with high-power lenses in the light microscope. The scanning electron microscope, however, produces higher-resolution images with greater depth of field than is possible with most light microscopes.

Further observations using transmission electron microscopy provide insight into the relationship of the cytoplasm to the skeleton. As a point of historical interest, it is most surprising that earlier theorists concluded that the alveoli interstices were the site of silica deposition given the limitations of light microscopy. The very thin cytokalymma surrounding the skeleton is sometimes situated within the interstices of the alveoli (e.g., Fig. 2-12D), and thus would be nearly impossible to resolve as a separate layer from the alveolar membranes. With electron optics, it is clearly separated from the surrounding alveoli, although it may be partially supported by them, and thus is a distinctly specialized structure mediating silica deposition. The question of how the silica is deposited within the cisterna of the cytokalymma remains to be answered. The inner membrane of the cytokalymma immediately surrounding the skeleton may act as a silicalemma to secrete silicate into the cisterna, as has been shown in silica-secreting vesicles of diatoms (Volcani, 1981; Sullivan and Volcani, 1981). The dense amorphous silica within the skeleton represents a considerable reduction in entropy compared to the low concentrations of dissolved silica in seawater, and must require some expenditure of metabolic energy to drive the process. Even if one hypothesizes that the cisternal milieu is at a low pH thus favoring silica deposition, metabolic energy would be required to maintain the strong pH gradient in relation to the surrounding seawater. Further research is needed on the kinetics of silica deposition and transport across the cytokalymma. It is of interest to examine also the molecular composition of the inner membrane, "silicalemma," of the cytokalymma to determine the ionic change properties of its molecules. If it is rich in amphipathic compounds with cationic polar groups, these may enhance silicate localization within the cisternae, presuming the cationic groups are oriented toward the cisternal inter-face.

Although much remains to be done on the biochemical aspects of silica deposition in radiolaria, some fine structure evidence has begun to elucidate some early events during silica deposition (Anderson, 1981). Early signs of silica deposition were observed in ultrathin sections of skeletons in regions where the silica was thin and evidenced signs of thickening, or where there was an incomplete bridge of silica between thicker regions. Sections were obtained from *Sphaerozoum punctatum*, *Collosphaera globularis* and a colonial spumellarian species with very early shell formation. In longitudinal sections passing through the skeleton and the growing edge, high magnification images reveal the presence of an osmiophilic, granular matrix (Figs. 2-18D and 2-25G, H) or a spongiose matrix with granular inclusions (arrow, Fig. 2-18D) at the surface of the growing skeleton. This granular deposit appears to consist of more widely spaced granules in younger regions of deposition and more closely spaced or fused aggregates of granules in the region where the silica has been more fully deposited. The granules within the matrix are ca. 80 nm in diameter. In some preparations, there appears to be fine filaments interconnecting them, thus perhaps contributing to the spongiose appearance of the early stages of deposition. The earliest deposited matrix may consist largely of a nonsiliceous deposit as hydrofluoric acid treatment of the sections failed to dissolve the earliest formed electron-dense matrix but removed the silica at more mature segments of the skeleton (Figs. 2-25 G, H). Hypothetically, the granular or spongiose matrix may serve as a nucleation site for deposition of the silica. The granules could become impregnated with or enclosed by the polymerizing silica, thus gradually increasing in size until they fused and formed the siliceous wall. This growth by accretion may also be supplemented by intussusceptive growth through polymerization of silica in between the nucleation sites thus gradually building up the amorphous silica until a solid mass is formed. Presently, it is not known how the solid silica phase emerges from the initial deposition sites. At the point of contact between the weakly silicified growing zone and the more strongly silicified segment, small, electron-dense granules can still be observed within the transition zone (arrow, Fig. 2-18D). In the heavily silicified zone, however, it is not possible to detect such fine substructure. Sectioning damage also tends to fracture and displace the larger masses of silica making interpretations difficult.

Evidence for a nonhomogenous infrastructure in radiolarian skeletons and other silica-secreting organisms (e.g., diatom frustules) has

been obtained from physical-chemical assays of shell porosity and transmission microscopic examination of thin sections (Hurd et al., 1981).
Fossil skeletons of varying ages from recent to 40×10^6 years before
present (million years B.P.) were examined to determine pore size,
volume, and distribution using a combination of gas adsorption techniques and electron microscopy. A layer of porous matter near the
surface of the skeleton was found in many fossil specimens. The more
central part of the skeleton, however, appeared to be considerably less
porous when observed in ultrathin sections using transmission electron
microscopy. To a fair degree, both gas adsorption and electron microscopic estimates yield comparable results. When the average number
of pores per unit area is plotted against the average radius, both techniques show a peak number of pores per unit area corresponding to a
radius of approximately 20–25Å. The transmission microscopic images
show that the pores in the outer layer of the fossil skeleton are open,
whereas those more deeply situated are closed and presumably may
enclose aqueous-filled spaces. It is not immediately clear how this
evidence of skeletal porosity relates to the granular matrix observed in
early stages of skeletogenesis; however, the two sets of data are complementary in suggesting a more complex infrastructure for radiolarian
skeletons than presumed heretofore.

The skeletons of Phaeodaria have long been recognized as highly
complex aggregates consisting of a mixture of organic and silica phases
often producing a decidedly "spongiose" quality to the shell wall (Haeckel, 1887). Modern scanning electron microscopic research (e.g., Ling
and Takahashi, 1977) has confirmed that the wall composition is complex in many phaeodarian species. For example, although the surface
of the shell in Protocystis thomsoni (Murray) is smooth and porcelaneous, images of a fractured edge show the presence of numerous elongated teardrop-shaped parietal cavities oriented approximately
perpendicular to the shell wall. They are surrounded by a hexagonal
meshwork or honeycomb structure previously observed by Ling (1966).
The magnification of the scanning electron micrographs is insufficient
to determine if there is a granular quality to the siliceous matrix of the
wall.

Additional evidence for the presence of silicification granules in
skeletal deposition has been obtained from scanning electron microscopic observations of the surface of mildly etched shells, using short
exposures to dilute HF solutions and from occasional shells exhibiting
pitted surfaces in the natural state that appear to have been incompletely
silicified (Fig. 2-20). Mildly etched shells of collosphaerid radiolaria
exhibit at occasional edges, fine granules of more HF-resistant silica.
These granules are about the same size as the larger silicified granules
observed in ultrathin sections of skeletons (Fig. 2-18D). Collosphaerid
shells with pitted surfaces (Fig. 2-20A), which appear to have an in-

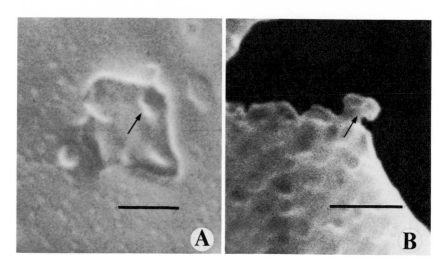

Fig. 2-20. Granular infrastructure (arrows) of a partially silicified skeleton of a collosphaerid radiolarian **(A)** and corresponding evidence obtained by mild etching of the skeleton with ammonium bifluoride solution **(B)**. Scales = 0.5 μm. (Anderson, 1981)

complete surface deposition of silica, exhibit siliceous granules projecting from the continuous silica phase in the wall. The surface surrounding the pit appears smooth as though it were deposited as a fine final layer or veneer. The pits appear as regions where this final coating is incomplete and thus the details of the underlying surface are revealed. It is not known whether these subsurface granules result from partial resorption through incomplete silicification. The edges of the pits, however, are smooth and contoured, suggesting that they result from incomplete deposition of silica rather than partial dissolution of the silica. The size of these subsurface granules corresponds to the size of granules observed in sections near the partially silicified wall.

Additional evidence for cytoplasmic deposition of granular silica at sites of skeletal growth has been presented by Cachon and Cachon (1972c). They observed a granular, stub-like spine called an axobate on the otherwise smooth surface of skeletons in certain sphaerellarian and nassellarian species. The granular composition and morphological distinctiveness of the axobate relative to the surrounding surface suggested that it was deposited by accretion of silica granules. Cachon and Cachon suggest that the axobate may be produced by cytoplasmic streaming carrying particles of silica to the site where it is assembled. They observed cleaned skeletons prepared for scanning electron microscopy; therefore it was not possible to determine the actual role of the cytoplasm in axobate formation.

Given the limited information available on the early events of silicification in radiolaria, it is not possible to generalize beyond the several instances cited here. There appears to be considerable difference in the porosity and granular composition of mature polycystine skeletons compared to phaeodarian skeletons based on scanning electron microscopic evidence of Petrushevskaya (1975b) and Takahashi (1981). The ontogenetic origins of these skeletal differences need to be examined in greater detail. Certainly skeletal morphology and physical composition are sufficiently different among Spumellaria, Nassellaria and Phaeodaria to warrant caution in generalizing across groups. Indeed, the early events of silica secretion may be variable among genera and more detailed examinations of early stages of silicogenesis are needed to clarify constant features compared to variable ones. It remains to be determined whether a silicification matrix is present in all species and to better characterize its composition. Indirect evidence from chemical microanalysis of cleaned radiolarian skeletons indicates the presence of organic matter, including amino acids, in the skeleton (King, 1974, 1975, 1977). Other silica-secreting organisms including diatoms (Volcani, 1981) and sponges (Garrone et al., 1981; Simpson, 1981) possess a silicification matrix of very fine constitution, thus further illustrating the generality of a silica-binding matrix in diverse organisms. In radiolaria it is unclear whether the cytokalymma secretes the earliest deposited substratum whereupon the skeleton is built or if vesicles within the cytoplasm perhaps fuse with the cytokalymma inner membrane and deposit substances into the cisterna. Although these molecular events are uncertain, considerable evidence points to a major role of the cytokalymma in skeletal morphogenesis and the origin of the unique, species-specific form of the skeleton.

Skeletal Morphogenesis

Solitary Spumellaria   Enriques (1931) was among the first to attempt to analyze the mode of skeletal growth among groups of radiolaria. He proposed three patterns of skeletal development: (1) pellicular development in the Collosphaerida, (2) simultaneous development, and (3) tangential progressive development; the latter two being particularly characteristic of the sphaerellarian radiolaria. By pellicular development, he apparently meant the skeletal deposition within the peripheral membranes surrounding the central capsule. He undoubtedly did not detect the earliest stage of the cytokalymma formation, which is the origin of the pellicular deposition of the skeleton, as it is very thin and becomes recognizable with Normarski or phase-contrast optics. By simultaneous development, he apparently meant the concurrent deposition of silica throughout the major part of the skeleton. Tangential progressive formation within the Sphaerellaria occurs as progressive shells are deposited and thickened tangentially as the protistan matures. Schwartz (1931) examined fossil radiolaria and attempted to infer growth patterns based on the preserved hard parts. He concluded that the secondary shells developed by centrifugal growth, but admitted the possibility that centripetal growth of the skeleton was also possible. In some of his preparations, growth appears to progress from the periphery toward the center. He also made observations on the relationship of the skeleton to the central capsule; however, as pointed out by Hollande and Enjumet (1960), it is questionable whether he was observing the central capsular wall, which is not likely to have been well preserved, or the spongiose material sometimes deposited on the surface of the skeleton.

Based on their light microscopic examinations of living and sectioned radiolaria, Hollande and Enjumet (1960) made some observations on skeletal growth in the Sphaerellaria. They observed that the "Medullary" shell in many species is deposited within the central capsule and is either endoplasmic or intranuclear. Based on more recent transmission electron microscopic observations (Anderson, 1976a), it would appear that they intended intranuclear to mean that the skeletal components were enmeshed with or enclosed by the elaborate nuclear lobes or reticula produced at its surface. Hollande and Enjumet also clarified the origin of the medullary shells in some species. They found convincing evidence in *Arachnosphaera* that the medullary shell is initially deposited by the immature protistan as a small cortical shell within its ectoplasm, presumably within the equivalent of the cytokalymma. Subsequently, as the radiolarian matures, additional cortical shells are deposited in the ever-enlarging ectoplasm. However, this is attended by increasing medullarization of the first-formed shells as they become enclosed within the endoplasm. A similar pattern was found in *Cladococcus*. The single shell is at first cortical, but is without exception medullary in secondary development. On the whole, Hollande and Enjumet conclude that skeletal growth is centrifugal in most species that they examined, although they recognize viewpoints to the contrary, based on fossil evidence and skeletal morphology (e.g., Schwartz, 1931; Deflandre, 1953).

Skeletal Structure of *Callimitra* sp.   *Callimitra* sp. (Nassellaria) are characterized by an elaborate skeleton composed of a net-like mesh supported by a framework connected to the tripodal elements of the skeleton. This is illustrated by the scanning electron micrographs of *Callimitra emmae* (Figs. 2-21–2-24). A corresponding light micrograph of a living cell (Fig. 1-2G) shows the central capsule filled with cytoplasm and the fine peripheral rhizopodia that radiate outward from the perimeter of the lattice. The fundamental architecture of the skeleton consists of three wing-like plates of mesh protruding from the side of the skeleton and three basal plates, each supported between two of the tripodal feet. The wings are formed by a framework suspended from the apical spine (arrow, Fig. 2-21A) and connected by a curved, undulating marginal rim to a basal foot. Additional reinforcing, lateral spars extend outward from the central capsule or from the surfaces of the feet. These appear to be thickened strands of the meshwork and provide reinforcement for the finer strands composing the latticed network of the skeleton. In a well-developed skeleton, the central capsule framework (cephalis region) is covered by a thin siliceous membrane deposited upon the bar-like framework of the cephalis (Fig. 2-21A). This very thin layer is not organic, but a remarkably delicate lamina of silica. A similar lamina has been observed in *Callimitra carolotae* (Robinson and Goll, 1978). A pore with a protruding lateral spine, originiating from the vertical rod of the central tripod, occurs on one side of the cephalis (Fig. 2-21B). It was not present in all specimens, but has been described also in *C. carolotae* by Robinson and Goll.

The geometry of the mesh-like lattice in *C. emmae* is exhibited more clearly in Fig. 2-22A and can be compared to the overall pattern in the lower magnification view of the whole skeleton. The strand-like elements of the lattice are arranged in a crisscross pattern that produces triangular openings in the mesh or, in less ordered regions, parallelograms, or irregular polygonal openings. The lattice elements consist of a set of laterally parallel strands running approximately horizontal to the basal plane. These can be seen most clearly in the basal plate (Fig. 2-21A) spanning the space between the tropodal feet. These approximately parallel strands, distributed like lines on a writing tablet, are intersected by oblique strands at an angle of approximately 60°. One set of these oblique strands runs from lower left toward upper right in the plane of the basal plate. The other set is oriented from the lower right to the upper left. These two sets of strands also intersect with one another at an angle of about 60°. Consequently, the intersection of the three produces equilateral triangular spaces. When one of the intersecting strands is incomplete in a region or deviates in its course so severely that it fails to transect the area enclosed by the other two strands, a parallelogram area is produced. The biological basis for the development of the triangular mesh pattern is not known. The pattern is clearly genetically determined. It is characteristic of this group of Nassellaria, and observed repeatedly in many specimens. The small or sometimes large deviations from a perfect triangular pattern are characteristic of epigenetic variability in biological systems. It is unlikely that the pattern is simply a result of static physical forces occurring within a network of "passively stretched" rhizopodia. The pattern in the lateral wings is consistently different from the pattern in the basal plates. In the wings, the nearly vertically oriented strands within the network are arcuate, not linear. Their curvature follows the contour of the outer margin of the wing and they undoubtedly represent earlier ontogenetic approximations to the mature curved magin occurring at the rim of the wing. The clearest instances of triangular networks formed by intersecting linear strands occur in the basal plates attached to the feet of the tripod. This generally triangular-mesh lattice is clearly a highly stable geometric configuration, and is further reinforced by the thickening of the strands to form the spar-like bars within the mesh. These are more clearly seen in the apical view (Fig. 2-21C). Numerous cleaned skeletons of C. emmae were examined to find evidence of the patterns of silica deposition that occur during morphogenesis. It appears that the tripodal element is the first formed structure. This is reinforced by the deposition of the fine strands spanning the spaces between major elements of the tripod and surrounding the central capsule where the helmet-shaped cephalis is formed. During the development of the meshwork on the lateral wings and basal plate, the major lateral strands destined to become the thickened spars are reinforced by additional silica deposition. This thickening occurs on the edge of the strand preferentially in a direction perpendicular to the plane of the mesh. Thus, these strands become a bar (Figs. 2-21D and 2-22B) that is oriented with its plane perpendicular to the plane of the lattice. The bar is all the more remarkable, as its lateral edge is also thickened and contoured (arrow, Figs. 2-21D and 2-22B) to yield an I-beam type construction. This undoubtedly increases the strength of the bar. The mother strand that was thickened to produce the bar is still clearly visible, embedded within the central shaft of the bar (asterisk, Fig. 2-22B). The orientation of these bars with relation to the lattice plane is also clearly exhibited in the apical view of the skeleton (Fig. 2-21C). During early stages of skeletal development when the bars are being

---

Fig. 2-21. *Callimitra* sp. (probably *C. emmae*). **(A)** Frontal view of a mature specimen showing the silicified cephalis (C), lateral wings (W) connected to the apical spine (arrow), basal plates (BP), and tripodal elements that are partially obscured by the lattice. Scale = 4 μm. **(B)** A pore and protruding spine commonly observed on the surface of the cephalis. Scale = 2 μm. **(C)** A top view of the specimen exhibiting the pyramidal shape of the basal plates and the three wing-like panels projecting laterally. Scale = 40 μm. **(D)** A segment of the lattice in higher-magnification view showing the configuration of the bar-like spars with thickened edges (arrow) and the cylindrical threads in the lattice. Scale = 10 μm. (I express appreciation to Horst Hooteman, technical manager of the scanning electron microscope laboratory, at Tübingen University where I pursued this research)

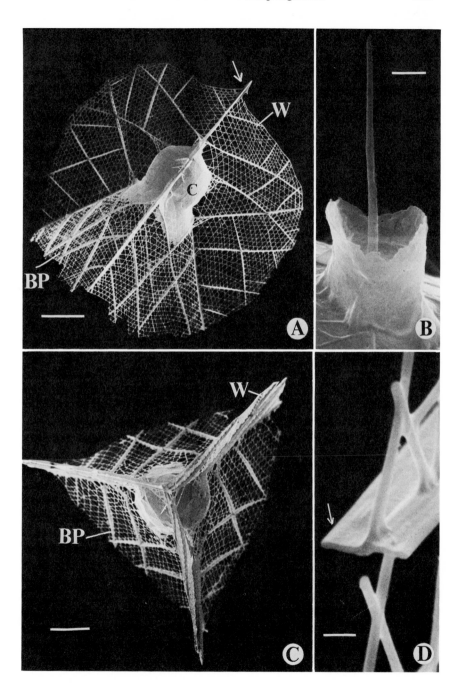

extended peripherally, the mesh is also built up concomitantly. Eventually it is terminated by a thickened border that delimits and frames the periphery of the lattice. The preferential thickening of the lateral strands that become the plate-like spars may result from a tendency for the rhizopodia to aggregate and stream outward along these developing siliceous strands, thus augmenting the cytokalymma and enhancing the amount of silica deposited. Light microscopic views of living *Callimitra* show that thickened strands of rhizopodia sometimes radiate outward along these flattened spars and extend into space surrounding the skeleton. In incompletely formed skeletons, the lateral bars extend outward beyond the lattice and serve as surfaces to which additional strands of the lattice mesh are attached.

High magnification views of the developing lamina on the cephalis show the remarkable sequence of events during its deposition (Figs. 2-22C, D, 2-23, and 2-24). An overall view of the cephalis region during an early stage of silica deposition shows a thin reticulated layer deposited in the interstices of the framework (Fig. 2-22C). At this stage, the cephalis framework is complete, but only localized patches of silica are being secreted on the surface of the framework. This indicates that the cytokalymma becomes active in localized regions at first, rather than making a uniform deposit of silica over the entire surface. A higher-magnification view (Fig. 2-22D) shows various stages of silica deposition within the areolae of the cephalic framework. During the earliest stages, a reticulated, lacy network of silica is suspended within the opening. The peripheral strands of the reticulum are clearly fused with the internal rims of the bars in the framework. At later stages, also visible in Fig. 2-22D, the reticulum becomes reinforced by additional deposition of silica producing a perforated siliceous septum. Further thickening results in a more substantial lamina with greater thickness than the earliest reticulate stages. During the early stages of thickening of the reticulum, the small irregular pores in the lacy network become cross-divided by bar growth, resembling the partitioning of the pores that occurs in the shells of collosphaerid colonial radiolaria (Anderson

---

Fig. 2-22. *Callimitra* sp. **(A)** Detail of the peripheral skeleton showing the laterally thickened bars and fine lattice meshwork. **(B)** A cleaved bar exhibits the I-beam type of construction (arrow) around the central shaft (asterisk). **(C)** An overview of a partially silicified cephalis and a higher-magnification view **(D)** shows the early stages of silica secretion of the cephalis wall proceeding from a thin reticulum suspended within the framework (1) to a more fully silicified, perforated lamina (2), and a nearly completely closed lamina (3). In the earliest stages (1), the web-like network appears to be segregated into increasingly smaller pores by cross-bridge growth (arrows). Scales = 3 μm.

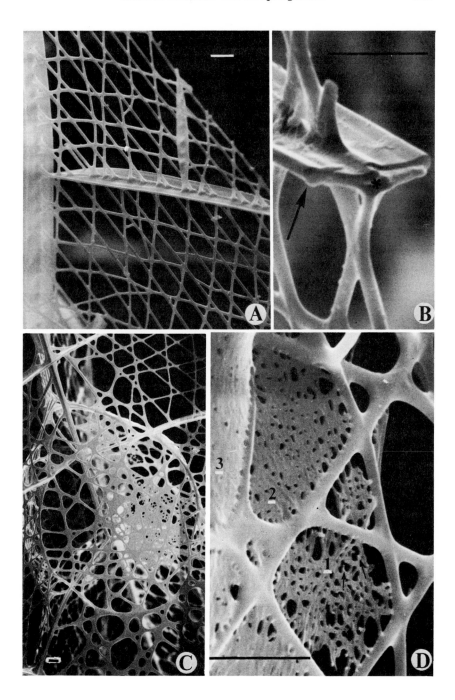

and Swanberg, 1981). The small bar-like cross-bridges originate on the inner surface of a pore rim and grow across toward the opposite side, until they fuse with the opposite rim of the pore or with an opposing bar that is also growing laterally across the pore. When the two bars meet and fuse, they create a partition that divides the pore into two smaller pores. By repeated partitioning accompanied by rim thickening and surface thickening, a substantial lamina with small irregular pores (ca. 0.2 μm diameter) is deposited on the framework (Figs. 2-22D and 2-23A). At some points, the early stages of lamina deposition form a lacy undulating layer attached to the surface of the framework (Fig. 2-23B). This clearly shows that the lamina is deposited upon the framework of the cephalis after it is formed. The delicate early stages of the lamina on the surface of the cephalis resemble a repoussé art decoration. A more complete stage of cephalis lamina development is shown in Fig. 2-23A. Much of the surface has been closed by the lamina on each of the three sides delimited by the lateral wings. A high-magnification view of a completed section of the lamina (Fig. 2-24A) shows the contour of the surface and the thickness (ca. 0.15 μm) along a fractured edge. In this portion of the framework, it is clear that the lamina has been deposited as a sheet attached to the inner rim of the bar-like frame surrounding each pore in the cephalis wall. The construction of the collar-like vertical pore in the cephalis wall (Fig. 2-21B) exhibits the delicate and irregular form of the pore rim surrounding the emergent, needle-like spine. The surface contour and texture of the surrounding cephalis wall are also visible in this view. At this stage, the small pores in the lamina have been closed by additional thickening during the last stages of silica deposition.

An interesting example of the remarkable adaptability of *Callimitra* to ameliorate errors in framework construction is exhibited in Fig. 2-24B. This is a portion of the lattice and associated lateral bar that apparently had been fractured or disrupted during an early stage of development. It is clear that this is not an artifact introduced during preparation of the specimen. The major fracture zone could not have been produced after growth without also shattering the triangulated framework spanning the fracture site. Hence this must have been an artifact of early growth. One of the most interesting aspects is the way in which the two surfaces in the fracture plane have been joined. The right-hand edge has been interleaved or dove-tailed with the left-hand edge. The plastic appearance of the edges where this overlap occurs suggests that the two halves were joined while the cytokalymma was depositing silica on the left hand edge of the fracture. The contact of the cytokalymma with the protruding right hand edge may have bifurcated the cytokalymma resulting in the apparent split-like fault on the front part of the left-hand edge. It is not possible that the solid amorphous silica could have been dislocated in this fashion and still yield

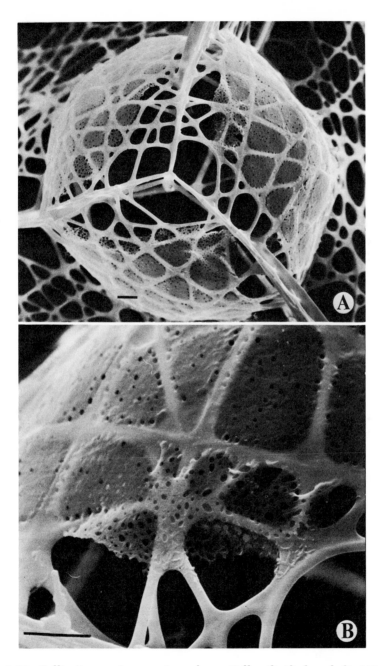

Fig. 2-23. *Callimitra* sp. An overview of a partially silicified cephalis **(A)** and a more detailed view **(B)** showing the lace-like conformation of the early stages of silicification, and the porous lamina produced at later stages when the lacy network has been partially filled in by further silicification. Scales = 4 μm.

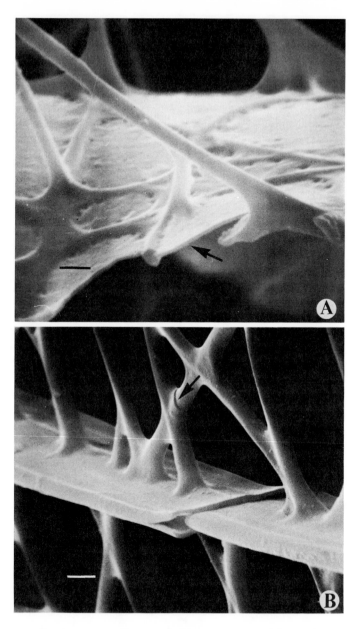

Fig. 2-24. *Callimitra* sp. **(A)** Fractured segment (arrow) of the cephalis showing the thin deposit of silica between the framework strands in a nearly completed wall. **(B)** A portion of the peripheral framework that was apparently damaged during development has been repaired by the radiolarian. The surface of one of the thread-like elements in the lattice has been partially coated (arrow) with a veneer of additional silica probably deposited during the repair process.
Scales = 1 μm.

such a plastic contour along the site of union. Additional evidence showing that this is a zone of new growth and surface repair is the veneer-like deposit (arrow, Fig. 2-24B) that coats the lattice strands. A distinct edge is visible where this secondary coating has been terminated. This suggests that the cytokalymma was active in this region redepositing silica as part of the major repair of the bar, but abruptly ceased the secondary deposition at the ridge-like edge on the upper set of strands. The remarkable differentiation of behavior and corrective response produced by these single-celled organisms are clearly exhibited in the complexity of their skeletal morphogenesis and the elaborate design of their skeletal structures.

Colonial Spumellaria   Combined evidence from high resolution light microscopy and scanning electron microscopy (Anderson and Swanberg, 1981) has provided additional evidence about the sequence of stages in skeletal development among some collosphaerid, colonial radiolaria. As some specimens were observed in the very earliest stages of deposition, when the shell was incomplete, it was not possible to make species identifications. Among the more mature specimens, five species were examined: 1) *Collosphaera huxleyi* (Müller, 1858a,b), (Fig. 1-15A), (2) *Acrosphaera cyrtodon* (Haeckel, 1887) (Fig. 1-15C), (3) *Acrosphaera spinosa* (Haeckel, 1887), (Fig. 1-15D), (4) *Siphonosphaera tubulosa* (Müller, 1858a,b) (Fig. 1-15G), and (5) *Siphonosphaera socialis* (Haeckel, 1887) (Fig. 1-15H). Species identifications were made on the best combined descriptions of Haeckel (1887) and Strelkov and Reshetnyak (1971). Light microscopic examination of colonies during early stages of shell deposition confirm the observations of Haeckel (1862) that many colonies do not develop shells until the colony has reached a certain critical size and that cells near the periphery are among the first to deposit skeletons. At the very earliest stages of development, the extracapsular cytoplasm forms a thin sheath (Figs. 2-25A–F) that is destined to become the cytokalymma. In some colonies exhibiting a discoidal shape, silicogenesis commenced in the most peripheral cells at the rim of the colony and progressed toward the center. During early development of the cytokalymma before substantial silica is visible, the central capsules may continue to divide. The peripheral cytoplasmic sheath also divides concurrently (Figs. 2-25C–D) thus each daughter cell is surrounded by a complete cytokalymma. At the earliest stages of skeletal secretion, the siliceous bars are delicate and reticulate (Fig. 2-25I). This confirms a finding of Hollande and Enjumet (1960) with sphaerellarian species that the earliest skeleton is a lattice and later produces circular pores.

In some of the very thin, newly deposited skeletons, the lattice is irregular, and incomplete growth of the lattice bars produces a spur-

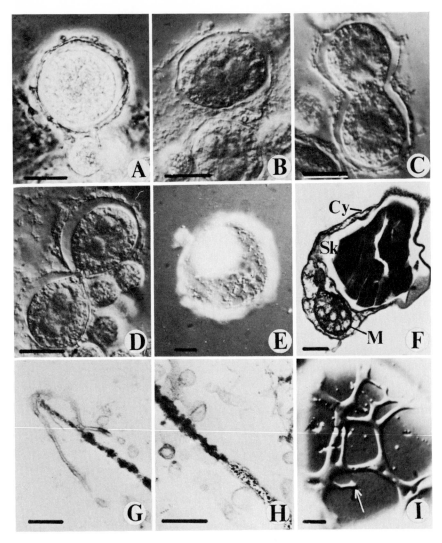

Fig. 2-25. (A) Phase-contrast light micrograph of a cell body surrounded by a cytokalymma (skeletal-forming cytoplasmic sheath). **(B–D)** Stages of cell division with concurrent cytokalymma division yielding two daughter cells, each with an entire cytokalymma (arrow). **(E)** Early stage of silica deposition within a cytokalymma exhibiting the reticulate morphology of the skeleton. Scales = 20 μm. **(F)** Transmission electron micrograph of a section through a cytokalymma (Cy) containing a skeletal segment (Sk). A mitochondrion (M) is in the cytoplasm of the cytokalymma. Scale = 0.5 μm. **(G and H)** Ultrathin sections of early shell deposition showing organic granules preceding silica desposition. HF-treated specimen **(H)** demonstrating HF resistance of the particles. Scales = 250 nm. **(I)** A light micrograph of a portion of a cleaned skeleton exhibiting early stages of pore development with incomplete bar growth. Scale = 10 μm. (Anderson and Swanberg, 1981)

like or finger-like projection from the rim of a lattice space (arrow, Fig. 2-25I).

Two major patterns of shell morphogenesis have been observed in varying degrees among the five species of colonial radiolaria examined: (1) a "bridge-growth" pattern consisting of bar-like skeletal components, which elongate and fuse to produce a reticulated skeleton exhibiting large polygonal or elliptical pores separated by relatively narrow, bar-like partitions (Figs. 2-26A–D). The pores are further subdivided by additional cross-bar partitions. These arise either by growth of a bridge from one side of a pore to the other side (Figs. 2-26A, B) or by two protuberances growing out from opposite sides of the pore and meeting each other (Figs. 2-26C, D). (2) The "rim-growth" pattern results in a perforated shell consisting of nearly circular pores of varying diameter separated by a wide partition between them. In some species (e.g., *Siphonosphaera* sp. or *Acrosphaera* sp.), the pores exhibit an elevated rim or become extremely ornamented by rim thickening to yield tube-like extensions (Figs. 2-27 E, F). During rim growth, the initially large and often polygonal pores become increasingly smaller and sometimes more curricular as additional silica is deposited in the wall. Concurrently, as the diameters of the pores decrease, the partitions increase in width. If the rim thickening proceeds uniformly around the circumference of the pore, the holes are nearly circular and, when carried to an extreme limit, may become so small as to be mere punctate perforations or nearly completely closed (Figs. 2-27B, C). In both types of growth patterns, the silica deposition is controlled by the cytoplasm which serves as a "living hollow template or mold" to determine the ultimate structure that is produced. In the case of bridge growth partitioning of pores, the cytokalymma may form a strand of cytoplasm across the pore as a guide for the silica secreted within it. Thus, even though we see in cleaned shells a pair of bars growing toward each other to produce by fusion a partition across the pore, these seemingly segregated growing points may in fact be growing across the same "hollow" strand of cytokalymma. Hence what may appear to be an unusually accurate trajectory for two opposed growing points that meet in midspace can be the result of their development within a common connecting cytoplasmic strand. These cytoplasmic strands are observable in scanning electron microscopic views of the cytokalymma adhering to the shell.

Evidence for bridge growth is presented in Figs. 2-26 to 2-27C. Various stages of bar growth range from early protuberance of a tooth-like form on the rim of the pore (Figs. 2-26A, B) to finger-like extensions spanning the pore (Fig. 2-26C) or beginning to fuse with protuberances emerging from the opposite rim (Fig. 2-26D). Further evidence of the mechanisms of pore development was obtained from a specimen exhibiting a fortuitous error in morphogenesis. At some point during the

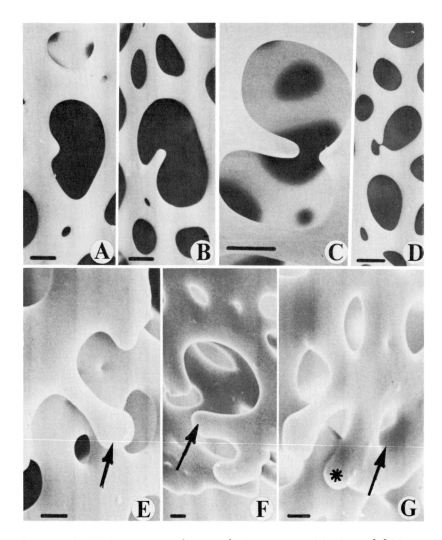

Fig. 2-26. (A–D) A sequence of stages during pore partitioning exhibiting an initial stage with a small protuberance on one edge of the pore and later stages forming bridge-like segments. Scales = 5 μm. (E–G) Stages of initial pore development on a growing edge of an incomplete shell. The first formed finger-like protrusion (arrow, **E**) gradually expands laterally (arrow, **F**) to form a closed pore (arrow, **G**), however, the incomplete segment (asterisk) exhibits an expanded tip characteristic of a bar termination step. Scales = 2 μm. (Anderson and Swanberg, 1981. I express appreciation to the staff of the Geological Institute of Tübingen University where I pursued research on skeletal morphogenesis)

development of the wall, the opposing two planes of development became misaligned and one plane partially overgrew the other one (Figs. 2-26E–G). We cannot determine from the skeleton alone what caused the misalignment of the cytokalymma during development, whether trauma at a very delicate stage, an error in coordination of cytoplasmic development, or other contributing factors in the environment. The fortuitous merit of this biological oddity is the unique information it yields about skeletal morphogenesis. Several stages of early pore development were arrested before completion as the two surfaces overlapped and contacted one another. Among the earliest stages, a bar-like protrusion (arrow, Fig. 2-26E) extends from the edge and terminates with an expanded tip that begins to approximate a contour of a pore. At a later stage (arrow, Fig. 2-26F), protrusions from the edge have expanded laterally to yield a partially completed pore; whereas at other more advanced stages, the pores have been fully formed by fusion of the lateral growth (arrow, Fig. 2-26G). The data also raise some interesting questions about cellular stereognosis during morphogenesis. What role does surface contact and membrane surface recognition factors serve, if any, in determining terminating steps in morphogenesis. One interpretation of the incomplete skeletal growth in the misaligned region is that "normal" processes of pore construction proceeded until the two surfaces of the cytokalymma overlapped and touched. Then, development ceased. Are surface tactile stimuli, primitive stereognostic cues to signal termination sequences in development? The finger-like incomplete bar (Fig. 2-26E) appears to have undergone a terminating expansion at the tip, upon contact with the underlying surface, that created the approximation to a semicircular edge that would have produced the rounded pore if the tip had in fact contacted an opposing coplanar surface. The abnormally abrupt ending may allow us to infer that the cellular mediating cues of membrane contact signaled a termination step of lateral expansion and protrusion as would produce the final union of a bar within a coplanar edge upon contact. Imagine, if you will, that the expanded tip in Fig. 2-26E had contacted an opposing rim instead of the underlying surface, the outward flaring of the edges and the forward protrusion of the tip against the opposed surface would have created a bridge bar delimiting on either side the semicircular edge of a pore. We see, moreover, that other pores partially completed (Fig. 2-26F) or nearly normally completed (Fig. 2-26G) as rounded pores have developed over regions where no contact was made with the underlying surface. As soon as contact is made, growth ceases. In Fig. 2-26G, an additional interesting anomaly appears to confirm that apposed contact induces a bar termination response, whereas lateral contact induces a pore closure response. Note that a mixed mode termination has occurred at the juncture of a nearly completed pore (arrow), where it made surface contact with the underlying developing layer of the shell. The right-hand part of the pore has closed in a nearly normal way, but the top and left-hand part of the rim exhibit the swollen profile characteristic of a bar-termination response, mimicking that of the free bar (asterisk), whose both sides have flared upon contact with the underlying surface. These bar termination responses can be compared to the light microscopic views in Fig. 2-25I. Here we also see the flaring of the tips of bars where they unite to form a curved rim of a pore.

Thus we can hypothesize the following model for early stages of skeletal morphogenesis. The cytokalymma commences development of the skeleton at various points and perhaps at varying rates throughout the surface of the incipient shell. We may think of it by analogy to building a jigsaw puzzle by assembling sections of it simultaneously until the parts unite. As the developing sections of the skeleton are elaborated, they proceed by pore-terminating steps; that is, they are deposited by arching arms that fuse with one another at their edges and give rise to closed pores, from whose edges yet additional arms are produced. This continues until opposing segments contact one another, as, for example, an arm contacting an opposing rim of another section. This induces bar-terminating responses, a flaring of the tip to produce the curved edges of the two adjacent pores completed on either side of the partitioning bar. Additional crossbars may form that also produce bar-terminating responses when they contact an opposing rim. These two morphogenetic units of pore-termination responses and bar-termination responses appear to be minimal essential steps in explaining the composite data presented in Fig. 2-26.

Among different species, there appears to be a range of contributions of bridge growth and rim growth as shown in Fig. 2-27. The shells in Figs. 2-27A–C exhibit substantial contribution of bridge growth as evidenced by the bar-like partitions between some of the pores. The shells in Figs. 2-27D–F exhibit increasing amounts of rim growth as evidenced by the more circular pores and tubular elongations on the pore rims. Further evidence for varying amounts of bridge and rim growth in shell morphogenesis among five species is presented in Table 2-3. Five biometric parameters were investigated: (1) mean shell diameter, (2) mean ratio of the bar width to hole diameter, (3) mean ratio of the major axis to minor axis for elliptical pores (the ratio approaches unity for nearly circular pores), (4) percentage of pores that have circular rims, and (5) mean hole diameter of the pores. The mean ratio of bar width to hole diameter is in part a convenient measure of the extent of rim thickening or silica deposition on the partitions between the pores. The measurements of bar width are made from rim to rim along the line of minimum distance between the pores. The diameter of the pores is measured along the same line projected across the pore. Of the five biometric parameters, the mean bar/hole ratio, mean major/minor axes ratio, and

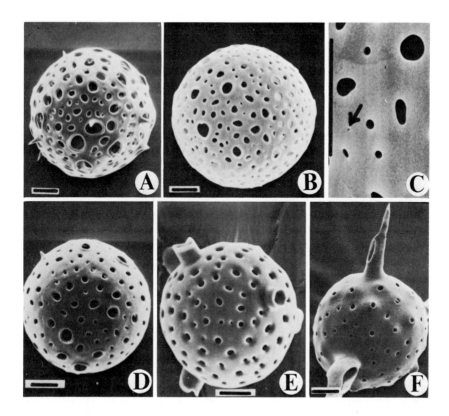

Fig. 2-27. A series of shells exhibiting increasing amounts of rim growth as evidenced by more circular pores and gradual accretion at the rims resulting in some of the pores becoming nearly closed (arrow, **C**). Scale = 2 μm. (Anderson and Swanberg, 1981)

percentage of holes that are circular were particularly useful in discriminating among shell growth patterns.

When the bar/hole ratio is small and the pores are elliptical (large major/minor axes ratio), the predominant form of growth is bridge growth. This is also reflected in a low percentage of pores with circular rims. Conversely, when the bar width or shell partition between pores is large compared to pore diameter and there is less ellipticity as shown by a lower major/minor axes ratio, then rim growth is more likely to predominate. The data in Table 2-3 show a predictable trend toward increasing contribution of rim growth progressing from *Collosphaera huxleyi* with skeletons exhibiting largely bridge growth to *Siphonosphaera socialis* with predominantly rim growth. This trend is particularly indicated by the consistent increase in mean bar/hole ratio and

Table 2-3. Shell morphology and pore geometry data for a range of developmental patterns: Predominately bridge growth to predominately rim growth[a]

Species	Mean shell diameter (μm)	Mean bar/ hole ratio	Mean major/ minor axes	Percentage circular	Mean hole diameter (μm)
Collosphaera huxleyi	143	0.51 (0.27)[b]	1.57 (0.35)	6	11.75
Acrosphaera cyrtodon	112	1.13 (0.52)	1.48 (0.31)	5	6.70
Acrosphaera spinosa	107	1.59 (0.67)	1.41 (0.39)	19	4.20
Siphonosphaera tubulosa	103	1.71 (0.51)	1.36 (0.29)	21	4.64
Siphonosphaera socialis	88	3.99 (0.90)	1.18 (0.20)	40	2.70

[a] Anderson and Swanberg (1981).
[b] Figures in parentheses are standard deviations.

a concomitant decrease in mean major/minor axes ratio. Simultaneously, the percentage of circular holes increases dramatically, especially between Collosphaera huxleyi and Acrosphaera spinosa and between A. spinosa and Siphonosphaera socialis. Likewise, the mean hole diameter decreases, as would be expected with increasing thickening of the rims. Circular pores, exhibiting varying degrees of rim thickening pore closure due to extreme deposition by rim growth, and tubular elongations of the pore rim are particularly evident in Siphonosphaera tubulosa and Siphonosphaera socialis. An analysis of variance of the mean differences among the major/ minor axes coefficients and the bar/ hole ratios was statistically significant ($p < 0.01$).

During early shell morphogenesis, if the final shape of the shell is spherical, then certain constraints occur on the length and/or degree of curvature of the initial lattice bars. The length of the lattice bars must be relatively small compared to the radius of the developing sphere, or they must be bent to follow the contour of the ultimate sphere. This is a necessary, but not sufficient, condition for sphericity. Clearly, if the initial bars are long relative to the radius of the developing sphere and they are relatively straight, then the skeletal lattice will be decidedly polyhedral and increasingly so as the bars have lengths approaching the dimensions of the radius. Given the slight tangential thickening that occurs, the wall could not be rendered spherical if the initial stage was decidedly polyhedral. Small bar lengths and even appropriate curvature, though necessary for sphericity, are not sufficient conditions. If the cytokalymma is distorted from a spherical shape either by normal control processes during morphogenesis or by trauma, then the ultimate shell will of course not be spherical. This is true, even though parts of the surface lattice meet the necessary conditions of short lattice bars and proper curvature. Thus, in addition to the basic principles of solid geometry that must be satisfied for sphericity, the construction of the cytokalymma as the guiding mold for ultimate shape becomes a critical factor in determining how the latticed body is eventually assembled into a final form. Examination of early stages of shell deposition (Fig. 2-25I) indicates that a typical ratio of bar length to shell radius is in the range of 0.1 to 0.33 for *Collosphaera huxleyi*; the former value is for one of the shortest bars and the larger value is for one of the longest bars. If these same bars are expressed as circumferential increments, that is, as a proportion of bar length/circumference of shell, we obtain the values of 1.6×10^{-2} and 4.3×10^{-2}, respectively. This is interpreted as each bar length being about 1/100 of the circumference of the shell. Even if these very small segments were nearly linear, the approximation to a sphere would be possible by their combined contribution in the lattice surface. Indeed, most scanning electron microscopic images suggest that the bars are linear or slightly circumflex. On the whole, however, a typical *Collosphaera huxleyi* shell is usually imperfectly spherical with indentations and ridges on its surface (Fig. 1-15A). This may result from the cytokalymma being distorted during silica deposition thus conferring a less than spherical form to the shell. Some of these indented impressions are more obvious on the sides of the shell than on the frontal surface.

In the species with thickened pore rims, particularly those with elongated tubular projections, the cytokalymma must be differentiated in these regions to secrete the elaborate surface flutes and cylinders that are added on the pore rim (Fig. 2-27E). In some cases the elongated pores are imperfect cylinders with deeply incised rims or they are partially closed at their distal ends. Again, we know very little about the cellular control mechanisms that regulate these morphogenetic processes. If the form of the cytokalymma is at least partially determined by cytoplasmic structural elements, such as microtubular cytoskeletons, then the addition of microtubule inhibitors to the culture medium while skeletogenesis is occurring should disrupt the normal development of the shell and produce teratological effects. Biochemical investigations and chemical probe analyses of this kind are limited at present.

Skeletal Composition   The presence of an organic and perhaps nonsiliceous inorganic infrastructure in radiolarian skeletons indicated by electron microscopy raises the interesting question of the chemical composition of radiolarian skeletons. It is widely recognized that radiolarian skeletons are composed of amorphous silica (opal) rather than crystalline silica. Kamatani (1971) examined the physical and chemical characteristics of diatom silica and concluded that there are many forms in biogenous silica ranging from silica gel to pseudo-opaline silica. It is not known to what extent his observations apply to radiolaria, although in both diatoms and radiolaria, the skeletal silica is deposited within a living membrane (silicalemma) and therefore may be expected to have similar biogenic properties. The chemical formula for silica is $SiO_2 \cdot nH_2O$, which is the hydrated form of amorphous, polymerized silicic acid, with undetermined molecular weight. The chemical composition of biogenic opal is such that divalent cations such as Mg^{2+} or Ca^{2+} may be present in the polymerized matrix of the skeletal wall; however, little is known about the distribution or composition of these inclusions in living radiolarian skeletons. Stanley (1973) reported the presence of Mg, Ca, Na, and Al in concentrations of 1–4% within the cleaned skeletons of fossil and recent species of radiolaria examined by electron microprobe assay. K and Sr were present as trace elements, but no Cl or Fe was found. A potentially significant area of research may exist in determining the role of divalent or polyvalent cations as binding centers for the anionic silica during early stages of silicogenesis.

Considerably more is known about the organic composition of radiolarian skeletons, particularly with reference to the amino acid content. King (1975, 1977) has examined the amino acid composition of a protein matrix in radiolarian skeletons obtained from core tops in sediment samples. He found a mean concentration of 1,465 nmole of amino acid/g of skeleton. Among the radiolarian samples that he examined, the major eight amino acids in decreasing order of abundance were glycine, aspartic acid, glutamic acid, alanine, valine, serine, leucine, and threonine. It is interesting to note that, with the exception of glycine, the most abundant amino acids (aspartic acid and glutamic acid) bear anionic side groups at physiological pH. The significance of these negatively charged groups in the molecular biology of skeletal deposition remains to be determined. A comparative analysis of amino acid composition in some spumellarian species (Spongoplegma antarcticum and Spongotrochus glacialis) and a nasselarian (Cyclampterium neatum) showed remarkably similar abundance patterns of amino acids in the two groups. Detailed analyses of amino acid composition revealed significant differences among the three species in the relative proportions of amino acids present. Approximately 25% of the total amino acid composition is free; however, none of the glutamic acid is in the free state (King, 1975). Comparative data on amino acid composition of diatoms (Volcani, 1981) and planktonic foraminifera (King, 1977) show striking differences with those obtained for radiolaria. Although glycine is also highly abundant in diatom thecae, threonine not aspartic acid is the next most abundant. In planktonic foraminifera, aspartic acid is most abundant, followed in decreasing order by glycine, glutamic acid, alanine and serine. Volcani (1981) has also found three new amino acid analogs of collagen residues in diatom thecae, i.e., a proline analog and two analogs of hydroxylysine.

Further research is warranted on the skeletons of living radiolarian species to determine if unique organic constituents are present within the organic matrix that may act as determinants of molecular structure. Quaternary ammonium groups with a stable positive charge are very likely functional groups to aid silicification, as they may bind the anionic silicic acid during early stages of deposition. Some of the cephalins, lecithins, and lysine are examples of these kinds of compounds that could be present in the organic matrix. The quantity and spatial distribution of the organic components must also be determined toward a more accurate understanding of the biomolecular bases of silicification in radiolarian skeletons.

Reproduction

Reproduction by binary or multiple fission, budding (in colonial forms), and flagellated spore formation was reported as early as the late 19th century (e.g., Cienkowski, 1871; Brandt, 1885, 1890; Haeckel, 1887). In some of the earlier accounts, the flagellated spores were interpreted as a sexual reproductive stage, and this view has been summarized in reviews of radiolarian literature (Gamble, 1909; Tregouboff, 1953). There is, however, no convincing evidence that the flagellated swarmers are gametes. They have not been reported to form zygotes or offspring. Consequently, to avoid confusion and perhaps a premature conclusion about the mode of reproduction by flagellated swarmers, they will be described only as reproductive swarmers.

Kling (1971) has suggested that some radiolaria may have sexual reproductive stages with dimorphic alternating generations analogous to the dimorphic, sexual–asexual, alternating generations in benthic foraminifera. He examined 19 species of radiolaria from the eastern North Pacific Ocean and found pairs of closely related forms. The "dimorphic" pairs are distinguished by similar modification of the same shell characteristics (size, shape, and pore structure). Members of each pair have essentially identical distributions; the two forms frequently occurred in the same sample. Hence their differences in morphology cannot be explained as regional variations of subspecies. An example of the description of morphological pairs observed by Kling (1971, p. 664) is as follows:

Castanidium apsteini Borgert, 1901
> Form A: Pores circular to subcircular, mouth large and circular; main-spines short and numerous.

Form B: Pores often subpolygonal, intervening bars with axial structure; mouth somewhat smaller and polygonal; main-spines longer and thinner.

Castanidium longispinum Haecker, 1908

Form A: Pores circular to subcircular, intervening bars relatively wide.

Form B: Pores subpolygonal, intervening bars narrow.

This very clever interpretation deserves additional empirical investigation; however, until actual syngamy is observed in radiolaria, it will be very difficult to state conclusively that variations in morphology are related to sexual reproduction. Inferential data of the kind reported by Kling are useful in guiding our search for species that are potentially suitable organisms for laboratory research on reproductive processes.

Fission

Reproduction by fission has been observed in Spumellaria and Phaeodaria. Among the Spumellaria, it is most commonly observed in colonial species, especially those that lack skeletons or possess only spicules (e.g., *Collozoum inerme* and *Sphaerozoum punctatum*). Fission of cells within the colony may occur repeatedly, thus increasing its mass. Spherical or spheroidal colonies become elongated, and the gelatin is constricted at intervals to form a segmented cylindrical colony, which eventually separates at the constricted regions into two or more daughter colonies. In spicule-bearing species, the enclosing layer of spicules becomes segregated around the daughter cells at the time of cell division. Multiple fission has been reported in the Thalassophysidae and may be the only means of increase in this family. The central capsule and nucleus become irregular-branching, vermicular, or radiating structures. The oil droplets and pigment granules are dispersed throughout the cytoplasm of the intracapsulum. The nucleoplasm separates into numerous distinct corpuscles followed by rapid division of the capsule and intracapsulum. The extracapsulum divides into numerous product fragments, each containing several nuclei surrounded by a layer of oily endoplasm and supported by the fragment of the extracapsulum. The subsequent development of these bodies is unknown. Multiple fission has been reported in *Thalassophysa spiculosa*, *Th. pelagica*, and *Th. sanguinolenta* (Brandt, 1902).

Brandt (1885) also reports asexual reproduction by budding of the extracapsular envelope in small colonies of *Collozoum inerme*, *C. radiosum*, *C. fulvum*, and *Sphaerozoum neapolitanum*. The buds appear as a highly refractive, lobate mass of endoplasm surrounding a cluster of fat globules with "modified muclei" at the periphery of the lobes. They cling to the jelly of the colony and may become part of the colony or give rise to flagellated microswarmers. Brandt emphasizes the latter.

These observations must be tempered in light of modern information on parasitism in radiolaria. Some of these earlier studies incorrectly identified escaping intracellular parasites as reproductive cells. This holds true also for some of the heterosporous swarmers observed in colonial radiolaria. I have not seen the budding process in colonies as described by Brandt, nor have I read recent reports of its occurrence. Hence it is advisable to treat these observations of budding with scholarly skepticism.

Fission in the Phaeodaria is nicely exemplified by *Aulacantha scolymantha*, a solitary radiolarian with a fine siliceous spherical skeleton composed of a peripheral veil of tangential needles penetrated by simple radial siliceous tubes; the distal third part of each tube is dentate bearing numerous (10 to 40) fine teeth. The polyploid nucleus is very large and centrally located (Fig. 2-28). During fission, the nucleus exhibits an exceptional binary division that differs significantly from a normal mitosis (Grell, 1973). The chromosomes do not align at a spindle equatorial plane, and there is no spindle apparatus. The chromosomes organize into a contorted "mother plate" (Figs. 2-28A, B), with their axes aligned parallel to the direction they will ultimately move (Fig. 2-28C) and not perpendicular to it, as is normally the case in mitotic metaphase. The daughter plates are also contorted at first (Fig. 2-28B), but eventually as they move apart, they form plane parallel discs (Fig. 2-28D). The daughter nuclei become segregated in separate masses of endoplasm. The central capsular wall disappears for a while, but reforms around the two masses of cytoplasm. Last of all, the extracapsulum, phaeodium, and spicules divide into two groups, and the entire organism divides into two. In Phaeodaria bearing a shell, various modifications of the process occur. In the Challengeridae, for example, the cell undergoes division within the helmet-shaped shell. One of the daughter cells escapes through the oral aperture and develops into a new individual.

Almost nothing is known about the factors regulating reproduction by fission or by swarmer production. Neither the rate at which cells divide, as, for example, within colonies, nor the frequency with which colonies separate to form daughter colonies has been determined. These processes may very well have seasonal rhythms or vary with environmental conditions of water depth and intensity of illumination, but little systematic research has been pursued in this field.

Swarmer Production

The release of flagellated swarmers has been reported in polycystine and phaeodarian species (e.g., Brandt, 1885; Borgert, 1909; Huth, 1913; Enriques, 1919; and Tregouboff, 1953). Much of this earlier work was based on light microscopy, and the details of swarmer production and release could not be resolved. Nonetheless, some very significant observations were made on the general pattern of swarmer production,

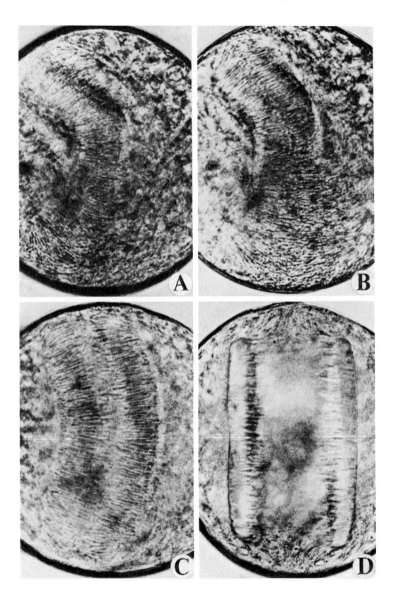

Fig. 2-28. Chromosomal organization during mitosis in the polyploid nucleus of the radiolarian *Aulacantha scolymantha*. (Grell, 1973)

the organization of the intracapsulum prior to swarmer release, and the organization of swarmer cytoplasm with respect to crystalline inclusions and the position of the nucleus. Some of the earlier literature contained reports that some radiolaria produce two kinds of swarmers, a large swarmer (macrospore) and a smaller swarmer (microspore). In some species, particularly in Collodaria, colonies were reported to yield only one type of swarm cell or in other cases both kinds of swarm cells. This led to the erroneous conclusion that radiolaria were either isosporous or anisosporous. The larger swarmers were thought to be female gametes and the smaller swarmers, male gametes. Subsequent cytological and fine structure research (Hovasse, 1924; Chatton, 1934; Hollande and Enjumet, 1953, 1955) has confirmed that the putative anisospores were in fact motile stages of parasitic peridinien dinoflagellates of different species probably belonging to the genus *Merodinium*. At least six infective species were described, and their developmental stages in the host were elucidated. The motile peridinien parasite enters the host cell and invades the nucleus, where it develops into a plasmodial stage. At maturity the plasmodium divides into numerous motile infective stages which escape from the host cytoplasm and once again invade another host, thus completing the life cycle. When multiple infection occurs, the motile stages may be of different sizes, thus accounting for the erroneous conclusion of anisosporogenesis in radiolaria. Among the conclusive lines of evidence confirming parasitism were the clear differences in nuclear organization between the host cell and the putative anisospores. In *Thalassicolla* sp., for example, at mitotic division there are at least 20 chromosomes (Belar, 1926), whereas the presumptive anisospores always possess much fewer (Pätau, 1937a,b). The parasite swarmers, moreover, possess a different shape and do not possess crystal inclusions observed in radiolarian reproductive swarmers. Consequently, modern researchers accept the conclusion that radiolaria produce only one size of swarmer (isospores); however, as these have not been observed to develop further, their role in reproduction remains undetermined.

Present knowledge of the temporal sequence of swarmer production and release is based largely on laboratory observations of specimens maintained in seawater culture (Hollande and Enjumet, 1953; Anderson, 1976c, 1978b, 1980, 1983). Representative events during swarmer production are presented for a solitary species *Thalassicolla nucleata* and a colonial species *Sphaerozoum punctatum* (Figs. 2-29 and 2-30). The earliest evidence of impending swarmer release in a laboratory cultured specimen is rejection of prey. This not an invariant sign, however, as some nonreproductive, recently fed specimens of *T. nucleata* also refuse prey. Confirmation of incipient swarmer release occurs within a matter of several hours when the alveolate extracapsulum is abruptly discarded. This is truly a remarkable event to witness, as

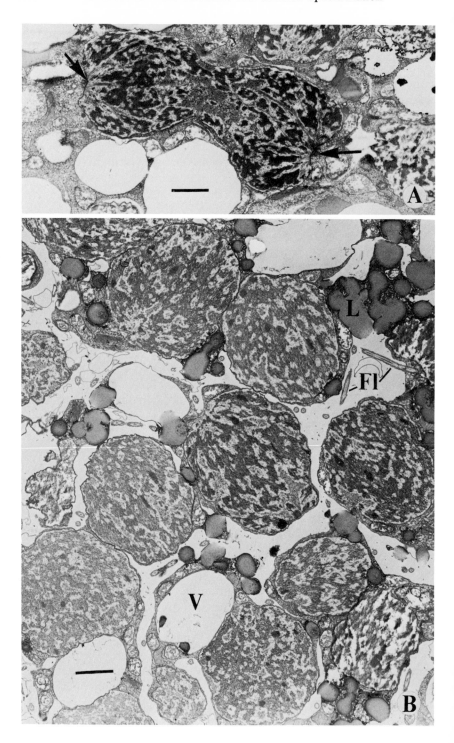

it occurs so rapidly and decisively. The normal-appearing extracapsular coat is suddenly jettisoned, and the large, dark central capsular sphere immediately sinks in the culture dish coming to rest on the bottom. There are no visible remains of the extracapsulum after it is shed. It appears to disintegrate into a very small mass once the alveoli have burst. The central capsule of some specimens may lose the dark pigment coat surrounding it and become a pearly-white sphere. In other cases, the pigment layer is retained, which facilitates observation of the next stage when a small white mass of cytoplasm protrudes from a pore that develops at one point in the capsular wall. The bulge of cytoplasm appears viscous and somewhat granular with fine cytoplasmic strands threaded throughout it. The bulge increases in size until it eventually fragments releasing numerous flagellated swarmers. Each swarmer observed with high power light optics possesses a crystalline inclusion appearing as a fine refractile rod of classical description (Tregobouff, 1953). The swarmers are slightly pear shaped and may have a fine conical projection of cytoplasm on the small end opposite the side bearing the flagella. The swarmers exhibit a sluggish, gyrating swimming motion, but usually move away from the mother cell along a fairly direct line rather than spiraling or gyrating near the mother cell. A considerable variation in behavior is observed among swarmers released in laboratory culture. Some are nonmotile and remain clustered around the mother cell. Others exhibit only weak motility and seldom swim very far from the point of release. It is not known whether this is normal variability in motility or an artifact of the culture conditions. Fine structure evidence amplifies the details of swarmer development and release (Anderson, 1978b). Before the extracapsulum is shed, the mother nucleus which is usually centrally located, undergoes repeated division to produce numerous daughter nuclei that proliferate throughout the intracapsulum (Figs. 2-30A, B). Groups of the nuclei become segregated into islands of cytoplasm, which possess typical vital organelles including mitochondria, Golgi bodies, endoplasmic reticulum, and lipid droplets that apparently serve as an energy source for the developing

---

Fig. 2-29. (A) A dividing nucleus of *Sphaerozoum punctatum* in late anaphase exhibits the persistent nuclear membrane and a constriction zone in the medial plane producing a dumb-bell shape. Microtubules radiate from attachment sites (arrows) on the inner side of the nuclear membrane at the poles of the dividing nucleus. Scale = 2 μm. **(B)** Multinucleated masses of cytoplasm during swarmer production in *S. punctatum* possess large vacuoles (V) where crystalline inclusions will be secreted, lipid droplets as food reserve (L), and flagella (Fl) protruding into the free space between the segregated nucleated masses of cytoplasm. Scale = 1.6 μm. (Anderson, 1976c)

swarmers. Before the cytoplasmic bulge emerges from the central capsule, the clumps of cytoplasm begin to form large vacuoles (Fig. 2-30A) that already contain the developing crystal (I, Fig. 2-30B) included within the swarmer. The multinucleated masses of cytoplasm undergo repeated division yielding the individual swarmers (Fig. 2-30B). They possess a prominent nucleus surrounded by a thin layer of cytoplasm. The small lipid inclusions are undoubtedly derived from the large lipid bodies found in the mother cytoplasm prior to nuclear division. The included crystal is clearly contained within a vacuole, but fractures during ultrathin sectioning and is lost, leaving only a clear space. The crystals are always enclosed in a vacuole, which is located slightly excentric to the nucleus and often creates a bulge in the thin layer of cytoplasm due to the bulky crystal. The swarmers are somewhat tapered on the side opposite the flagella and sometimes exhibit a pronounced prolongation of the cytoplasm from the tapered end.

The fine structure events during swarmer production in two colonial species (*Sphaerozoum neapolitanum, Collozoum pelagicum*) and in *Plegmosphaera* sp. was examined by Hollande (1974). He reported the fundamental features of the swarmers including the presence of a strontium inclusion in the cytoplasm, nuclei with cord-like masses of chromatin, and cytoplasmic lipid inclusions (presumably food reserves) mixed among the membranous organelles. He did not investigate the sequence of events during swarmer production. Some recent research at our laboratory has elucidated the events accompanying swarmer release in the colonial species *Sphaerozoum punctatum* (Anderson, 1976c).

Production of swarmers in *Sphaerozoum punctatum* in the main resembles the sequence in *Thalassicolla nucleata* with modifications unique to its colonial form. As with other radiolaria observed thus far, one of the early signs of reproduction is prey rejection followed within hours by a clearly perceptible shrinkage of the colony attributable to the contraction of the rhizopodia in the extracapsular network. Concomitantly, the gelatinous sheath becomes less firm and aids the ensuing contraction of the colony into a mass of closely adhering central capsules. As the jelly sheath is contacted, the colony loses buoyancy and settles in the water. This process of descent is much slower than that observed in *Thalassicolla nucleata*; however, when the cells have been completely aggregated, the "condensed" mass sinks rapidly and in the open ocean may reach a depth of 400 to 600 m by the time the swarmers are released. These estimates are based on sinking rates observed in our laboratory and those previously published (Gamble, 1909). The closely aggregated mass of cells often becomes tightly compacted and bound within a dense network of actively streaming cytoplasm. Within a period of several hours after settling, the large lipid droplet at the center of the cell becomes dispersed in the cytoplasm as smaller lipid

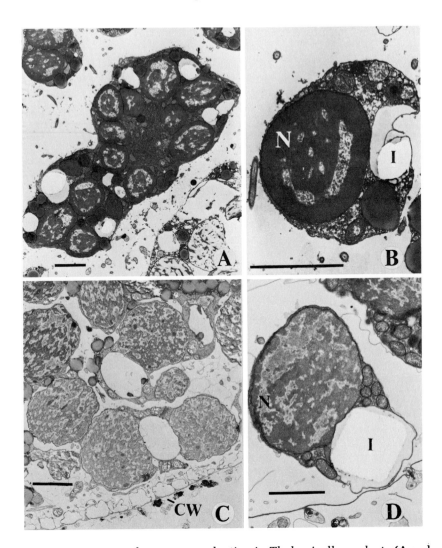

Fig. 2-30. Late stages of swarmer production in *Thalassicolla nucleata* (**A** and **B**) and *Sphaerozoum punctatum* (**C** and **D**). Multinucleated lobes of cytoplasm (**A**) in *T. nucleata* gradually segregate into flagellated swarmers (**B**) containing a large nucleus (N) and a crystalline inclusion (I) that has fractured and fallen out during sectioning leaving only its profile. (Anderson, 1978b) Similar stages in *S. punctatum* exhibit the delicate capsular wall (CW) enclosing incipient swarmers (**C**) that eventually are released within the intracapsulum (**D**) before rupture of the capsular wall and release. Scales = 3 μm. (Anderson, 1976c)

droplets, which presumably supply the developing swarmers. The numerous nuclei that have proliferated well in advance of the onset of colony shrinkage (Fig. 2-29A) can be detected in living cells with the light microscope, owing to the very thin, translucent capsular membrane. Within an hour or two before swarmer release, the distinctive crystalline inclusions are clearly visible and resemble small refractile rods nearly equally spaced from one another and apparently randomly oriented. Electron microscopic examination of ultrathin sections of central capsules at the earliest stage, when crystalline inclusions can be detected by light optics, shows the presence of interconnected, multinucleated masses of cytoplasm (Figs. 2-29B and 2-30C) containing large vacuoles interspersed among the nuclei. The vacuoles enclose developing crystalline inclusions that appear before the individual swarmers are released from the mother cytoplasm. The very thin central capsular membrane surrounds a loosely connected layer of vacuoles that have become segregated from the internal nucleated masses of cytoplasm. Shortly before swarmer release, the capsular membrane disintegrates, apparently by separation at the fine slit-like junctures where the component segments of the capsular membrane join to one another. The individual swarmers (Fig. 2-30D) escape profusely from the disrupted membrane of the mother cell. They sometimes settle on the surface of the culture dish and become attached by a thin prolongation of cytoplasm issuing from their tapered end. They remain attached for some time and rotate slowly around the point of attachment. The significance of this settling phenomenon is not known, nor do we know whether they become attached to any surface in the open ocean. The swarmers have never been observed to develop further in the laboratory.

Based on reports in the early literature (Brandt, 1885; Tregouboff, 1953) and present fine structure research, it appears many solitary and colonial radiolaria exhibit a similar general pattern of events in swarmer production: (1) proliferation of nuclei by repeated division of the mother nucleus sometimes well in advance of swarmer production, (2) segregation of the cytoplasm into multinucleated masses containing dispersed lipid droplets apparently derived from the large lipid reserves in the mother cell, (3) progressive vacuolarization of the cytoplasm and the appearance of crystalline inclusions within the vacuoles prior to swarmer release, and (4) separation of individual swarmers from the intracapsular masses of cytoplasm concomitant with rupture of the capsular membranes releasing the motile swarmers. Apparently most species reported in the literature lose buoyancy at some point prior to or concurrent with the second stage of cytoplasmic segregation; however, the mechanisms for settling vary. Among the solitary species, it is common for the extracapsular cytoplasm to be shed or resorbed in the multinucleated stage, whereas colonial radiolaria appear to increase their density by contraction of the colony and concurrent collapse of their alveoli when present. It is not known to what extent the crystals in the swarmers increase their density, but clearly they must contribute substantially to the weight of the organism. The chemical composition of the crystals has long been of interest. Brandt (1885) suggested that the crystals may be strontium sulfate (the same substance as in acantharian skeletons) based on the symmetry of the crystal. His hypothesis has been strengthened by recent electron probe analyses, showing that the crystals in *Sphaerozoum neapolitanum* are rich in strontium (Hollande and Martoja, 1974), and observations in our laboratory using electron dispersion X-ray analysis, showing abundant strontium in crystals of swarmers from *Sphaerozoum punctatum*. The presence of strontium sulfate within reproductive swarmers in radiolaria may be an indication of an ancestral trait, thus perhaps linking radiolaria and Acantharia to a common primitive phylogenetic ancestor with a strontium sulfate skeleton (Anderson, 1981).

The proliferation of nuclei prior to swarmer production affords an excellent opportunity to observe the cytology and fine structure of nuclei in various stages of division. Hollande and Enjumet (1953) made cytological observations on the staining properties of nuclei in collodarian radiolaria at various stages of the reproductive cycle during swarmer production. They noted the curious observation that the chromosomes are Feulgen negative, regardless of several standard preparative techniques, until the stage of early swarmer production when the nuclei begin to divide. Then, the chromosomes become faintly Feulgen positive. At this point the chromosomes appear very thin. The fuchsin response increases through their sequential development as they become encased once again in a matrix. The nuclei of the swarmers (spores) exhibit maximum chromaticity. Hollande and Enjumet discuss the possible reasons for this chromogenic variablility including (1) changes in DNA composition or content and (2) the variations in chromosomal coiling. They favor the second hypothesis as being very probable and explain the faint staining of Feulgen negative chromosomes as a result of their fine dispersement compared to swarmer nuclei. Electron microscopic evidence (Anderson, 1976a,c; 1978a) also supports the conclusion that the multiple nuclei preceding swarmer production possess very condensed masses of chromatin (cords) that are much more evident than the finely dispersed chromatin found in the solitary mother nucleus of some species. Additional research is needed to clarify the variations in chromatin organization as a function of varying developmental stages and variations in physiology of the radiolaria.

The mechanism of nuclear division has been observed in some spumellarian species (Anderson, 1976a,c). There is not a normal mitosis as the nuclear envelope is persistent during karyokinesis and does not disintegrate as in many cells. An electron micrograph (Fig. 2-29A) of a late anaphase stage of nuclear division in *Sphaerozoum punctatum* exhibits the dumb-bell shape profile of the nucleus. The chromosomes, enclosed within the persistent nuclear envelope, have begun to move to the poles of the nucleus. A constriction ring (apparently composed of a circumferential band of contracitle microfilaments) has begun to contract at the equator, thus rendering the bilobed profile. Eventually, the constriction ring contracts until the nucleus is separated into two daughter nuclei, each containing an entire complement of the duplicated chomosomes.

The mechanism of chromosomal attachment to the nuclear membrane is shown in Fig. 2-29A (arrows). A thickened osmiophilic plaque on the inner membrane of the nuclear envelope serves as an attachment surface for the intranuclear microtubules that converge toward the central part of the nucleus. A small osmiophilic granule lies immediately outside the nuclear envelope and may serve as a microtubule organizing center or as a primitive centrosome. The microtubules intersect with electron dense masses of chromatin, which appear to be attachment sites on the chromosomes. It appears that the persistent nuclear membrane and attached microtubules pull the chromosomes apart during elongation of the nuclear envelope, which is in contrast to normal mitoses where the spindle fibers appear to provide the translational motion to separate the chromosomes. Variations in attachment of the chromosomes to the inner membrane of the nucleus are observed among species. In *Collosphaera globularis* (Fig. 2-13A), the chromsomes attach to a flattened segment of the nuclear envelope, which imparts a decidedly aspherical shape to the nucleus. In *Sphaerozoum punctatum*, the attachment site is slightly convex on the inner side, thus rendering a distinct depression on one side of the nucleus where the chromosomes are attached. Thickened attachment sites on persistent nuclear membranes are not unique to radiolaria, however. They have also been reported in such diverse organisms as fungi (Robinow and Marak, 1966; Zickler, 1970), amoeboflagellates (Schuster, 1975), and benthic foraminifera (Schwab, 1975). In the foraminifera, the plaque lies outside the nuclear envelope rather than attached to the inner membrane as in some radiolaria. Additional research may be profitably directed toward examining nuclear fine structure with the aim of possibly elucidating phylogenetic affinities among species and clarifying taxonomic categories.

Fine Structure Synopsis

Based on the literature currently available, some polycystine, solitary, and colonial radiolaria can be organized into general categories representing their major fine structure features. These are not intended as taxonomic groupings. They are convenient groups of taxa possessing similar fine structure features. It is likely, however, that these groups may overlap with or refine currently established taxonomic categories as both systems are further enlarged and refined. Current fine structure data are limited to a few major species among the hundreds believed to exist in the oceans. The categories presented here are a convenient synopsis of the general fine structure features of several representative groups of species and are clearly not exhaustive.

Each group of radiolaria possessing similar fine structure features is placed under a heading designated by a modified generic name of one of the organisms typical of the group. For example, those radiolaria with fine structure features resembling *Thalassicolla* sp. are placed under the general heading of Thalassicollid Type. Those resembling *Sphaerozoum* sp. are placed under the heading Sphaerozoid Type, and so forth. After each generalized description, some additional examples of species are cited that have not been described in the foregoing sections on fine structure. The categories of radiolaria are grouped under two major divisions: (1) Solitary Forms and (2) Colonial Forms. Within these major divisions, categories are listed in sequence from skeletonless forms to skeleton-bearing forms.

Solitary Radiolaria

Thalassicollid Type   The species included here possess a relatively large central capsule (diameter in millimeter range) surrounded by a thick organic capsular wall, perforated by long strands of cytoplasm forming the fusules passing through circular or nearly circular pores in the capsular wall. The nucleus is usually large and centrally located in the intracapsulum and is surrounded by a dense deposit of cytoplasm organized into closely spaced lobes radially arranged or contorted and interconnected in a nearly radial arrangement. The lobes are separated by narrow cisternae extending from the nuclear region to the periphery of the intracapsulum. Lipid deposits and vacuoles may be arranged in nearly radial rows or more diversely organized in patterns of concentric patches within the cytoplasm. The voluminous extracapsulum is organized as a massive corona of rhizopodia interspersed with digestive vacuoles in the pericapsular cytoplasm, and in some species it is associated with membranous alveoli in the more peripheral zones of the extracapsulum. Symbionts when present are enclosed within thin envelopes of rhizopodial cytoplasm. The fine structure of *Thalassicolla* sp. (Fig. 2-4B) exemplifies this group. During reproduction, in *T. nucleata* at least, the centrally located nucleus disperses into the peripheral cytoplasmic lobes which become widely spaced. In *Lampoxanthium* sp. (Fig. 1-2C), the closely spaced lobes are much more interdigitated than those in *T. nucleata* and they possess scattered vacuoles that tend to be more abundant in the region immediately surrounding the nucleus. The nucleus is centrally located, surrounded by a thin cytoplasmic layer, and delimited from the surrounding intracapsular lobes by a narrow cisterna. The capsular wall is thick and perforated as in *T. nucleata*. All symbionts examined thus far were dinoflagellates.

Thalassolampid Type   The central capsule is very large (several millimeter diameter range), with the thin capsular wall composed of a membranous envelope (0.08–0.1 µm thick) filled with an organic, electron-opaque deposit. Fusules are formed by an elevated collar-like opening (as in *Spongodrymus* sp., Fig. 2-14B) through which the fusule strand passes into the surrounding extracapsulum. Microtubules occur within the shaft of the strand, and a dense, osmiophilic plug partitions the proximal part of the fusule strand from the distal part. The fusule strand is continuous with widely spaced intracapsular lobes separated by large interlobular spaces. Large alveoli, surrounded by thin membranous septa connected within the intracapsular lobes, are frequently observed in the interlobular space. Mitochondria, Golgi bodies, and endoplasmic reticulum occur abundantly within the cytoplasm of the lobes. The large nucleus located near the center of the central capsule is peculiarly constructed of a nearly spherical central mass surrounded by a thick fibrous coat, perforated with pores through which lobes of the nucleus protrude and expand peripherally. The overall organization of the nucleus, therefore, is a central sphere enclosed in a porous organic capsule with lobose radially arranged protrusions of the nucleus covering its surface. The lobes are surrounded by a typical nuclear, double membrane envelope. The nuclear lobes are continuous with the surrounding loosely organized cytoplasmic lobes extending toward the central capsular membrane. In some places, the lobes are reduced to mere strands of cytoplasm and elsewhere become expanded. These expanded regions are well supplied with granular reserve substances surrounded by closely aggregated mitochondria. It is not uncommon for the periphery of the intracapsulum to contain numerous alveoli, distributed as a loose layer near the capsular wall membrane and interspersed with the narrow extensions of the lobes that terminate as usual strands passing through the capsular wall pores. During reproduction, the extruded nuclear lobes apparently generate daughter nuclei that become the nuclei of the swarmers. The thin gelatinous extracapsular envelope is penetrated by rhizopodial strands and, in some species, encloses siliceous spicules of varying shapes including curved needles, hook-shaped spicules, and those with a C-shaped structure.

Species of *Thalassolampe* and *Physematium* are presently included in these groups. The symbionts if present are dinoflagellates or sometimes prasinophytes (Cachon and Caram, 1979) based on current fine structure evidence.

Cladococcid Type   This group, especially, may require additional refinement and perhaps separation into subgroups as more data are accumulated. Included here are some of the anaxoplastidies (Cachon and Cachon, 1972b) exemplified by the genus *Cladococcus* and related genera including *Diplosphaera, Rhizoplegma, Lychnosphaera,* and *Arachnosphaera*. The nucleus is centrally located and may be surrounded by a thickened organic lamella (nucleotheca), through which small protrusions of the nuclear membranes bulge out into the surrounding endoplasm. The vesiculated endoplasm is compact and not so lobulate as in the Spongodrymid-type species. The axopodia originate as rods of microtubules inserted on the surface of the nucleotheca and radiate outward through the dense endoplasm into the fusule strands passing through the pores in the caspular wall. The slightly elevated rim of the fusule cylinder is in continuity with the cytoplasm of the axopodial sheath. An osmiophilic, dense plug of organic matter surrounds the microtubule bundle in the base of the axopodium, where the capsular wall fuses with the axopodial sheath.

Cyrtidosphaerid Type   This group is assembled from among the representatives of the Periaxoplastidies (Cachon and Cachon, 1972b). The nucleus contains an invagination enclosing the axoplast, from which originate radially arranged rods of microtubules extending through the endoplasm to the fusules in the capsular wall. The axoflaglellum, when present, also originates in the axoplast and emerges within a densely granular sheath of cytoplasm surrounding the axoflagellum. The shafts of microtubules pass through the fusules in the capsular wall and continue distally within the axopodia. A slightly granular plug of cytoplasm encloses the microtubule shaft in the proximal part of the fusule, where it joins with the base of the axopodium. A thin collar-like projection from this cytoplasmic deposit projects outward into the axoplasm. The capsular wall is usually a thin membranous envelope enclosing a dense deposit of organic matter forming the capsular wall.

Actinosphaerid Type   This group comprises the radiolaria assigned to the Centroaxoplastidies by Cachon and Cachon (1972a). The main distinguishing feature is the presence of the axoplast within a centronuclear position. The nucleus consists of a centrally located lacuna formed by the invagination of the nuclear envelope. This lacuna is lined by the nuclear outer membrane and is continuous with the surrounding endoplasm by way of radial canals. The rods of microtubules that originate in the axoplast pass outward through the radial canals into the endoplasm, where they continue radially toward the fusules in the caspular wall. The conical outward directed pore of the capsular wall is connected distally to the base of the axopodium. The fusule strand, passing from the endoplasm into the conical pore of the capsular wall, contains a dense osmiophilic deposit surrounding the microtubule bundle. In this region of the microtubule shaft, there is an osmiophilic fibrillar deposit between the microtubules. A dense disc of organic matter occurs at the level of the fusule where it enters the base of the axopodium. The genera included in this group are *Actinosphaera*, and members of the Families: Spongosphaeridae, Rhizosphaeridae, and Thecosphaeridae.

Spongodrymid Type   The spongiose, skeleton-bearing species exemplified by *Spongodrymus* sp. possess a central capsulum surrounded by a thin capsular wall containing numerous fine, complexly organized fusules, consisting of a central cytoplasmic strand surrounded by a perforated, outwardly directed collar originating from the capsular wall membrane. The nucleus is located in the center of the intracapsulum and in some cases is surrounded by a complex lacy network of membranes formed by the nuclear envelope. The surrounding cytoplasm is organized into radial lobes, usually widely spaced, sometimes differentiated into inner and outer segments with differentiated intracytoplasmic organization. The extracapsular cytoplasm consists of a corona of axopodia exhibiting surface contact with the symbionts and containing copious deposits of internal microfilaments lying parallel to the axopodial axis. Symbiont fine structure reveals that there are three kinds of symbionts that are present in *Spongodrymus* sp.: (1) a dinoflagellate, (2) a prasinophyte, and (3) a brown-pigmented alga. Only one type of symbiont, however, is associated with a given host.

Colonial Radiolaria

Collozoid Type   The central capsule is delimited by a thin boundary of capsular wall membranes containing fusules that in some cases emerge from an elevated cylindrical portion of the capsular wall. The intracapsular cytoplasm is organized into closely spaced radial or nearly radial lobes, or in some species radially arranged, contorted lobes, forming an interlaced pattern when viewed in tangential section. Intracapsular lobes are closely associated at the periphery with the capsular wall membranes and possess either a single large peripheral vacuole or a group of smaller vacuoles distributed in the distal part of the lobe near the capsular wall. The peripheral vacuoles may be electron lucent or possess an osmiophilic granule as in *Collozoum caudatum*. Nuclei possessing fine filamentous chromatin are widely dispersed within the intracapsular cytoplasm. Intracapsular vacuoles may be sparse or abundant and arranged in approximately concentric layers and varying in electron opacity and granularity of their contents. The extracapsulum consists of a pericapsular layer of vacuoles surrounded by thin, membranous, cytoplasmic boundaries connected to the fusules and the peripheral network of rhizopodia. Symbionts are enclosed within perialgal vacuoles within the rhizopodia that project as fine strands from the periphery of the colony. In *C. amoeboides*, possessing irregularly shaped central capsules, the lobes are not as regularly arranged as those in C. inerme and consist of interconnected and intertwined lobes of varying widths. These terminate with connections to the capsular wall membranes and are interspersed with large vacuoles occurring sporadically around the periphery of the intracapsulum. These vacuoles are not as numerous nor as regularly arranged as those in C. inerme or C. caudatum. They sometimes contain a dense spongiose, granular material. The periphery of the central capsule is surrounded by a layer of cytoplasm containing digestive vacuoles. One contained a partially digested alga resembling a symbiont as has been observed in C. inerme. The symbionts in all cases were dinoflagellates.

Sphaerozoid Type   These colonial radiolaria possess spiculate skeletons. The central capsule is surrounded by a thin membranous capsular wall, perforated by short fusules that connect rather directly with the extracapsular frothy layer of vacuoles in the gelatinous envelope. The intracapsulum frequently possesses a large, central lipid droplet and one or more occasional peripheral lipid bodies in the cytoplasm, which is organized as a network of closely spaced, interconnected lobes. A cisterna of varying widths and meandering configuration separates the intracapsular lobes. At the periphery of the intracapsulum near the capsular wall, the lobate cytoplasm may contain a single layer of vacuoles separated by a thin layer of cytoplasm or produce a more complex array of closely packed vacuoles. These may be interspersed in a staggered fashion when viewed in ultrathin sections. Numerous nuclei with cord-like masses of chromatin are scattered throughout the cytoplasm and are surrounded by a thin layer of cytoplasm containing mitochondria, Golgi bodies, endoplasmic reticulum, and other membranous organelles. Belanazoum sp. and Raphidozoum sp. are included in this group. Belanozoum atlanticum (?) possesses a loosely spaced network of interconnected lobes resembling S. punctatum, but with wider cisternae between the lobes. The periphery of the intracapsular cytoplasm is elaborated to form a layer of widely spaced vacuoles surrounded by a thin layer of cytoplasm connected to the intracapsular lobes. Nuclei possess cord-like masses of chromatin and are numerous. The pericapsular cytoplasm contains vacuoles filled with dense matter, which appears to be late stages of digested food particles. Raphidozoum pandora is similar to Belanozoum sp., but has more numerous closely packed vacuoles in the periphery of the intracapsulum forming a multiple layer of irregularly shaped, closely packed vacuoles separated by a thin cytoplasmic septum. All specimens examined thus far contained dinoflagellate symbionts.

Collosphaerid Type   Each central capsule is surrounded by a shell varying in complexity from an irregular latticed sphere to more complexly ornamented spherical shells. The intracapsulum is dominated by a large lipid droplet and the surrounding closely spaced, interconnected lobes form a compact mass of cytoplasm extending to the perimeter of the intracapsulum. In the region near the capsular wall membranes, the intracapsular cytoplasm may become segregated into a thin peripheral layer segmented by intervening cisternae (e.g., Fig. 2-13B) or variously segmented into peripheral thin lobes possessing vacuoles of varying dimensions. On the whole, however, the intracapsulum is much less vacuolated than in foregoing groups. Numerous nuclei with cord-like masses of chromatin are closely spaced within the cytoplasmic lobes in the intracapsulum. The fusules are connected with a delicate layer of pericapsular vacuoles separated by a thin cytoplasmic septum. The shell and surrounding cytokalymma lie immediately distal to this layer of ectoplasm.

Collosphaera globularis and C. huxleyi are characteristic of this group. They differ mainly in the organization of the peripheral part of the intracapsular cytoplasm. C. globularis often has a thin peripheral layer of cytoplasm segregated by a cisterna from the main central mass. But in C. huxleyi, the peripheral cytoplasm is much more irregular, consisting of a thin somehwat convoluted or sometimes vacuolated layer closely appressed to the numerous nuclei filling the intracapsulum. On the whole, the cytoplasm is much more compact with narrower interlobular cisternae than observed in C. globularis. Both species contain dinoflagellate symbionts.

Chapter Three
Physiology and Ecology

The remarkable complexity and highly specialized organization of radiolaria, as exhibited by their cytology and fine structure, suggest that their physiological differentiation may be equally elegant. Our current knowledge, however, is meager in comparison to the challenge that these very complex protists present. The long lapse of programmatic research between the late 19th century and the mid-20th century has been especially critical with respect to new knowledge of radiolarian physiological ecology. Only sporadic studies were reported between the beginning of this century and the post World War II period, and much of the present research is located in only a few laboratories including our own. We are beginning, however, to discover some of the functional correlates of cellular structure and to relate these findings to the broader ecological questions of interest to a wide interdisciplinary audience including biologists and micropaleontologists. The physiology and ecology of radiolaria are combined in this chapter to create a common epistemological framework for these usually separate disciplines, and also as a logical consequence of the significant interrelationship that exists between these delicate holoplanktonic protozoa and their environment. The open-ocean environment is remarkably consistent over vast regions of the earth's surface. Although variations in productivity and temperature as well as other physical and chemical factors may differ appreciably from one major geographical region to another, or within the water column, there are remarkably large regions of rather consistent environmental quality in the open ocean, and it appears radiolaria have made a close adaptation to their oceanic environment. They are, for example, exclusively holoplanktonic and seldom survive for long when carried into near-shore water, where there is little replenishment by open-ocean water. In regions where open-ocean currents come near to the land, or in inland deep water masses such as the fjords, the radiolaria and other holoplanktonic organisms are sometimes as abundant as in pelagic environments. We do not know at present what factors in many coastal waters are deleterious to radiolaria; however, their mortality in these regions clearly highlights the close link between radiolarian physiology and pelagic environment.

In the first part of this chapter, information is presented largely under physiological headings; however, reference to ecological questions and pertinent ecological principles will be interwoven with the presentation. The latter part of the chapter focuses more particularly on topics directly related to ecology. Some pertinent review articles and research reports on various aspects of physiology (Anderson, 1980, 1981, 1983) and ecology (e.g., Casey, 1966, 1971a,b; Casey et al., 1979a; Kling, 1966; Swanberg, 1979; Petrushevskaya, 1981; and Boltovskoy, 1981) may provide supplementary information on topics of interest.

Trophic Activity and Nutrition

Prey Diversity

Haeckel (1887) mentions quite incidentally that radiolaria ingest diatoms, tintinnids, and other calcareous monothalamic and polythalamic organisms. He does not describe the variety of prey observed in various groups. Many of his samples were undoubtedly improperly preserved to make these discriminations, and most were collected in nets, thus making it difficult to separate prey from adhering organisms introduced in the net. Recent studies on specimens collected in jars by SCUBA divers and laboratory studies on prey selectivity of spumellarian radiolaria have confirmed that many species are omnivorous (Anderson, 1978b, 1980, 1983; Swanberg, 1979). Current knowledge of prey consumed by some groups of radiolaria is summarized in Table 3-1. Little detailed information is available on diversity of prey among the three major groups of radiolaria (Spumellaria, Nassellaria, and Phaeodaria). The variation, if any, in predation by radiolaria in diverse geographical regions or at varying depths in the water column has not been determined. The smaller radiolaria as occur among the Nassellaria and some Spumellaria are a particular challenge in laboratory studies of predation as their small size makes it difficult to monitor their behavior over extended periods of time. Moreover, the prey they can accept is clearly very small. Some of the radiolaria are only 30–80 µm themselves, which suggests a priori that they ingest microplankton and perhaps even bacteria. Electron microscopic evidence (Fig. 2-8B) has confirmed that some Nassellaria consume bacteria (Anderson, 1977a); however, their predation pressure on microbial populations has not been investigated. Special radioactive labeling techniques and sophisticated light microscopic observations are required to investigate the predatory behavior of many of these small species.

Table 3-1. Prey observed in some Spumellaria and Nassellaria freshly collected from the open ocean and prey accepted in laboratory culture

Radiolarian	Natural Prey	Prey Accepted in Culture
COLLODARIA *Thalassicolla* sp.	Crustacean larvae, diatoms, thecate algae	Calanoid and harpacticoid copepods, *Artemia* nauplii, colorless flagellates (*Crypthecodinium cohnii*), diatoms (*Thalassiosira* sp.), Coccolithophora (*Coccolithus huxleyi*), dinoflagellata (*Amphidinium carteri*)
Collozoum inerme and *Sphaerozoum punctatum*	Copepods, ciliates	Copepods, *Artemia* nauplii Phytoplankton predation not examined, except *C. inerme* ingests algal symbionts occasionally.
SPHAERELLARIA *Spongodrymus* sp., *Hexastylus* sp., and *Diplosphaera* sp.	Copepods, tintinnids, nonthecate ciliates, copepod larvae, Pteropoda (205 μm), *Sticholonche* sp., trochophore larvae, Larvacea, Acantharia, silicoflagellates, diatoms	Copepod nauplii, *Artemia* nauplii, tintinnids, colorless flagellates (*Crypthecodinium cohnii*), dinoflagellata (*Amphidinium carteri*), diatoms (*thalassiosira* sp.), Coccolithophora (*Coccolithus* sp.)
NASSELLARIA	Bacteria	

A convenient method of simultaneously obtaining data on the amount of phytoplankton and zooplankton consumed by a predator employs a radioactive isotopic labeling technique. This method that I have previously applied to planktonic foraminifera (Table 3-1, comparative data) yields an estimate of the amount of phytoplankton prey consumed as a ratio of the amount of zooplankton prey consumed per unit of time. This ratio symbolized as "P/Z" is a coefficient of omnivorous trophic activity, and can be applied to predatory organisms of diverse size. Thus, it is a convenient technique to use with a wide variety of zooplankton to make comparative estimates of their predatory behavior. If an organism consumes a very small quantity of phytoplankton compared to zooplankton, the P/Z coefficient will be very small, whereas a larger proportion is obtained when phytoplankton consumption is large compared to zooplankton consumption. It is clearly inappropriate to compute the P/Z coefficient when either P or Z is zero, as then the predator is either exclusively carnivorous ($P = 0$) or herbivorous ($Z = 0$). The quantity of prey consumed is expressed either as biomass weight units or as protein weight determined by chemical analysis of a representative aliquot of the prey sample.

The fundamental procedure in obtaining the P/Z ratio is to label potential prey with a radioisotope (e.g., ^{35}S, ^{32}P, or ^{14}C where appropriate). The relative activity of the isotope per unit of prey biomass or protein is determined by assaying a representative aliquot for isotope activity. The potential prey is offered to the predator for a specified interval of time (usually 2 hr or less to avoid biases that may be introduced if the predator completely digests and excretes prey substance). The phytoplankton prey is offered to one set of randomly selected predator organisms (in this case, the radiolaria) and the zooplankton prey is offered to an equivalent sample of the predator organisms. At the end of the prescribed feeding interval, the predators are gently washed free of noningested prey by passing them through three washes of culture medium. The washed preparations are immediately collected and prepared for isotopic assay. Based on the isotopic assay of the washed predators, the amount of prey consumed can be determined using the estimate of isotope relative activity as obtained in a previous step. The mass of phytoplankton prey consumed divided by the mass of zooplankton yields the P/Z ratio. The total quantity of phytoplankton and zooplankton consumed is also obtained as part of the procedure. This technique has been applied to radiolaria (Anderson and Swanberg, 1982) to determine their degree of omnivorous trophic activity (Table 3-2). The solitary Spumellaria yield P/Z ratios of approximately 6.2 × 10^{-2}, whereas the planktonic foraminifera cited yield values ranging from 3.6 × 10^{-3} to 8.5 × 10^{-3}. The fractional P/Z ratios indicate that planktonic foraminifera and the spumellarian radiolarian species consume more zooplankton protein per unit time than the amount of phy-

Table 3-2. Comparative data on phytoplankton and zooplankton predation by some planktonic foraminifera and spumellarian radiolaria (*Spongodrymus* sp.)

Species	P/Z[a]	Prey protein consumed (μg)[b]	
		Phytoplankton	Zooplankton
Pl. Foraminifera[c]			
Globigerinella aequilateralis	3.6×10^{-3}	2.0×10^{-2}	5.56
Globigerinoides sacculifer	5.4×10^{-3}	2.6×10^{-2}	4.82
Globigerinoides ruber	8.5×10^{-3}	3.2×10^{-2}	3.74
Radiolaria			
Spongodrymus sp.	6.2×10^{-2}	5.5×10^{-2}	0.89

[a] Ratio of phytoplankton protein to zooplankton protein consumed during a period of 2 hr exposure to prey.
[b] Based on a standard aliquot of *Amphidinium carteri* as phytoplankton prey and 1-day-old *Artemia nauplii* as zooplankton prey offered in laboratory culture. See text for explanation of method.
[c] Anderson and Bé (unpublished data).

toplankton protein. This is based on the use of the dinoflagellate, *Amphidinium carteri*, as phytoplankton prey. Under these circumstances, it appears that the planktonic foraminifera and the spumellarian radiolaria are proportionately more carnivorous than herbivorous in their omnivorous predation. The radiolaria, however, comsume more *Amphidinium carteri* than the planktonic foraminifera, suggesting that these radiolaria consume slightly more phytoplankton prey than do some spinose planktonic foraminifera. Electron microscopic and light microscopic examination of snared prey tends to confirm this generalization. The food vacuoles of the larger, spongiose skeletal Spumellaria frequently have algal remains in them. Occasional algal prey are also observed in the vacuoles of spinose planktonic foraminifera.

These findings must be interpreted in the perspective of the limited species of radiolaria examined. The *Spongodrymus* sp. are large radiolaria (0.5–0.8 mm diameter) which is comparable to the size of mature spinose planktonic foraminifera cited in Table 3-2. The *Spongodrymus* sp., however, are much larger than many other abundant Spumellaria in pelagic environments; therefore, additional research with other small species and appropriate sized prey will be required to determine the prey P/Z ratio for a variety of radiolarian species.

There are several assumptions and limitations in the method that must also be considered in interpreting the data. The P/Z ratio is calculated for a feeding interval of 2 hr. Predatory behavior may change with varying environments and nutritional states of the predator. In this research, only freshly collected specimens free of adhering prey were used. Thus we assumed they were fairly representative of the natural state of the organism when in the open ocean. The kind of prey offered may influence the predatory behavior. Hence it is useful to try several kinds of phytoplankton and zooplankton prey to assess the generality of the results. Clearly, the method is only a first approximation to quantifying the selective feeding behavior of a predator and should be combined with light and/or electron microscopic observations to provide qualitative supplementary data.

Prey Capture

Radiolaria snare their prey on the axopodia or within the peripheral network of rhizopodia. Many of the solitary Spumellaria are surrounded by a layer of jelly, varying in thickness and viscosity among species, which appears to aid prey capture. The sticky jelly substance augments the holding capacity of the rhizopodia by adhering to the appendages and broad surfaces of the prey, thus hindering their escape. It is not known if the radiolaria possess a means to "narcotize" or disable the prey before major disruption of the prey has occurred. There is no evidence of stinging organelles in the extracapsulum. Further research on the possible subtle interactions between radiolarian-feeding rhizopodia and prey capture is needed in addition to the basic fine structure research reported here. In large measure, it appears that the radiolarian depends on the holding capacity of its rhizopodia and the massive flowing activity of the extracapsular cytoplasm to engulf and subdue the prey. This is particularly evident in large radiolarian species where the extracapsular cytoplasmic layer may be a millimeter or more in thickness. In the smaller, skeleton-bearing radiolaria, the long spines on the skeleton of some species appear to aid prey capture. The rhizopodia stream along the spines and anchor themselves to the surface, thereby attaining a mechanical advantage in holding the prey. The radial configuration of the spines in some radiolarian species enhances rhizopodial holding capacity by providing solid surfaces of attachment at some distance from the central capsule. Thus they permit prey apprehension in the peripheral extracapsular network of rhizopodia without sacrificing rhizopodial tensile strength, due to the very long strands of cytoplasm that would be required otherwise. In this sense, the siliceous framework is truly a skeleton as opposed to merely being a protective "shell." The silica framework provides anchorage for the rhizopodia, hence providing enhanced stability and greater mechanical advantage analogous to the skeletomuscular system in higher animals.

With respect to prey capture and skeletal morphology, it appears that among the spumellarian radiolaria belonging to the Collodaria and

Sphaerellaria, the larger species lack a skeleton, whereas the smaller species show increasing evidence of skeletal development and the formation of radially arranged spines. The larger collodarian species of *Actissa, Thalassolampe, Thalassicolla, Thalassophysa,* and *Physematium,* for example, bearing central capsules with diameters in the millimeter range and total cell diameters of commonly 1–15 mm, lack skeletons or possess simple spicules. They readily catch copepod prey and retain them largely through their sheer mass of rhizopodia, and the stickiness of the surrounding gelatin. The smaller sphaerellarian species with diameters typically of 0.07–0.2 mm exhibit increasing evidence of long spine development to augment the centrally located, smaller-diameter shells or lattices. These spines as in species of *Octodendron, Hexastylus, Cladococcus,* and *Centrocubus* may not only provide protection against predation, but also facilitate prey capture. Table 3-3 presents illustrative data on the length of spines in relation to the size of some Spumellaria and the structure of their skeleton, when present. There may be several adaptive advantages of radiolarian skeletons including protection, support of peripheral cytoplasm, mechan-

Table 3-3. Relationship between spine length and radiolarian size among some spumellaria (Collodaria and Sphaerellaria)[a]

Genus	Maximum diameter of jelly mass or central cell mass (mm)	Skeletal structure	Spine or spicule length (mm)
Thalassolampe	15.0 Jelly sphere	None	—
Physematium	12.0 Jelly sphere	None	—
Thalassophysa	8.0 Jelly sphere	None	—
Thalassicolla	6.0 Jelly sphere	None	—
Thalassopila	5.0 Jelly sphere	None	—
Actissa	1.5 Jelly sphere	None	—
Thalassoxanthium	0.5 Central capsule	Spicules	
Thalassosphaera	0.2 Central capsule	Spicules	
Plegmosphaera	0.32 Central shell	Spongy sphere	—
Cenosphaera	0.30 Central shell	Spongy sphere	—
Xiphosphaera	0.35 Central shell	Lattice sphere with spines	0.2
Cladococcus	0.20 Central shell	Lattice sphere with spines	0.3
Xiphostylus	0.17 Central shell	Lattice sphere with spines	0.2
Sphaerostylus	0.14 Central shell	Lattice sphere with spines	0.3
Octodendron	0.02 Central shell	Lattice sphere with spines	1.2

[a] Based on data from Haeckel (1887).

ical advantage in snaring and subduing prey, and perhaps as a kind of ballast to enhance vertical positioning in the water column and to allow rapid descent when disturbed.

We will need to examine additional evidence of prey capture mechanisms and prey preferences to determine if the complexity of radiolarian skeletons and the length of spines, particularly in smaller species, correlate with the size and motile strength of prey consumed. Hypothetically, we would expect species with elaborate spines and large surface areas for anchorage to be more efficient in capturing and retaining vigorously motile prey. Small species of radiolaria (e.g., Prunoidea, Discoidea, and Larcoidea) that have nonspinose or spongy skeletons may consume only microzoa and possibly algae. They do not appear to be able to retain larger microzooplankton and crustacea when offered in laboratory culture; however, these observations are limited to a few species.

Among the larger spumellarian species, there are considerable differences in the mechanisms of prey apprehension between very small prey (e.g., microzooplankton and algae) and relatively large, vigorously motile prey (e.g., copepods, copepod larvae, small crab zoea, and Larvacea). Apprehension and ingestion of small prey will be described first, followed by a description of larger prey engulfment.

Algal and Infusorial Predation   Light and electron microscopic observations of prey capture in *Thalassicolla nucleata* (Anderson, 1978b; 1980) illustrate the diverse responses exhibited by spumellarian radiolaria during prey capture. Small prey such as algae are immediately immobilized upon contact with the surface of the rhizopodia. There appears to be a complex process of prey selection, involving at least two stages or types of information processing during acceptance or rejection of prey. The first type appears to be a rapid tactile response (Type I). A potential prey organism upon contacting the surface of the rhizopodium is either immediately apprehended or is rejected and released. It is not engulfed within the rhizopodial cytoplasm. The second type of response is more delayed (Type II response). An alga, for example, may be apprehended and enclosed within a thin layer of rhizopodial cytoplasm, apparently within a vacuole. After several seconds to several minutes, the prey may be rejected and shed copiously from the rhizopodia. This delayed rejection response has been observed with *Dunaliella* sp. offered as prey. It is interesting to note that the *Dunaliella* sp. though motile when captured is no longer motile when shed. This is in contrast to a Type I radiolarian response where the rejected alga is usually still motile.

We do not know the molecular basis for the difference between these two response modes. It appears that some form of complicated cellular coordination or physiochemical processing of information is taking place in the delayed response. It requires longer time intervals to complete than with the rapid rejection occurring in the Type I response. Further research is also needed to determine whether the type of response to potential prey is related to the quantity apprehended. A Type I response may occur with smaller numbers of some prey algae; however, a Type II response may occur when large quantities of potential prey have been apprehended and a delay occurs as the radiolarian accumulates information on the total load of food apprehended. If the quantity exceeds an amount that can be efficiently utilized, the excess can be discarded. There is some evidence to support this hypothesis based on unpublished light microscope observations with *Thalassicolla nucleata*.

When a small prey such as an alga is accepted, it is firmly enclosed within a vacuole surrounded by a substantial envelope of cytoplasm in the rhizopodium (arrow, Fig. 3-1A). This is in sharp contrast to the rhizopodial attachment of the symbionts (Sy, Fig. 3-1A), which are only thinly enclosed within the host cytoplasm. The algal prey is carried by rhizopodial streaming into the pericapsular, cytoplasmic envelope (sarcomatrix), where it is consumed within a digestive vacuole (Fig. 3-1B). In the larger skeletal-bearing species (e.g., *Spongodrymus* sp.), with numerous axopodia surrounding the central capsule, a very remarkable contractile response of the shorter axopodia is observed during small prey capture. Upon contact of the prey with the surface of the axopodium, it is immediately snared and the axopodium swiftly contracts carrying the prey into the pericapsular cytoplasm. As the axopodium is contracted, it thickens and appears ruffled on the surface, apparently due to the accumulation of cytoplasm beneath the axopodial membrane. It gives the appearance of a ruffled shirt sleeve. Eventually, the axopodium elongates. It is also interesting to note that in these large spongiose skeletal species, the halo of fine axopodia continually undergoes rhythmic contractions, clearly visible with oblique illumination in a dissecting microscope. The constant rhythmic waves of contraction give a shimmering appearance to the extracapsular halo of axopodia. These continuous cycles of contraction never appear as vigorous or as decisive as when a microscopically visible prey is snared and withdrawn into the digestive layer of cytoplasm. The function of the rhythmic contractions of the axopodial corona is unknown. It is possible that it creates movement of water near the surface of the large central capsule, thus enhancing exchange processes across the capsular wall membranes. As a consequence, metabolic wastes could be eliminated more efficiently and fresh supplies of oxygenated water would be provided near the surface of the central capsule. It is also possible that these rhythmic contractions are associated with the transport of very small food particles into the digestive layer near the capsular wall; however, this needs to be examined more thoroughly by use of more sophisticated optical techniques than conventional light microscopy.

Crustacean Predation

The capture of copepods by *Thalassicolla nucleata* illustrates the major events during ensnarement and ingestion of "larger motile prey." When an acceptable prey encounters the halo of rhizopodia surrounding the gelatinous envelope, it is immediately entangled within the cytoplasmic network Fig. 3-1C). The motion of the prey as it struggles to free itself usually results in it becoming more entangled within the "sticky" rhizopodia and the viscous jelly surrounding them. Concurrently, the rhizopodia become organized into massive bands of aggregated rhizopodial strands that stream over toward the prey and engulf it (Fig. 3-1D). Within several minutes or less than an hour, the prey is drawn into the inner layers of the extracapsulum, where it eventually succumbs and is digested. The rate of prey ingestion is more swift when the radiolarian has not been recently fed.

Electron microscopic observations of the fine structure events during prey capture and digestion elucidate the mechanism of rhizopodial action. During the earliest stages of capture, the rhizopodia stream onto the appendages of the prey, enclosing them in a thickened envelope of cytoplasm (arrow, Fig. 3-2A). These enclosing envelopes of cytoplasm are clearly differentiated forms of pseudopodia that serve the function of prey capture and ingestion. To distinguish them from the strand-like rhizopodial precursors, they have been named *coelopodia* (Anderson, 1978b). This term is used to signify their appearance as hollow, engulfing pseudopodia, when observed in electron microscopic ultrathin sections. A dense deposit of microfilaments (contractile protein fibrils) is clearly visible in the ultrathin sections, and appears as a slightly electron dense fibrous layer enclosed with the hyaloplasm of the coelopod. It is particularly dense on the inner surface of membranes

---

Fig. 3-1. Prey capture by *Thalassicolla nucleata*. **(A)** A dinoflagellate symbiont (Sy) loosely enclosed within the extracapsular rhizopodia is clearly differentiated from a prey alga (Pr) that is encapsulated in a thickened cytoplasmic sheath (arrow). Scale = 20 μm. **(B)** A prey alga drawn into the digestive matrix near the central capsular wall of *Thalassicolla nucleata* is partially digested and enclosed within a large digestive vacuole (V) surrounded by a frothy layer of extracapsulum. Scale = 1 μm. **(C)** Crustacean prey (harpacticoid copepod) is snared in the peripheral rhizopodia of *T. nucleata*. Scale = 200 μm. **(D)** At a later stage, engulfing rhizopodia (arrows) arch over and stream onto the prey to further snare and subdue it. Scale = 50 μm. (Anderson, 1978b)

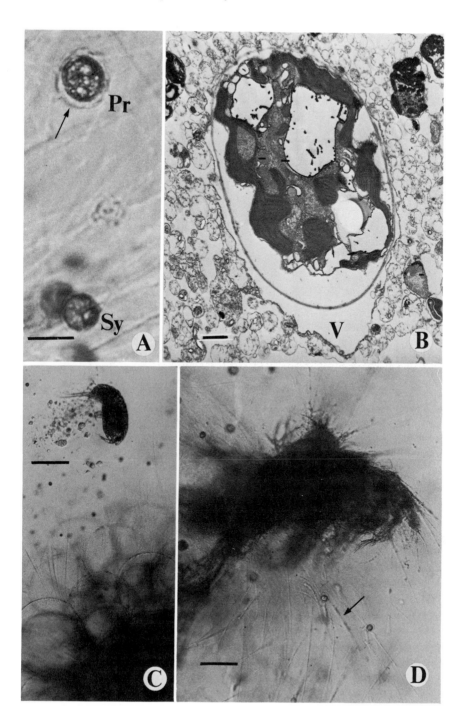

that are in contact with the prey surface. Vacuoles, mitochondria, and other subcellular organelles, but not Golgi bodies, are located in the cytoplasmic region opposite the dense mass of microfilaments. The microfilaments may mediate cytoplasmic streaming and enhance the tensile strength of rhizopodia during prey capture. Hence it is not unreasonable to find copious amounts of microfilaments within the coelopodia. When the prey has been enclosed within the rhizopodial network, contractile strands of cytoplasm attach to the carapace of the prey and fracture the exoskeleton at weakened sites (Fig. 3-2B), thus opening the prey to invasion by the rhizopodia. In Fig. 3-2B, the contractile rhizopodia attached to either side of a fracture in the prey exoskeleton (arrows) exhibit dense deposits of microfilaments lying parallel to the long axis of the rhizopodial strand. Other, more vacuolated rhizopodia (asterisk) have invaded the opening and begun to engulf prey tissue. A higher-magnification view of an invading tip of a rhizopodium (arrow, Fig. 3-3) shows a layer of prey tissue being prised from the surface of the exoskeleton as the rhizopodium penetrates deeply into the prey body cavity. A second lobe of radiolarian cytoplasm (asterisk) has begun to engulf the detached prey tissue. The radiolarian cytoplasm is readily distinguished from prey tissue by the osmiophilic granules in the rhizopodia and the presence of mitochondria with tubular cristae that are characteristic of the radiolaria and most protozoa.

When the prey tissue has been engulfed within the rhizopodia, it is carried by cytoplasmic streaming into the pericapsular digestive region and sequestered in digestive vacuoles (Fig. 3-4A). The products of digestion may be temporarily stored in the extracapsulum, which occasionally possesses oil droplets when the specimen is well nourished. However, much of the digested tissue products must be carried into the central capsule where the major storage depots are found. Presumably, these products pass from the extracapsulum into the intracapsulum by way of the fusules and, perhaps to some extent, through the fine cytoplasmic strands passing through the slits in the capsular wall. The major events during prey capture, rhizopodial engulfment, and digestion are diagramed in Fig. 3-5. A similar sequence of events occurs

---

Fig. 3-2. (A) During early stages of crustacean prey apprehension, the rhizopodia of the radiolarian (T. nucleata) engulf the appendages of the prey (An) and enclose it within specialized structures called "coelopods" (arrow) possessing a rich deposit of microfilaments (Mf) within the cytoplasm. Scale = 2 μm. **(B)** At later stages, the microfilament-rich rhizopodia exert tension upon the exoskeleton and rupture it (arrows) at weakened sites opening the body cavity to invasion by other rhizopodia (asterisk). Scale = 1 μm. (Anderson, 1978b)

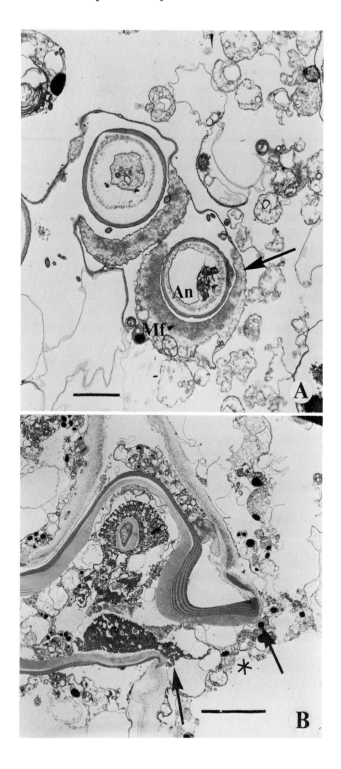

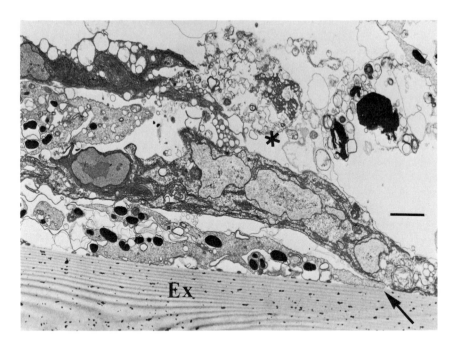

Fig. 3-3. An invading rhizopodium (arrow) of *Thalassicolla nucleata* prises prey tissue from the surface of the ruptured exoskeleton (Ex) while nearby more vacuolated rhizopodia (asterisk) surround and begin to engulf prey tissue. Scale = 2 μm. (Anderson, 1978b)

during copepod capture by large skeleton-bearing Sphaerellaria. The prey is eventually drawn into the rhizopodial network near the central capsule and anchored betwen the spines (Fig. 3-4B), where it is fully invaded by feeding rhizopodia and digested.

The feeding behavior of the larger radiolaria exhibits some of the remarkable diversity of responses that single-celled pelagic organisms possess to maximize survival in an uncertain existence imposed by a floating habit. Food must be apprehended as it floats by or wanders within contact of the rhizopodia. Hence prey capture must be swift and certain. The coordinated activity of rhizopodia in apprehending the prey, rupturing its exoskeleton, and simultaneously sending engulfing coelopodia into the prey tissue illustrates the sophistication of differentiated locomotory response in these single-celled organisms. This is further illustrated by the differentiation of responses to algal symbionts as compared to algal prey. As an example, while algal symbionts are being moved toward the periphery of the rhizopodial network, algal prey captured in a nearby rhizopodial strand are carried centri-

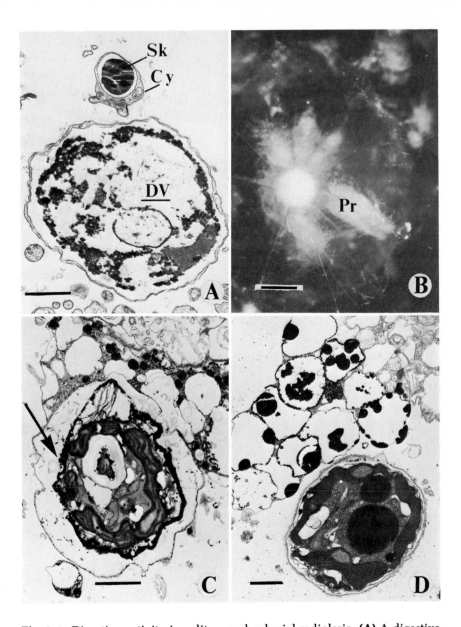

Fig. 3-4. Digestive activity in solitary and colonial radiolaria. **(A)** A digestive vacuole (DV) in the extracapsulum of a solitary skeletal-bearing species contains prey in a late stage of digestion. A segment of the skeleton (Sk) is enclosed in the cytokalymma (Cy) produced from rhizopodia in the extracapsulum. Scale = 1 μm. (Anderson, 1976a) **(B)** A copepod prey has been drawn close to the central capsule of a sphaerellarian radiolarian where digestion occurs. It is firmly secured by rhizopodia within the lattice work of the skeleton. Scale = 200 μm. **(C)** A dinoflagellate symbiont in a *Collozoum inerme* colony is partially digested within a digestive vacuole containing a dense deposit of acid phosphatase reaction product (arrow) indicating the presence of lysosomal enzymes. Scale = 2 μm. (Anderson, 1976b) **(D)** A healthy symbiont is enclosed within a perialgal vacuole near the capsular wall and may be destined to be digested by the host, *Collozoum inerme*. Scale = 2 μm.

petally toward the pericapsular digestive matrix. The thickened cytoplasmic envelope, surrounding the prey in the rhizopodial strand, compared to the thin perialgal vacuolar envelope, surrounding symbionts, may be evidence of a fundamental discriminatory mechanism employed by the radiolarian to signal an engulfed prey as differentiated from a symbiont.

Radiolarian predation is markedly different from planktonic foraminiferan predation (Anderson and Bé, 1976). Although both organisms are holoplanktonic and snare prey by rhizopodial contact, the planktonic foraminifera observed so far do not form the extensive coelopodia observed in *Thalassicolla nucleata*. They invade the prey by rupturing it and penetrating it with rhizopodia bearing finger-like protrusions that attach to the prey tissue. The segments of disrupted tissue are carried by cytoplasmic streaming to the digestive vacuoles that are scattered throughout the intrashell cytoplasm and within the vacuolated mass of cytoplasm protruding from the aperture.

The pericapsular sarcomatrix of the radiolaria is remarkably differentiated. Vacuoles enclosing algal symbionts occur in close proximity to the digestive vacuoles and are often separated by only thin cytoplasmic boundaries. Nonetheless, the symbionts are not digested, whereas the nearby prey are fully degraded. The fine structure of the radiolarian feeding pseudopodia also exhibit differentiation. The pseudopodia invading prey tissue possess small vacuoles containing osmiophilic granules (ca. 0.8 μm) distributed within the cytoplasm (Anderson, 1978b). At some points on the surface of the feeding pseudopodia, the vacuoles have opened, and the dense granules appear to be released by exocytosis. These granules look remarkably like the extrusomes reported in Heliozoa (Bardele, 1976; Patterson, 1979) that are believed to provide the ensnaring substances surrounding prey during early stages of engulfment (Hausmann and Patterson, 1982). In *Thalassicolla nucleata*, the number of these dense granules appears greatest in the peripheral feeding pseudopodia, although some are also observed within the cytoplasmic septa separating the large vacuoles in the sarcomatrix surrounding the capsular wall. No granules were observed in the intracapsulum, which further supports the hypothesis that the osmiophilic granules are differentiated structures in the extracapsulum. Their similarity to dense bodies in the axopods of Heliozoa may indicate that these are characteristic of several Actinopoda, and represent a fundamental mechanism of prey capture in the peripheral cytoplasm of these axopod-bearing protists.

Although little is known about the development of axopodia during radiolarian ontogeny, Cachon and Cachon (1977a) have examined the fine structure origin of axopodial membranes during regeneration after axopod removal by merotomy. They compared these events among a wide variety of polycystine radiolaria, including *Thalassicolla* sp.,

Thalassophysa sp., *Spongosphaera* sp., and *Litharachnium* sp. Numerous coated vesicles were produced by the endoplasmic reticulum in the vicinity of the fusules. These vesicles (ca. 130–200 nm diameter) are budded off from the surface of the endoplasmic reticulum. They exhibit a distinctive fine structure in ultrathin sections. The outer surface is coated with a layer of bristle-like protrusions (up to 30 nm long), and the inner side of the membrane bears a deposit of fuzzy material. An accumulation of small membranous inclusions may also occur inside the vesicles. Once free from the surface of the endoplasmic reticulum, the coated vesicles migrate through the fusules to the ectoplasm where the axopods are forming. The origin of the vesicles varies depending on the thickness of the capsular wall. In thick-walled species, the coated vesicles originate on the surface of the endoplasmic reticulum near the base of the fusule, migrate along the microtubule axis, and pass through the dense substance into the proximal part of the axopodium. In thin-walled species, the endoplasmic reticulum projects distally into the proximal part of the developing axopod. The coated vesicles arise directly from the endoplasmic reticulum projecting through the dense layer and migrate into the axopodial cytoplasm. Although these vesicles are always present in limited numbers in the intracapsular cytoplasm, they become much more numerous during axopod regeneration. There is substantial evidence from diverse organisms to suggest that this type of vesicle acts principally in intracellular transfer of membrane (Pearse, 1975; Franke *et al.*, 1976a,b).

Several different kinds of vesicles have been observed in the cytoplasm of the central capsulum. Some originate from the Golgi and others, from the endoplasmic reticulum. The latter are apparently the major vehicle for membrane transfer. The variety of membranous vesicles secreted by the intracapsulum and their varied functions in membrane transport in the radiolarian cell yields further evidence of the subcellular differentiation characteristic of these protists. The origin of the axopodial membranes within the intracapsulum and their transport through the fusules to the regenerating surfaces in the extracapsulum further indicate the fine structure specialization of the intracapsulum as a source of membrane biosynthesis.

Intracapsular and Extracapsular Metabolic Differentiation

The fine structural differentiation between intracapsulum and extracapsulum observed in many radiolaria suggests a high level of metabolic differentiation in these organisms. As early as 1887, Haeckel recognized that very different physiological functions may occur in the intracapsulum and extracapsulum. Without the aid of modern knowledge of

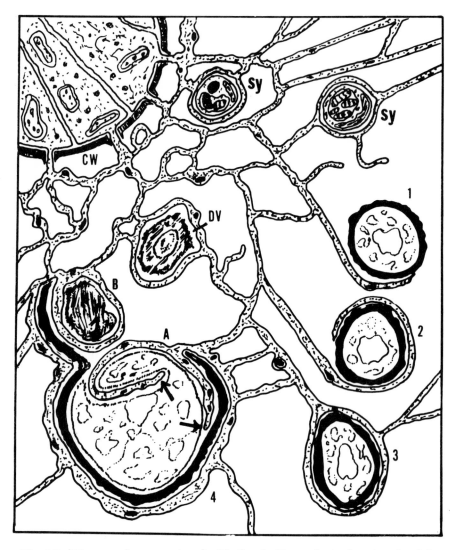

Fig. 3-5. Diagram of prey capture by *Thalassicolla nucleata* showing the elaborate extracapsular network of rhizopodia surrounding the capsular wall (CW). During crustacean prey capture, the appendages are engulfed by coelopods (1–3). The exoskeleton of the prey is ruptured (4) and invading rhizopodia (A) enclose prey tissue (arrows) which is subsequently enclosed in a food vacuole (B). The vacuole is eventually transformed into a digestive vacuole (DV) by fusion of the food vacuole membrane with lysosomes carrying digestive enzymes.
(Anderson, 1978b)

the molecular basis of life, the interpretation of these functions sometimes was phrased in rather mystical constructs as exemplified by the
following statement (Haeckel, 1887, pp. cxxviii–cxxix).

   The distribution of the functions among the various parts of the unicellular
organism of the Radiolaria corresponds directly to their anatomical composition, so that physiologically as well as morphologically the central capsule and
the extracapsulum appear as the two coordinated main components. On the
one hand the *central capsule* with its endoplasm and enclosed nucleus is the
central organ of the 'cell-soul' (Zellseele), the unit regulating its animal and
vegetative functions, and the special organ of reproduction and inheritance.
The *extracapsulum* forms, on the other hand, by its calymma the protective
envelope of the central capsule, the support of the soft pseudopodia and the
substratum of the skeleton; the calymma acts also as a hydrostatic apparatus,
whilst the radiating pseudopodia are of the greatest importance both as organs
of nutrition and adaptation as well as of motion and sensation. If, however,
the vital functions as a whole be divided in accordance with the usual convention into the two great groups of *vegetative* (nutrition and reproduction)
and *animal* (motion and senseation), then the central capsule would be mainly
the organ of reproduction and sensation, and the extracapsulum the organ of
nutrition and motion.

Haeckel concluded that the layer of vacuolated cytoplasm immediately
surrounding the capsular wall was a digestive region and that prey was
captured largely in the more distal portions of the pseudopodia. He,
therefore, recognized the fundamental principle of physiological differentiation in radiolaria.

   Modern biochemical and cytochemical research provides additional
molecular evidence for a marked metabolic differentiation between intracapsulum and extracapsulum. The distribution of enzyme activity
between the intracapsulum and the extracapsulum of *T. nucleata* was
determined for cytochrome *c* oxidase, a marker enzyme for respiratory
activity and acid aryl phosphatase, a marker enzyme for lysosomal
digestive activity (Table 3-4). The total cytochrome oxidase activity in
the intracapsulum is 1.2 times as great as that in the extracapsulum,
whereas, the total acid aryl phosphatase activity in the intracapsulum
is only about one-third as much as that in the extracapsulum. These
distributional differences are also observed when the enzyme activity
is expressed as enzyme units per microgram of protein, but to a lesser
extent for acid phosphatase than for cytochrome oxidase. These data
indicate a marked specialization for lysosomal activity in the extracapsulum and a comparatively greater respiratory activity in the intracapsulum. To determine further the localization of acid phosphatase in *T.
nucleata*, cytochemical light microscopic and electron microscopic
techniques were employed. Light microscopic evidence demonstrated
the presence of cytochemical reaction product in large vacuoles in the
extracapsulum; however, the deposits in the intracapsulum were too
fine to be detected. Electron microscopy, however, clearly demonstrated acid phosphatase reaction product in the Golgi bodies and pe-

Table 3-4. Comparative activity of acid phosphatase and cytochrome c oxidase in the intracapsulum and extracapsulum of T. nucleata[a]

	Acid phosphatase	Cytochrome oxidase	Total protein (μg)
Intracapsulum	7.38 (0.37×10^{-4})	440 (47×10^{-4})	2.3
Extracapsulum	20.72 (0.54×10^{-4})	360 (9.5×10^{-4})	4.3
Total	28.1 (0.91×10^{-4})	800 (56.5×10^{-4})	6.6

Activity is expressed as units per total cytoplasmic volume in each compartment and in parentheses as units per microgram of protein, based on a homogenate from 95 specimens.

[a] Anderson and Botfield (1983).

ripheral secretory vesicles within the intracapsulum. These vesicles were approximately 60 to 90 nm in diameter, thus explaining their invisibility with light optics. Dense deposits of acid phosphatase were found in the large vacuoles near the capsular wall, in the spaces between alveoli in the extracapsulum, and within small vesicles (ca. 1 μm) in the cytoplasmic partitions between the alveoli and the vacuoles in the extracapsulum. The stain was specific to the small vesicles. Nearby mitochondria were free of the dense reaction product. This evidence suggests that the digestive enzymes are largely synthesized within the endoplasm of the central capsule and become sequestered in Golgi-derived vesicles which pass from the intracapsulum to the extracapsulum through the fusules. These primary lysosomes (enzyme-rich vesicles) move through the extracapsular cytoplasm to sites of digestive activity where they fuse with the membranes surrounding the food-containing vacuole, thus emptying their digestive enzymes into the vacuole. This would explain the accumulation of large amounts of lysosomal marker enzyme activity within extracapsular digestive vacuoles even though the source of the enzymes is within the intracapsulum.

In total, these data indicate that the extracapsulum is a major site of prey capture and digestion, whereas the intracapsulum is a significant region of respiration and anabolism. To further test this hypothesis, the relative number and size of mitochondria (respiratory organelles) and Golgi bodies (secretory organelles) in the intracapsulum and extracapsulum were determined from representative ultrathin sections observed with the electron microscope. There was a mean of 26 mitochondria/100 μm² in the intracapsulum compared to a mean of 10 mitochondria/100 μm² in the extracapsulum. The mean size of the mitochondria in the extracapsulum was 0.92 μm compared to a mean size of 1.4 μm in the intracapsulum. There were no Golgi bodies observed in the extracapsulum sections; however, their mean density per unit area in the intracapsulum was 4.9/100 μm². Moreover, the endoplasm of the intracapsulum is rich in large vacuoles containing food reserve substances, further supporting the hypothesis that this is a region of active anabolism. It appears therefore that there is a marked specialization of function within these two major cytoplasmic regions in *T. nucleata*. Based on similarities in fine structure among the larger Spumellaria, this pattern of differentiation may be a general phenomenon among many of the larger species in this group.

The success of larger spumellarian radiolaria in the pelagic environment, particularly in regions of lower productivity, may be attributed in part to their highly differentiated physiological functions localized within different major morphological regions of the organism. This is reflected in their effective exploitation of algal symbiosis as a source of nourishment to supplement prey-based nutrition and more especially the remarkably efficient compartmentalization of metabolic activity. The abundant if not exclusive occurrence of Golgi bodies in the central capsule and the cytochemical evidence for major secretory activity of digestive enzymes in Golgi-derived vesicles indicate that the central capsule is a major site of anabolism. The large number of mitochondria and significant amounts of cytochrome *c* oxidase activity within the intracapsulum suggest that it is also a major region of respiration. It clearly serves as a major site of food reserve storage, as shown by the rich deposits of lipid and organic matter in the large vacuoles observed by light and electron microscopy.

The extracapsulum, however, is rich in lysosomal enzyme activity as demonstrated by the high acid phosphatase activity. Not all of the acid phosphatase activity occurs in the extracapsulum, as the primary lysosomal vesicles carrying the newly synthesized digestive enzymes originate in the intracapsulum. The major site of acid phosphatase reaction product is within the large vacuoles of the pericapsular envelope (sarcomatrix) and in membrane-bound regions among the alveoli. These data confirm Haeckel's conclusion that the sarcomatrix is a site of catabolism. The clear separation of anabolic activity in the intracapsulum from catabolic activity in the extracapsulum undoubtedly enhances the efficiency of these unicellular organisms. By establishing the major digestive functions in the pericapsular cytoplasm, an efficient centripetal sequence of prey capture, digestion and food product storage is achieved. The prey apprehended at the periphery of the extracapsulum is partially degraded by coelopod invasion and enclosure of prey tissue. The small fragments of food are carried to the pericapsular sarcomatrix, where they are digested and transformed into small molecules that are transported into the central capsule, presumably by way of the fusules and stored in food reserve vacuoles. These food products are also utilized for major respiratory and biosynthetic activity in the intracapsulum. By clearly segregating digestive functions from biosynthetic functions, there is less likelihood that these counteractive processes are likely to be confused, as could occur if the major digestive functions were comingled with biosynthetic organelles. We do not know at present how much or what kind of biosynthetic activity occurs in the extracapsulum, and future research clearly needs to be addressed to this question. Since the symbionts are located exclusively in the extracapsulum, it is feasible that some of their secretory products are immediately fixed into larger molecules in the extracapsulum before translocation to the central capsule. Evidence for carbohydrate, lipid, and amino acid synthesizing enzymes should also be investigated in the extracapsulum and compared to the intracapsulum. Among the significant areas of physiology that are not well investigated are (1) the rate and kind of excretory products released into the environment, (2) rates of metabolism as assessed by appropriate marker enzyme activities, oxygen consumption, and utilization of food reserves; and (3) changes in physiology during ontogeny both in solitary radiolaria of varying sizes and in colonial forms. Many of these physiological variables should also be examined in relation to changes in environmental variables commonly observed in the open ocean.

On the whole, the marked structural and metabolic–functional differentiation in radiolaria suggests that they are highly evolved, unicellular organisms, and should not be considered primitive in their phylogenetic development, though derived from an ancient ancestral line. Their efficient metabolic organization implies that they may be very efficient in mobilizing reserves for rapid growth. Additional research is needed on their chemical composition and rate of growth as part of our total understanding of the productivity of the oceans.

Chemical Composition

Little information is available on the chemical composition of solitary and colonial radiolaria, particularly with reference to the major biochemical components: carbohydrate, lipid, and protein. Hence, we are not able to make good estimates about their contribution to the productivity of the oceans nor to assess their nutritional value as prey for other organisms. Some preliminary assessment of chemical composition was presented in an earlier publication (Anderson, 1980) and is further expanded here (Table 3-5).

Thalassicolla sp. (probably *T. nucleata*) was prepared for analysis by removing the extracapsulum so that only the intracapsulum was assayed. The whole organism was assayed for the remaining solitary

Table 3-5. Chemical composition of some solitary and colonial radiolaria[a]

Specimen	Cell or colony mean diameter (mm)	Expressed as μg/cell or as μg/colony		
		Carbohydrate	Lipid	Protein
Solitary species				
Spongodrymus sp. and *Hexastylus sp.* (total cell)	0.3	57×10^{-3}	—	3.4×10^{-1}
Thalassicolla sp. (central capsule)	3.0	1.6×10^{-1}	1.5	6.5
Colonial species (total colony)				
Collozoum inerme	15.0	43.4	81.7	85.1
Collozoum pelagicum	8.0	45.8	138.3	101.1
Collosphaera sp.	10.0	91.2	92.3	260.0
Acrosphaera sp.	8.0	33.7	32.1	134.0

[a] Based on analytical methods published by Strickland and Parsons (1972).

and colonial radiolaria. These contained the normal complement of jelly mass, extracapsular cytoplasm plus symbionts, and the central capsule(s). Among the solitary radiolaria, lipid was determined only for *Thalassicolla* sp., which was sufficiently abundant to yield a reliable estimate of lipid content. This large species is relatively rich in lipid, which probably represents much of the reserve substances stored within the large vacuoles in the endoplasm of the central capsule. Due to its large size, it clearly contains substantially more carbohydrate and protein than the small skeleton-bearing species (e.g., *Spongodrymus* sp. and *Hexastylus* sp.). Among the colonial species which were collected in May and June from the Sargasso Sea near Bermuda, *Collozoum inerme* possessed small central capsules, widely spaced in the thin gelatinous sheath, and interconnected by fine strands of rhizopodia. Even though the colonies were large compared to the others in Table 3-4, they contain less protein than the other species. This is probably attributable to the sparse density of central capsules. *Collosphaera* sp. (largely *C. huxleyi*) is rich in carbohydrate, lipid, and protein. They are substantial in appearance when freshly collected. They possess a heavy gelatinous sheath and numerous, closely spaced central capsules, which sometimes contain a large centrally located oil droplet. Their extracapsulum harbors numerous symbionts. It is not surprising to find that these robust colonies contain relatively substantial amounts of protein and carbohydrate. The large amount of lipid in *C. pelagicum* is probably explained by the abundance of lipid storage droplets in the intracapsulum and particularly within the rhizopodial network.

The carbon and nitrogen content of *Collozoum inerme*, *Collozoum radiosum*, and *Acrosphaera spinosa* was determined by Swanberg (1979). He found a strong correlation between carbon content and colony size expressed as number of central capsules. The carbon content for *C. inerme*, *C. radiosum*, and *A. spinosa* was about 50, 100, and 200 ng carbon per central capsule, respectively. The carbon/nitrogen ratios were 11.0 for *C. inerme*, 8.3 for *A. spinosa*, and 8.4 for *C. radiosum*.

Additional research is needed to characterize the carbohydrate and lipid components of radiolaria better, to determine the possible metabolic pools where they are located, and to establish more closely the nutritional value of radiolaria as potential prey in marine food webs. Present knowledge about their predators is meager; however, it appears they are not heavily predated. Nonetheless, their sheer abundance in surface water and clear potential for contribution to flocculent detrital matter, "marine snow" (Silver and Alldredge, 1981) settling in the depths of the water column, suggest that they may be a significant part of benthic food webs in the open ocean. Hence, further research on their chemical composition and changes occurring during settling into deep water (perhaps as simulated in the laboratory), both for vegetative and for reproductive stages, may contribute to our knowledge of their potential role as food sources in benthic communities. Additional research on the enzyme composition and major metabolic pathways in radiolaria is needed to characterize more fully their chemical composition and mechanisms of food transformation and storage.

Predators

Little research has been done on the predators of radiolaria. Ostensibly, fish could consume radiolaria, but there is no clear evidence to support this view. It is possible that the dinoflagellate symbionts occurring abundantly in some solitary and colonial Spumellaria confer a disagreeable taste to the radiolaria, thus protecting them against predation by vertebrates. Dinoflagellates are known to synthesize substantial amounts of sterol (Withers et al., 1979), and therefore radiolarian symbionts may secrete some of these products into the surrounding jelly mass, where it could become sufficiently concentrated to make the radiolaria unpalatable. Indeed, among the other benefits provided by the symbionts, protection against predation may also contribute to the explanation for the close association between radiolaria and the dinoflagellate symbionts. Swanberg (1979) tried feeding various colonial radiolaria to several species of fish including *Fundulus* sp., *Poecilius* sp. (guppies), and *Tetragonurus cuvieri*. None of the fish ingested the

colonies after mouthing them. During the several years that research
has been underway on microzooplankton at our laboratory, we have
examined prey captured by radiolaria (Anderson, 1980) and planktonic
foraminifera (Anderson and Bé, 1976; Bé et al., 1977; Anderson et al.,
1979). We have found occasional instances where planktonic foraminifera contained solitary radiolaria or Acantharia snared in their rhizopodial net. Since these foraminifera were collected in glass jars by
SCUBA divers, it is unlikely that these prey were artificially introduced.
Thus it appears radiolaria may be consumed at least by planktonic
foraminifera and perhaps other pelagic invertebrates. It is not possible
at present to determine how heavily planktonic foraminifera prey on
radiolaria.

One of the best-documented and extraordinary instances of predation
by a crustacean on colonial radiolaria was reported by Swanberg (1979).
He found amphipods of the Family Hyperiidae embedded within the
gelatin of the colony or firmly attached on the outer surface. The amphipods clearly were not radiolarian prey as they moved freely within
the colony. Postjuvenile stages were observed crawling around within
a hollowed space inside the colony, and juveniles (0.53 to 1.9 mm)
were found embedded in colonies in groups of as many as 23 (Swanberg,
1979, p. 75). Usually, a hole was observed in the colony wall where
the amphipod invaded. Swanberg also reports seeing the amphipods
feeding on the radiolarian central capsules. Predatory or parasitic relationships of amphipods with colonial radiolaria were observed in 143
colonies collected at 92 of 224 stations scattered throughout the ocean
without apparent geographical pattern. Among other invertebrates found
attached to the exterior of colonial radiolaria, Swanberg reports observing harpacticoid copepods (Miracia efferata) probing into the gelatinous coat. Some of these may be ectocommensals obtaining food
particles from the gelatin. Juvenile decapods, phyllosome larvae, and
other Harpacticoida (Sapphirina sp.) were also observed on the surface
of the colonies. Interestingly, Swanberg also observed flatworms (Tur-
bellaria) browsing on the surface of some colonies; however, he did not
find conclusive evidence that they were radiolarian predators. His observations were carefully documented with photographs and detailed
accounts, thus offering convincing evidence that these organisms were
not prey of the radiolaria. Great care must be taken in observing potential predatory relations between other invertebrates and colonial
radiolaria, as they are known to snare and engulf crustacean prey, which
become drawn into the gelatinous sheath. These snared prey could be
mistaken for predators during early stages of capture. There is no question, however, when these are prey, as careful observation will show
they are eventually immobilized and digested.

Among the various mechanisms that hypothetically may protect radiolaria against predation are (1) the unpalatable taste previously men-

tioned, (2) spines or spicules when present, (3) the large size, as particularly among some colonial forms, and (4) bioluminescence observed widely in larger solitary and colonial spumellarian radiolaria which may deter or confuse potential predators.

Bioluminescence

Luminescence in solitary and colonial radiolaria was reported as early as 1885 by Brandt, and mention was made of it by Haeckel (1887). The organisms emit a distinctly bluish flash of light, which may be extended by repetitive emissions when disturbed. The source of the emission has been variously described as intracapsular or extracapsular and perhaps associated with the dinoflagellate symbionts. Some free-living species of dinoflagellates are known to exhibit bioluminescence (Esaias and Curl, 1973). Harvey (1926) reported that the luminescence of *Collozoum inerme* and *Thalassicolla nucleata* was not inhibited by illumination or by the absence of oxygen. The latter is unusual and is known to occur only in certain coelenterates in addition to radiolaria. Moreover, Harvey found no evidence of a classical luciferin–luciferase reaction or any cross-reactions with the *Cypridina* sp. reaction system. Nicol (1958) also reported recordings of low-intensity flashes from the large solitary radiolaria, *Aulosphaera triodon* and *Cytocladus major*, upon mechanical stimulation. The physical quality of the emitted light in *Thalassicolla* sp. has been analyzed by Herring (1979), who reported that the specimens gave a unimodal emission spectrum, λ_{max} at 446 ± 4 nm. Electrical stimulation of the extracapsulum elicited a variety of response modes, including a single train of repetitive flashes following each stimulus or repeated flash trains containing pulses of light with a frequency of 2, 5, or 10 per second. The bioluminescence was emitted by the outer gelatinous extracapsulum and apparently not by the intracapsulum according to Herring. Crude homogenates of the extracapsulum were activated to luminesce by the addition of calcium ions, but luminescence was suppressed when calcium-binding agents such as tetrasodiumethylenediaminetetracetate (EDTA) was added. Based on the available physical and chemical evidence, Herring hypothesizes that a calcium-activated photoprotein may be involved. A group of these photoproteins has been identified in certain luminous hydrozoans and ctenophores. He proposes that the flashes deter predators thus explaining their adaptive value. Similar flashing in dinoflagellates has been experimentally shown to diminish grazing by predators (Esaias and Curl, 1973). At present, it is not known whether the dinoflagellate symbionts or the radiolarian cytoplasm is the source of luminescence in *T. nucleata*.

Symbiosis

The presence of algal cells in radiolaria was noted in the Collodaria as early as 1851 by Huxley, who described them as "yellow cells." Müller (1858) also described them in many Spumellaria and Nassellaria. At first, they were thought to be part of the radiolarian, and a prevailing opinion developed that the radiolaria might be multicellular organisms. The independent existence of the yellow cells was established by Cienkowski (1871), who showed that the "yellow cells" were capable of dividing outside of the radiolaria. But he incorrectly concluded that they were algal parasites. With the discovery of symbiosis in fungi (lichens), Brandt (1881) examined the role of the yellow cells in the perspective of a symbiotic association and concluded they were indeed algal symbionts. He gave them the name *Zooxanthella nutricola* and subsequently explained more fully their vital functions in relation to the host. Geddes (1882), who almost simultaneously came to the same conclusion as Brandt, showed experimentally that the yellow algae produce oxygen when illuminated with sunlight. For some time, it was thought that the relationship was obligate for the radiolarian, and that the main function of the symbiont was to provide oxygen for the host. This view, however, was dispelled by Haeckel (1887), who noted that not all radiolaria possess algal symbionts and some with symbionts descend to great depths in the water column, where there is insufficient sunlight to support substantial amounts of oxygen production. The possibility that the algal symbionts supply nourishment to the host was apparently already being discussed at the time Haeckel wrote the Challenger report (1887). He mentions the presence of starch in the yellow cells and suggests a nutritive role for the symbionts. This view was also previously suggested by Brandt (1882), who showed, moreover, that the yellow cells could definitely serve as a source of nourishment. He did a very simple, but dramatic, experiment. The survival time of two comparable samples of *Thalassicolla nucleata* containing symbionts was determined when one of the samples was placed in the dark and the other in sunlight. The specimens in the light survived much longer than those in the dark, thus indicating that the symbionts provide photosynthate nourishment for the radiolaria. He also reported that illuminated *Thalassicolla* sp. could be maintained in filtered seawater for several months without additional sources of food. We do not know how effective his filtering system was, and hence prey microorganisms may have been added that could supplement the symbiont-derived nutrition.

In most of the early literature, it was presumed that the symbionts were of one type, i.e., yellow-pigmented zooxanthellae. It was not until recently (Anderson, 1976a) that the presence of green-pigmented prasinophyte algal symbionts was reported in solitary radiolaria. The early physiological studies on radiolarian symbiosis, though provocative, were not conclusive, since they were unable to unequivocably demonstrate host assimilation of symbiont photosynthates nor to evaluate how much of the host nutrition may be supplied by occasional cropping (ingestion) of symbionts. Indeed, during the early 20th century, there appeared to be a vigorous debate as to whether the algal symbionts were ingested by the host or merely secreted photosynthates that were assimilated by the host. They were unable to resolve the polemic, given the limitations of their research methods; nor did they entertain the idea that both methods of nutrition were possible depending on the species. The latter appears to be the case based on modern cytochemical and biochemical isotopic tracer techniques. A summary of current knowledge about host–symbiont physiology and its ecological implications is presented for some spumellarian species.

The colonial radiolaria are particularly useful experimental subjects, as they consist of numerous central capsules and abundant algal symbionts enclosed within a gelatinous envelope. Consequently, they provide sufficient biomass to make them effective tools in biochemical analyses of host–symbiont physiology. *Collosphaera* sp. are especially useful in the investigation of host assimilation of symbiont-derived photosynthates. The central capsule is surrounded by a nearly spherical shell perforated by pores that are usually too small to permit the algal symbionts to be drawn into the pericapsular cytoplasm. Consequently, the central capsules can be effectively separated from the symbionts by gently disrupting the colony, washing the central capsules repeatedly in filtered seawater to remove adhering symbionts, and collecting algal-free samples with a micropipet while inspecting them with a light microsocope. Hence, it is possible to analyze the central capsules separate from the algal fraction (Anderson, 1978a). To assess the assimilation of symbiont photosynthates by the host, the total colony is illuminated in the presense of $[^{14}C]$-HCO_3 added to the seawater culture medium and after a period of 1 to several hr, the colony is washed free of externally adhering isotopic label. The colony is disrupted by drawing it back and forth repeatedly through a wide tip of a Pasteur pipet. The separated, washed central capsules are collected, counted and assayed for ^{14}C-containing compounds using a liquid scintillation technique. To control for spurious transfer of isotopic label during disruption, a dye-stained nonradioactively labeled colony is added with the isotopically labeled colony. Both are disrupted in the standard way but only the dye-stained central capsules are collected for assay. These serve as a control for cross-contamination during disruption. Colonies incubated in the dark serve as controls for nonphotosynthetic uptake of ^{14}C. The results of such an experiment are presented in Table 3-6. These results, presented as radioactive decays per minute (DPM), are

Table 3-6. [14]C isotopic evidence for host assimilation of symbiont-derived organic compounds in *C. globularis*[a]

Incubation (hr)	Sample	Radioactivity minus blank (dpm)
1.5	Host cells (100)	21.7
3.5	Host cells (100)	51.6
1.5	Symbionts (300)	447.0
1.5	Dark control cells (100)	1.3
—	Adsorption control cells	<1.0

[a] Anderson (1978a).

expressed on the basis of ^{14}C incorporation per 100 cells. As the radiolarian cells are small, the counts are low, but they are significantly different among the treatments. The colonies were incubated for either 1.5 hr. and immediately assayed or washed and allowed to translocate photosynthates for an additional 2 hr before isotopic analysis. As shown in the table, the amount of isotopically labeled photosynthates assimilated by the radiolarian cells incubated for a total of 3.5 hr (51 DPM) is over twice the amount assimilated by the colonies incubated for 1.5 hr (21 DPM). Both the dark controls and the dye-stained adsorption controls show negligible uptake of [14]C. The symbiont fraction, however, shows substantial uptake as is required if they serve as the source of the translocated ^{14}C-labeled compounds. The total carbon translocated to the host based on the ^{14}C isotopic assay is calculated to be 54×10^{-5} g/hr/100 cells.

The number of symbionts associated with a radiolarian host varies among species. The colonial radiolaria are usually rich in symbionts as are many collodarian solitary species. Many of the larger spongiose skeletal species (e.g., *Spongodrymus* sp.) are also rich with symbionts, although there is considerable variability in the number of symbionts among the specimens within a species. The sphaerellarian species, however, vary in symbiont abundance. The reasons for inter- and intraspecific symbiont variability are unknown. We know very little about the factors involved in the regulation of symbiont numbers within a radiolarian cell or colony. The radiolarian cell may control their abundance in relation to the nutritional state of the host, incident light intensity, or other environmental physical or chemical factors impinging on the host–algal association. There may be a complex set of genetic and environmental factors that mutually determine symbiont population density. The number of symbionts in a colony may be as great as

3×10^3, assuming there are approximately 50 radiolarian cells in a small typical colony, and 30 to 50 algal symbionts associated with each radiolarian cell. Solitary, large spumellarian radiolaria possess up to 5×10^3 algal cells per organism (Anderson, 1980). Very frequently, their halo of extracapsular cytoplasm appears distinctly yellow or bright green depending on the kind of symbiont present. Since some of the carbon compounds assimilated by the radiolaria are derived from symbiont photosynthesis, the primary productivity of the symbionts was assessed. Solitary radiolaria were used as their symbionts are distributed in the peripheral extracapsulum and therefore are in close contact with the ^{14}C-labeled seawater. The mean primary productivity of the symbionts (15–20 μm diameter) based on two experiments was 350×10^{-9} μg C/cell/min (Anderson, 1980). This result compares favorably to productivity of free living flagellates when differences in cell volume are taken into account. The dinoflagellate Gonyaulax polyhedra Stein (30–40 μm), with a cell volume approximately 10 times that of the radiolarian symbionts, has a primary productivity about an order of magnitude greater, 301×10^{-8} μg C/cell/min (calculated from data published by Prezelin and Sweeney (1978). The primary productivity of the small green alga, Dunaliella primolecta Butcher (5–7 μm), is 44×10^{-9} μg C/cell/min (Thomas, 1964). Additional research is required to determine how much variation exists in primary productivity as a function of the symbiont species harbored by the radiolaria and in relationship to the physiological state of the host.

The primary productivity of symbionts in a colonial-radiolarian species, Collozoum longiforme, was reported by Swanberg and Harbison (1979) to be 0.72–7.7 nmole CO_2/mm colony length/hr at a temperature of 24°C and with an illumination of 1,000–44,000 μW/cm². Since they calculate that there are approximately 7.5×10^5 algae/ml of colony volume, the primary productivity expressed on a cellular basis is 378×10^{-9} μg C/cell/min. This compares favorably with the primary productivity found for symbionts in solitary radiolaria, particularly when differences in illumination, incubation, temperature, and possible symbiont species differences are considered. Khmeleva (1967) reported that Collozoum inerme colonies in the Gulf of Aden were three times as productive in fixing carbon per cubic meter than the surrounding phytoplankton.

The primary productivity of symbiont-bearing colonies has also been studied by Swanberg (1979). He found a net incorporation rate of 3.3 pmoles CO_2 per central capsule per hour with C. inerme at low-level intensities of artificial light (10^3 μW/cm²). Further studies using sunlight as a source of illumination at varying intensities yielded some interesting, contrasting results among the several species examined. For example, net incorporation of carbon for C. radiosum under full sunlight (1.7×10^4 μW/cm²) was 51.4 pmoles CO_2 per central capsule per hour, while at a lower intensity of sunlight (0.6×10^4 uW/cm²) later in the day, the rate of incorporation was higher, 76.1 pmole CO_2 per central capsule per hour. By contrast, *Acrosphaera spinosa* in general showed increasing net incorporation rates with increasing light intensity. For example, at sunlight intensity of 4.2×10^3 μW/cm², the net incorporation rate was 10.4 pmole/cell/hr, while at high intensities of 14×10^3 and 19×10^3 μW/cm², the respective net incorporation rates were 14.3 and 20.9 pmole/cell/hr. This represents a substantial primary productivity within a small volume of living substance in the open ocean and clearly indicates the potential photosynthetic food source available to the radiolarian from its symbionts.

The translocation of carbon-containing compounds to the host from the symbiont undoubtedly represents a substantial part of the benefit that the symbiont renders to the host; however, there is the additional possible contribution of nutrition through host ingestion of algal symbionts. Although there is little evidence to suggest that the collosphaerid radiolaria (e.g., *Collosphaera* sp.) ingest symbionts, substantial light microscopic and electron microscopic cytochemical evidence indicates a moderate cropping of the symbionts in the collodarian radiolarian *Collozoum inerme* (Anderson, 1976b). Colonies of *C. inerme*, maintained in Millipore-filtered seawater and illuminated with fluorescent lights in a room with temperature equivalent to ambient seawater, will live for several weeks. No external food need be introduced, though the colonies will accept small calanoid copepods as prey. The colony floats normally, and the symbiotic algae remain abundant. Light microscopic examination of the colonies, maintained in a 12-hr light/12-hr dark cycle, shows that some of the radiolarian central capsules contain symbionts enclosed deeply within the cytoplasm of the pericapsular envelope. These symbionts often appear crenated and denser brown in color than the symbionts distributed in the extracapsular rhizopodial netork. Electron microscopic examination of ultrathin sections of *Collozoum inerme*, prepared with a cytochemical stain to indicate sites of lysosomal marker enzyme (acid phosphatase), demonstrated the presence of partially digested dinoflagellate symbionts within lysosomal vacuoles (Fig. 3-4C). The electron-dense reaction product surrounding the partially digested alga indicates the presence of the lysosomal enzyme. The symbionts enclosed in perialgal vacuoles within the rhizopodial network or near to the pericapsular digestive layer appeared normal. Indeed some were clearly dividing (Fig. 2-11B), thus confirming their robust and healthy condition. Since the number of symbionts in a colony in an illuminated laboratory culture appears to remain fairly constant, it is hypothesized that the radiolarian ingests algae at a rate that is concomitant with symbiont proliferation. Thus a fairly steady-state population of symbionts is maintained in the colony.

Further evidence for a steady state cropping of symbionts was obtained by placing colonies in the dark to inhibit algal division and the change in symbiont abundance was noted at regular intervals. After 4 days, there was a 10–15% decrease in the number of symbionts (Anderson, 1976b). Many of the symbionts were withdrawn into the digestive matrix near the capsular wall, and these appeared to be undergoing digestion. The remaining symbionts in the peripheral extracapsulum appeared normal, thus the reduction in numbers of algae is attributed to radiolarian ingestion and not to algal death caused by light deprivation. Undoubtedly, the algal symbionts were under stress, however, and these data are meaningful only as a supplement to the experiments with cultures maintained in a regular light/dark cycle simulating the natural environment. It is not known whether the rate of symbiont ingestion changes with variations in external food source. The colonies examined were kept in filtered seawater; thus they had little exogenous prey. The time of day when the symbionts are most likely to be cropped is at evening, when they are withdrawn by rhizopodial streaming in close proximity to the capsular wall (Fig. 2-17). Most of the symbionts remain in the peripheral vacuolar system surrounding the central capsule; however, a few are engulfed deeply within the sarcomatrix. These, apparently, are the ones destined to be digested. How the radiolarian selects the cells to be ingested is unknown. Generally, it is believed that decrepid or senescent cells are ingested; however, we do not know the frequency with which such cells occur in the colony nor what factors might promote the occurrence of decrepid symbionts.

The general question of what factors promote symbiont vitality in the host–symbiont association has not been explored extensively. The fact that so many symbionts can be accommodated within such a small volume of host cytoplasm, both in solitary and colonial radiolaria, suggests that the radiolarian provides a particularly favorable environment. It is clear that in such a densely populated space, both an adequate algal nutrient supply and a metabolic waste removal must be provided to sustain the algal growth. It is generally presumed that the host provides excretory products such as ammonia, other small nitrogen-containing compounds, carbon dioxide from its respiration, and perhaps other products that can be used by the algae as nutrients. Correspondingly, the symbiont assimilation of these potentially toxic substances to the host may be an aid in host survival. The assimilation of the radiolarian waste products by the algae may also help the radiolarian conserve energy by providing a metabolic "sink" for the wastes rather than requiring the radiolarian to transfer them to the exterior of the colony or large cell. In this respect, it is interesting to note that the largest solitary radiolaria and the colonial species consistently possess symbionts. There are only a few exceptions. Whether this co-association is particularly beneficial to the large radiolarian because of the symbiont waste removal role remains to be determined. It is possible that the large radiolaria have become dependent on the symbiotic algae for a number of vitalizing functions. Among these are (1) the constant source of nourishment from the algal primary productivity that may support the host during periods of limited food supply or indeed sustain the host during periods of severe food shortage; (2) a light sensing function to establish the diel rhythms of cytoplasmic streaming and perhaps other fundamental metabolic systems (Anderson, 1983); and (3) waste removal by assimilation of host excretory products.

The beneficial activities of the host in sustaining the algae may also derive from nutrient sources supplied from prey. The massive assault of the radiolarian in rupturing the prey exoskeleton and degrading large quantities of its tissue may release small-molecular-weight organic compounds and minerals that can be utilized by the symbionts. Whether this process has immediate benefit for the algae or whether long-term, secondary released products sustain the algae needs to be investigated. Swanberg (1979) has found some preliminary evidence that well-nourished colonial radiolaria may possess more symbionts or more chlorophyll-rich symbionts than less well-nourished radiolaria. He found a direct relationship between the number of tintinnid loricas in the colony (an indicator of prey abundance) and the amount of chlorophyll per colony. This suggests that the better-nourished colonies sustained enhanced symbiont vitality.

It appears theoretically that the radiolarian–symbiont association is a remarkably balanced micro-ecosystem. In the large colonies, the translucent gelatinous envelope and the very thin perialgal cytoplasmic envelope of the host rhizopodium facilitate light reception by the symbiont, whose primary productivity is a significant organic energy source for the colony. Likewise, a complex system of waste removal and nutrient exchange including dissolved gases and soluble molecules may occur in the dynamic interrelationship between symbiont and host.

This rather "self-sufficient" micro-ecosystem is remarkably adapted for survival in pelagic, oligotrophic environments. Indeed, colonial radiolaria with abundant symbionts are characteristically found in high densities in low productivity regions such as the Sargasso Sea. On a calm day, when the sea surface is unusually smooth, they often rise into the near-surface layer of water, where they appear in copious numbers. SCUBA divers report that on these occasions their abundance in the water column is reminiscent of a flocculent snow storm, so great is their number in the passing current. The phenomenon of symbiosis in radiolaria is undoubtedly a profound aspect of their survival, their remarkable adaptiveness, abundance, and their widespread occurrence. Research into the dynamics of these micro-ecosystems promises to reveal much about the complex factors contributing to their exploitation of open ocean environments and more generally to elucidate some ecological principles of stability in pelagic microzooplankton communities.

In this endeavor, we see the clear merit of combining modern methods of cellular biological research with emerging techniques in ecological investigations toward a synthesis of knowledge at the molecular and ecosystems levels. The challenge of microzooplankton, physiological, and ecological research is to find the broader theoretical principles that unite cellular and ecosystems research paradigms into a coherent set of explanatory constructs. Much potential for advance of scientific knowledge will be lost if the traditionally separate disciplines of cellular and ecological research are not united in the exploration of these remarkably complex single-celled marine organisms.

Motility and Cytoplasmic Streaming

On the whole, most radiolaria are passively floating pelagic organisms largely dependent on ocean currents to move them from one location to another. They possess limited ability to move vertically in the water column, and this appears to be mediated in some species by the expansion and contraction of alveoli in the extracapsulum; however, the source of buoyancy and its possible regulation in many species, particularly among the small skeletal-bearing radiolaria, are largely hypothetical. The lipoidal composition of intracapsular vacuoles and perhaps the secretion of buoyancy-enhancing fluids, such as ammonium chloride into the vacuoles, have been suggested as possible flotation mechanisms (Anderson, 1980). The heavy skeletons may also serve as a kind of ballast to permit rapid descent when the radiolarian escapes noxious or threatening conditions in the surface water. *Thalassicolla* sp. can escape noxious conditions by shedding much of the alveolated extracapsulum, thus losing buoyancy and sinking in the water column. Any substantial abrasive contact with the surface of the extracapsulum or compression, as might occur when a predator encounters *Thalassicolla*, elicits the escape response. Even very forceful repetitive or agitated water movement can elicit the sinking response. The central capsule devoid of alveoli sinks quite rapidly. Subsequently, a new extracapsulum is generated from the cytoplasm within the intracapsulum. If the loss of extracapsulum during escape is heavy, then, considerable shrinkage of the central capsule may occur when the extracapsulum is regenerated.

Colonial radiolaria also exhibit vertical movements. These changes in vertical position are mediated by contraction and expansion of the fluid-filled alveoli within the gelatinous envelope and also perhaps by accompanying changes in the volume of the gelatinous matrix. The mechanism of regulating alveolar volume may be rather sophisticated, involving specialized regions of the membranous alveolar wall that serve as regulatory sites for release of the intraalveolar fluid. Fine structure evidence (Anderson, 1976c) shows the presence of thin, membranous regions in the thickened cytoplasmic septum surrounding the alveoli of *Sphaerozoum punctatum*. These sites may serve as a kind of regulatory valve to permit either partial release of intraalveolar fluid during fine adjustments in buoyancy or massive release when rapid descent occurs during avoidance of noxious or turbulent conditions in the upper regions of the water column. When the surface water is turbulent, they sink to more quiescent regions; however, under calm conditions, they occupy the near surface. In regions such as the Sargasso Sea, they are very abundant in the surface water during the summer months. The intensity of sunlight in the near-surface layer of the ocean at midday is quite high; thus the symbionts have an ample source of illumination to sustain photosynthesis. When light intensity is low as sometimes occurs in laboratory culture, the symbionts are distributed near the periphery of the colony in the fine rhizopodial network immediately beneath the surface of the gelatin. Presumably, if the light intensity becomes too high, the symbionts may be withdrawn by rhizopodial streaming into the inner portion of the rhizopodial network where illumination may be less intense. A fine halo of rhizopodia radiates from the surface of the colony. These rhizopodia are retractile and are withdrawn into the gelatinous envelope if the colony is disturbed. When the rhizopodia are extended, they serve as feeding rhizopodia to ensnare prey and withdraw it into the rhizopodial network within the gelatinous sheath.

The phenomenon of cytoplasmic streaming and the transport of vacuole-enclosed particles are of profound interest in cellular biology. Most radiolaria exhibit various modes of cytoplasmic flow including bidirectional streaming, fusion of separate cytoplasmic strands to form lamina, cords or networks of strands, and expansion of bubble-like surfaces. These activities are particularly evident among the rhizopodia where symbionts, food particles, vacuoles, and other cytoplasmic organelles are transported from one part of the cell to another. According to current theory, the structure and organization of the rhizopodia are determined largely by the intracytoplasmic cytoskeletal system including the microtubular arrays (Spooner, 1975). These intracellular, long, thin, tube-like structures may provide the guiding surfaces that determine the path of streaming particles in the cytoplasm. The motile force, however, is generally attributed to microfilaments that exhibit contractile properties. Light and electron microscopic investigations of cytoplasmic transport phenomena in rhizopodia show that the larger particles such as vacuolar-bound prey particles are carried near the periphery of a rhizopodial strand and often produce a distinct bulge on the surface, particularly in fine strands. Electron microscopic views of longitudinal sections through rhizopodia from a variety of rhizopod- and axopod-bearing protists, including planktonic foraminifera and proteomyxida (e.g., Anderson, 1976a; Anderson and Bé, 1976; Anderson and Hoeffler, 1979), exhibit the presence of a central shaft of microtubules running parallel to the long axis of the rhizopodium. Fine microfilaments are attached to the cytoskeletal microtubules and may serve to locomote the cytoplasmic particles moving through the rhizopodium. The vacuole-bound particles are located in the near vicinity of the microtubule shaft and its associated microfilaments, thus causing them to protrude from the surface of the thin layer of cytoplasm surrounding the cytoskeletal shaft (Fig. 3-6A).

The presence of microfilaments in actively streaming cytoplasm in coelopods and prey capturing rhizopodia has already been noted. To determine whether the microfilaments in the radiolarian *Thalassicolla nucleata* are significantly associated with cytoplasmic streaming, specimens with recently captured prey were treated with cytochalasin B, a microfilament inhibitor, to determine its effects on rhizopodial streaming and prey apprehension (Anderson, 1978b). Within minutes after applying 19 μg of cytochalasin B in a vehicle of dimethyl sulfoxide (DMSO) in 100 ml seawater, the radiolarian ceased rhizopodial streaming. The rhizopodia became aggregated and granular in appearance, joined lengthwise and produced cord-like masses. Control specimens treated only with DMSO exhibited normal cytoplasmic streaming. When the treated *Thalassicolla* were transferred to new seawater, they recovered within 24 hr and exhibited typical rhizopodial motility. These experiments, combined with the electron microscopic evidence of abundant microfilaments within streaming rhizopodia and the presence of axial arrays of microtubules, supports the concept of microtubule-guided flow, mediated by microfilament action.

Additional evidence for microfilament-mediated motility was obtained by Cachon et al. (1977), in a clever study of the structural and macromolecular basis of axopod oaring action in the protist *Sticholonche zanclea*. Although the taxonomic status of this organism is uncertain as explained in Chapter 1, and no other radiolarian is known to exhibit the oar-like locomotion of *Sticholonche*, the basic principles of contractile action discovered by Cachon et al. may provide insight into the more general role and activity of microfilaments in a broad range of radiolarian species. The oar-like axopods of *Sticholonche* contain a central shaft of microtubules surrounded by a thin layer of cytoplasm. The axopod is situated in a well-like depression formed by the infolding of the plasma membrane. At the base of the microtubule bundle, there is a thickened, osmiophilic cap that articulates with a concavity in the nuclear membrane resembling a hip joint (Fig. 3-6B). The axopod swivels about this joint when it exhibits its rhythmic beating action. Microfilaments, 20–30Å in diameter, connect with the dense

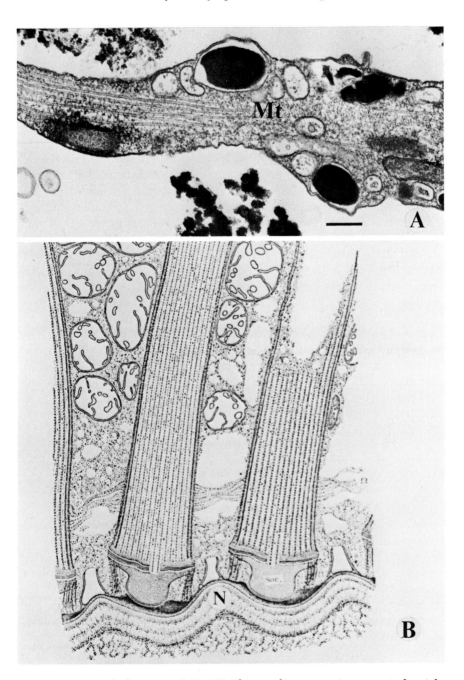

Fig. 3-6. Microtubule arrays (Mt). **(A)** Rhizopodia possessing a central axial array of microtubules and peripheral vacuoles with pigment granules. Scale = 0.5 μm. (From Anderson and Hoeffler, 1979) **(B)** Axopodia of *Sticholonche zanclea* that are articulated with socket-like depressions in the nuclear membrane (N). (From Hollande *et al.*, 1967)

cap and, by their cyclical contraction, produce the rocking motion of the axopods. Cachon *et al.* report that the microfilaments are not composed of the protein, actin, believed to be present in some microfilaments of many cells. They make this conclusion based on the following evidence: "(a) the small diameter of the filaments, (b) the lack of decoration with heavy meromyosin, and (c) their ability to coil, spiral, or fold during contraction." From cytochemical tests and the use of calcium-binding ionophores and chelators, they present evidence that the motile process may be controlled by calcium. Given the specialized quality of this contractile process, further research is needed to determine whether translational motion of cytoplasmic particles along the rhizopodium of other radiolaria can be explained by patterns of microfilament contraction similar to those in the basal apparatus of axopods in *Sticholonche*.

Although most radiolaria are believed not to propel themselves through the water in the open ocean, some interesting contradictions are observed in laboratory culture when the radiolaria contact the glass surface of a vessel. Some Nassellaria, for example, adhere to the surface of the glass with their rhizopodia. Observations with the light microscope show that the nassellarian is able to move across the surface by a kind of "walking motion." The rhizopodia are extended and attached to the surface, then, the central cell body is drawn forward. Once again, the rhizopodia are attached at a place ahead of the forward motion, and the radiolarian is translated by an additional forward increment. Similar translational motion has been observed in nonspinose planktonic foraminifera (e.g., *Globorotalia* sp.) when maintained in laboratory culture. The significance of this crawling motion for the existence of these protists in the pelagic environment is unknown. It may be a unique response to the artificial surfaces encountered in the laboratory. The fact, however, that they are able to make such use of surfaces for locomotion bears further testimony to the remarkable adaptability of these single-celled organisms. It may also suggest from a very speculative viewpoint that these pelagic organisms arose from not too distant phylogenetic ancestors that were benthic or inhabited surfaces of floating matter. The attachment and locomotion on solid surfaces may be a persistent ancestral trait. On the whole, it is not too surprising that the rhizopodia attach to solid surfaces as they commonly adhere to and stream over the surfaces of the siliceous shell. This is particularly evident during prey capture and transport of prey substances toward the central capsule. It may also represent part of the complex locomotory activities used by radiolaria to free themselves from large particles that sometimes adhere to their rhizopodial network. Such "cleaning action" is observed in laboratory cultured specimens that become fouled with large particles of *Trichodesmium* or other planktonic debris introduced when the specimen was initially collected in a plankton net. Hence, such locomotory activity may contribute to the radiolarian's survival.

Radiolarian Longevity

Our knowledge of the life span of radiolaria is limited due to the difficulty of maintaining continuous cultures in the laboratory. Mature radiolaria collected from the open ocean will release reproductive swarmers in culture dishes; however, they do not produce offspring. Consequently, we do not have precise data on the length of the life span from zygote stage to reproductive maturity. Moreover, it is almost impossible to collect very young juvenile stages in the open ocean, as they are far too small to be seen by SCUBA divers, and net collected specimens are often disabled by agglomeration with other gelatinous plankton specimens concentrated in the cod end. This can be partially avoided by using shorter periods of collection, thus reducing the total amount of plankton in the cod end, or by employing large cod ends as with the Reeve net (Reeve, 1981). Research on juvenile stages is at present very limited. Most of our knowledge of radiolarian longevity comes from estimates based on observations of maximum longevity of mature or nearly mature organisms maintained in laboratory culture (Anderson, 1980) or by estimating mean longevity based on a comparison of population densities in the surface water and accumulation rate of dead specimens in the sediments. The latter approach has been employed by Casey et al. (1970) in a very clever mathematical analysis of radiolarian longevity in populations in near-shore waters off the coast of California. Their method is to collect plankton samples in the near surface water (0 to 480 m depth) and determine the population density of each species per unit volume of seawater sampled. These data are extrapolated to yield the number of individuals per unit volume of water per year occurring in the water column (N/m^3/year pl). Sediment samples are taken in locations where a fairly accurate chronological record can be determined. From these samples, the number of individuals in each species accumulating in the sediment per unit volume of seawater per year is computed (N/m^3/year sed). By dividing the plankton sample value by the sediment value (N/m^3/year pl)/(N/m^3/year sed.) and multiplying by 365, the mean survival time of the species in the locality is determined in days.

Two species, *Theoconus zancleus* and *Eucyrtidium hexagonatum*, are considered to be equatorial shallow water faunal members and, therefore, may be assumed to be within a suitable reproductive environment in surface waters near the coast of California. The mean life span for these species estimated by Casey *et al.* was about 1 month (37 days). This compares remarkably well with laboratory observations of life spans of radiolaria. Mature specimens of skeletal-bearing Spumellaria including *Spongodrymus* sp., *Hexastylus* sp., and *Diplosphaera* sp. live on the average of about 14 days in laboratory culture before releasing reproductive swarmers. Assuming they were already quite mature when collected and may have developed to about half their full maturity, this also yields longevity of several weeks to a month. *Thalassicolla* sp. in laboratory culture often live for well over a month. They also were fairly mature when collected, which suggests that the life span of these skeletonless Spumellaria may be considerably longer than for some skeletal-bearing species (Anderson, 1980). Clearly much additional information is needed on the effects of such factors as nutrition, temperature, and quality of the water on radiolarian maturation and reproduction to determine what variations in these parameters may lead to variability in life span culminating in normal reproductive maturity. These data should contribute to our theoretical understanding of environmental factors that may modulate longevity (mean life span of the species) within a normal reproductive pattern of maturation (Anderson, 1983), as opposed to those factors that result in mass mortality, and may, therefore, govern in part the spatial and temporal distribution of radiolaria.

Abundance and Geographical Distribution

Two major sources of information are used in mapping radiolarian abundance and geographical distribution: (1) plankton sampling data, obtained in open ocean locations at various depths in the water column using nets, glass jars in the hands of SCUBA divers, sediment traps, or various forms of filter-collecting devices connected to pumps; and (2) sediment sample data obtained with coring devices or other forms of bottom sampling mechanisms.

The former provides information exlcusively about extant species present at the location during the sampling program. The latter data, obtained from sediment samples, permits a broader time span of analysis, depending on the age of the sediment stratum that is examined. It is limited, however, to those species that leave a fossil record. Thus only skeleton-bearing species that have relatively stable shells are preserved in the sedimentary record. The large gelatinous collodaria and some spicule-bearing colonial radiolaria are not represented in sediment samples. Many Phaeodaria do not leave a fossil record as their organic-rich skeletons are porous or granular in texture and do not persist. Other Phaeodaria possess skeletons that are formed of rod-like elements joined with organic connectors. These usually fall apart due to decay of the organic matter. A thorough analysis of radiolarian distribution in space and time requires a combined approach of water column sampling and sediment analyses.

One of the earliest programs of open-ocean sampling was the Challenger Expedition conducted in the late 19th century. The results of the analysis of radiolarian samples were reported by Haeckel (1887). He concluded that radiolaria occur in all of the seas, in all climatic zones, and from the surface layers to great depths in the water column. His samples were taken in plankton nets and from sediment cores. Among the approximately 4,000 species that he examined, many were undoubtedly fossil forms obtained from the sediments, and perhaps only 400 or 500 of the more common polycystine radiolaria are living in the oceans today (Casey et al., 1979a). Haeckel determined that some radiolarian species are limited to certain bathymetric faunal zones. He recognized three zones: Pelagic Faunal, Zonarial Faunal, and Abyssal Faunal. The Pelagic Faunal zone occurs from the surface to about 46 m. The radiolaria found here consist largely of the Spumellaria and Acantharia with a few members of the Nassellaria and Phaeodaria. The Zonarial Fauna occur in strata at various bathymetric depths between the Pelagic Fauna and the Abyssal Fauna. In the upper portions of the Zonarial Faunal zone (46 to 3,656 m), Spumellaria predominate, but are replaced gradually by Nassellaria and Phaeodaria in the deeper strata (3,656 m to just above the ocean floor). The Abyssal Fauna encompass largely Phaeodaria and Nassellaria that float very near to the ocean floor. Geographically, Haeckel reported that the greatest diversity and largest number of radiolarian species occur in the tropics. The abundance of species gradually declines toward the poles. This gradient is more steep in the northern hemisphere than in the southern hemisphere, and the southern hemisphere appears to have more species than the northern hemisphere. The richest diversity and greatest abundance of radiolaria were found in the Pacific.

Subsequent research in the early 20th century on general patterns of radiolarian distribution has been concisely reviewed by Casey (1971a,b, 1977) who proposed a convenient model of biogeographical zones suitable for use with polycystine radiolaria. A summary of his model is presented as a conceptual view for the more detailed information presented on solitary and colonial radiolarian abundance and distribution. Several of the sources used by Casey in constructing his model are cited as background for the model. Lo-Bianco (1903) proposed four depth zones based on his studies in the Mediterranean Sea. These zones may be correlated with degrees of illumination:

1. 0–50 m—Phaeoplankton (illuminated zone).
2. 50–400 m—Knephoplankton (partially illuminated zone).
3. 400–1,500 m—Skotoplankton (barely illuminated zone).

4. 1,500–5000 m—Nyctoplankton (nonilluminated zone).

Haecker (1908a) correlated radiolarian population zones with those of Lo-Bianco, and defined the following regions based largely on phaeodarian species distribution: Colloid zone (0–50 m), Challengerid zone (50–350 or 400 m), a Tuscarorid zone (350 or 400–1,000 or 1,500 m), and a Pharyngellid zone (1,000 or 1,500–5,000 m). Haecker (1908b) hypothesizes that these stratified populations are related to environmental conditions. More recently, Reshetnyak (1955) proposed the following scheme:

1. Surface radiolaria, 0–50 m.
2. Subsurface forms, 50–200 m.
3. Moderately deep forms, 200–1,000 m.
4. Bathypelagic forms, 1,000–2,000 m.
5. Abyssal forms, 4,000–8,000 m.
6. A transitional fauna, 50–1,000 m.

Reshetnyak reported that radiolarian species are most abundant and exhibit greatest diversity between 200 and 2,000 m. Casey (1966) also noted a distinct vertical zonation for polycystine radiolaria in waters off Southern California. There are three major zones: From 0 to 200 m, there is a distinct zone with a fluctuating fauna throughout the year; two other distinct zones occur at 200 to 400 or 500 m and from 400 or 500 to 1,000 m. During the summer months, Casey reports that additional zones appear with different assemblages of radiolaria in the 0- to 200-m zone. Beneath a 0- to 25-m zone, there is a 25- to 50-m stratum, which correlates with the thermocline; from 50- to 125-m, a zone is found that is delimited at 125 m by a weak pycnocline; the next zone extending to 200 m correlates with surface water overlying Pacific Central water.

The vertical distribution of polycystine radiolaria based on plankton and sediment samples was examined by Petrushevskaya (1966) in a wide range of geographical locations. In the central Pacific Ocean, the species composition is nearly homogeneous from the near surface to a depth of 100–300 m. At greater depths, however, the fauna is relatively impoverished and new species occur that are not found in the upper water strata. Polycystines occur in all layers from 0- to 5,000 m and reach concentrations as high as 16,000 specimens/m^3 of water. Almost no radiolaria were found in the surface layers of the Antarctic waters south of the Antarctic Convergence. Nassellaria and Spumellaria are encountered in greatest abundance at 200–400 m in the Antarctic depths; however, below this their density decreases.

Casey (1971a) proposes a synthesis that encompasses seven geographical zones distributed from north to south and further refined on the basis of depth into shallow zones (surface to 100 or 200 m, and perhaps 400 m in the lower latitudes) and deeper zones (Fig. 3-7). The hydrographic conditions correlated with each of the seven shallow-water zones are reproduced as follows from Casey (1971a, p. 156).

1. Subarctic faunal zone: waters north of the North Pacific Drift and the Subarctic Convergence (also sometimes termed Arctic or Polar Convergence).
2. Transition faunal zone: the North Pacific Drift waters bounded on the north by the Subarctic Convergence and on the south by the Subtropical Convergence.
3. North Central 'shallow' faunal zone: waters within the large anticyclonic circulation pattern of the North Pacific, which could be divided easily into two parts (east and west) as the circulation and Central Water Masses are.
4. Equatorial faunal zone: the regions occupied by the North and South Equatorial Current systems.
5. South Central 'shallow' faunal zone: waters within the large anticyclonic circulation pattern of the South Pacific, which could be divided easily into two parts (east and west) as the circulation and Central Water Masses are.
6. Subantarctic faunal zone: waters bounded to the north by the Subtropical Convergence (the Subantarctic Convergence) and to the south by the Polar Convergence (the Antarctic Convergence).
7. Antarctic faunal zone: waters bounded by the Polar Convergence on the north and the Antarctic Continent on the south.

Some species of radiolaria may be endemic to or characteristic of certain water masses below the shallow water zone. Casey suggests the following classification for these indicator fauna:

1. The North Transition-Central Fauna, which is endemic to waters of the North Pacific Central Water Mass (or masses, east and west), is shallow at the areas of formation and dives with the water mass at the North Pacific Subtropical Convergence.
2. The Subarctic-Intermediate fauna endemic to waters of the North Pacific Intermediate Water Mass, which is shallow north of the Subarctic Convergence and dives with the waters at the Convergence (Casey, 1970).

A general distributional scheme for polycystine radiolaria that is applicable to the World Ocean and adjacent seas and incorporates the foregoing hydrographic and biogeographic data has been proposed by Casey (1971a, p. 156) as follows:

I. Shallow Water Faunal Zones (approximately 0–200 m)
   1. Polar (60°N or S and greater)
      (1) Subarctic
      (2) Antarctic

    2. Subpolar (50°–60°N or S)
      (1) Transition (North Pacific and North Atlantic)
      (2) Subantarctic
    3. Central (10°-50°N or S)
      (1) North (may be divided into east and west where appropriate)
      (2) South (may be divided into east and west where appropriate)
    4. Equatorial (0°-10°N or S)
    5. Special (adjacent seas such as Arctic and Mediterranean shallow-water faunas)
II. Deep-Water Faunal Zones (diving below or existing below Shallow-Water Faunas)
    1. Central
      (1) Transition-Central (0°–40°S, 200–700 m)
      (2) Subantarctic-Central (0°–40°N, 200–700 m)
    2. Intermediate
      (1) Subarctic-Intermediate (Equator > 900 m to Arctic > 200 m)
      (2) Antarctic-Intermediate (Equator > 900 m to Antarctic > 200 m)
    3. Also Common, Deep and Bottom, and Bottom Faunal zones (in oceans where appropriate)
    4. Special (adjacent seas, such as Arctic and Mediterranean deep-water faunas)
III. Cosmopolitan Faunal Zone (cutting across other faunal zones)
    1. Shallow
    2. Deep
IV. Narrow Endemics (within a part of any previous zone)

The intermediate water zones in Section II-2 are water masses that occur at great depths (greater than 900 m) at the Equator, but gradually extend toward the surface near the poles, where they encompass all of the water columns beneath 200 m, as illustrated for the Pacific Ocean in Fig. 3-7.

It is essential to recognize that the deep water zones are marked by cosmopolitan species dwelling within this rather homogeneous environment. Transfer of species from one major geographical region to another within this zone is very likely. Thus, high-latitude forms are also tropically submergent, resulting in an exchange of deep-water forms beneath the tropics (Casey et al., 1970). An example is the exchange of fauna between the North Pacific Intermediate water and the Antarctic Intermediate water masses. An enlarged model appropriate for paleoecological analyses has been proposed by Casey as presented in Chapter 4.

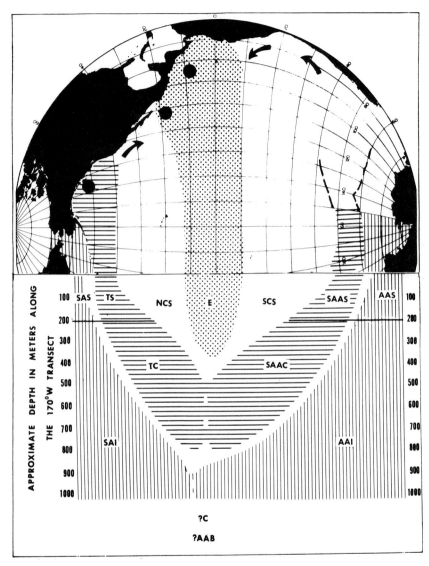

Fig. 3-7. Biogeographic zones for polycystine radiolarians in the Pacific Ocean. AAS, Antarctic Shallow Faunal Zone; AAI, Antarctic-Intermediate Faunal Zone; SAAS, Subantarctic Shallow Faunal Zone; SAAC, Subantarctic-Central Faunal Zone; SCS, South Central Shallow Faunal Zone; E, Equatorial Faunal Zone; NCS, North Central Shallow Faunal Zone; TC, Transition-Central Faunal Zone; TS, Transition Shallow Faunal Zone; SAI, Subarctic-Intermediate Faunal Zone; SAS, Subarctic Shallow Faunal Zone. Perhaps a Common Faunal Zone (C) may exist below the Intermediate Zones, and an Antarctic Bottom Faunal Zone (AAB) below the Common Zone. (Casey, 1971a)

The research on biogeography and ecology by Casey (1971a,b; 1977; Casey et al., 1979a,b), Kling (1966, 1976), Petrushevskaya (1971a,b), Nigrini (1967), Boltovskoy and Riedel (1979), and others suggests that there is merit in examining the distribution of radiolaria in relation to hydrographic and ecological parameters. Some specific information of this kind is presented for solitary and colonial radiolarian species.

Solitary Radiolaria

The major geographical distribution of some commonly occurring extant polycystine and phaeodarian radiolaria published by Haeckel (1887) is presented in Table 3-7. These data have been synthesized from several sections of Haeckel's report, including distributional data presented at the end of species descriptions and in the narrative. Where possible, the habitats have been coded to correspond to Casey's biogeographical model (1971a). Among these few species chosen from each of the major radiolarian groups, there are clearly diverse faunal zones. Some dwell in near-surface waters, whereas others occur at great depths. They occur in all major oceans and some species are cosmopolitan, widely distributed throughout the world's oceans. More recent studies of solitary polycystine radiolarian distributions have begun to elucidate abundance patterns in relation to physical, chemical, and biological factors in the environment.

Some of these studies are summarized for each of the major oceans and restricted geographical regions beginning with the Atlantic Ocean.

Atlantic Ocean   Goll and Bjørklund (1971) examined polycystine radiolarian skeletons in the surface sediments of the North Atlantic Ocean, and determined their abundance (as radiolaria per gram bulk sediment) and quality of preservation based on 334 samples taken at locations between latitudes 11°S and 63°N. A summary of their results is presented as Fig. 3-8. The maximum concentration of radiolaria known in North Atlantic sediments is 118,000 specimens/g or about 5% of the dry-bulk sediment weight. Several regions in the North Atlantic accumulate specimens in a concentration greater than 10,000 radiolaria/ g. In general, the sedimentary accumulation of radiolaria coincides with current systems. Radiolaria are most abundant in sediments from east of the Mid-Atlantic Ridge and from the Caribbean Sea. A high abundance of radiolarian skeletons (species 44 μm or larger) was found in a zone occurring approximately between 15°N and 10°S and extending as a tongue into the Caribbean Sea. A similarly high density region occurred further north approximately between 30 and 60°N, which was near the northern limit of their sampling range. Very large quantities of radiolarian skeletons (>10,000 rad/g) occurred near the Equator and in the northeastern Atlantic approximately between 45 and 60°N and 15 and 35°W. Sediment samples, underlying the central circulation gyre of the North Atlantic (approximately east of Florida, Fig. 3-8), were low in abundance or barren of skeletons. Sediments underlying strong currents (Equatorial, Gulf Stream, and North Atlantic Currents) generally yielded samples with more than 1000 radiolaria per gram, although abundance fluctuated greatly. Goll and Bjørklund recognize that abundance patterns of this kind cannot be explained simply; however, they suggest three possible categories of causal factors: (1) variations in production of living radiolaria, (2) the masking effect of other sedimentary constituents, and (3) opal solution. Unfortunately, they did not find that the sediment abundance of radiolarian skeletons could be related to such biologically significant factors as primary productivity or even to estimates of surface water abundance of radiolaria as assessed by other researchers. They also eliminated the masking effect of other constituents as a contributing factor and conclude that opal solution is probably a major factor in explaining variations in sediment abundance. Although low abundances may be attributed to opal solution, regions of high abundance of skeletons probably represent localized high productivity if current effects can be neglected.

Goll and Bjørklund have also mapped the distribution of several species of solitary radiolaria. Interestingly, they find a substantial difference between the species distribution of radiolaria and those of planktonic foraminifera. Most notably is the restricted nature of radiolarian species distribution compared to the broader distribution of most planktonic foraminiferal species. Some species of radiolaria were truly cosmopolitan, e.g., *Theocalyptra davisiana*. However, the majority of the species exhibited a much more limited distribution. Eight species of radiolaria were examined as typical representatives of solitary radiolarian distribution. These were *Hexacontium hostile, Corocalyptra craspedota, Tholospyris devexa, T. scaphipes, Octopyle stenozona, Ceratospyris hyperborea, Pterocanium praetextum,* and *Anthocyrtidium zanguebaricum.* A distinct radiolarian faunal boundary occurs at latitude 45°N, which marks the southern limit of species restricted to the North Atlantic Current and the northern limit of species occupying the central gyre current system. This is the latitude where the Gulf Stream divides into the North Atlantic and Canary Currents. Species of the North Atlantic Current were confined to sediment samples from the east side of the Reykjanes Ridge. Both *Hexacontium hostile* and *Corocalyptra craspedota* are representative of these species. *C. craspedota* occurs with high frequencies in samples north of latitude 50°N from the eastern Atlantic Basin. Small numbers of this species were found in eight of their samples west of the Reykjanes Ridge. Ocean currents apparently sweep small numbers of this species around the southern tip of Greenland, and the Labrador Current could carry *C. craspedota* to as far as latitude 40°N. Some species have modified bisubpolar distributions occurring in both the northern North Atlantic

Table 3-7. Distribution of some solitary radiolaria[a]

Species	Atlantic Ocean	Location Pacific Ocean	Indian Ocean
SPUMELLARIA			
Actissa princeps	Med., Canary Is.—surface [E]	Surface [E]	Ceylon, Surface [E]
Thalassolampe margarodes	Mediterranean and		
Thalassicolla nucleata	cosmopolitan in all oceans between 40°N and 40°S, Surface [NCS, E, SCS]		
Cenosphaera primordialis		Central, surface [E]	Surface [E]
C. coronata		Central, 4,800 m	
C. gigantea		Central, 5,300 m	
C. reticulata	Med. (Messina), surface [E]		
C. tenerrima		Central, surface [E]	
Carposphaera infundibulum	North		
C. melissa		Central, 5,300 m	
C. prunulum	South, surface [SCS]		
C. nobilis	Cosmopolitan in all oceans at various depths		
Plegmosphaera pachyplegma		Central, surface [E]	
P. leptoplegma	North, surface [NCS]		
Saturnalis circularis	South, 4,000 m		
S. circoides			Zanzibar, 4,000 m [AAB]
S. rotula	Tropical, 4,100 m	North, surface [NCS]	
Stylosphaera musa	Cosmopolitan in all oceans,		
S. polyhymnia	surface		

Species			
Hexastylus phoenaxonius			
H. sapientum	North, surface [NCS]	Central, 4,700 m	
H. maximus		Central, 5,300 m	
H. marginatus		South, 2,700 m	
Hexacontium hexagonale			
Cladococcus arborescens	Med., North, Canary Isl.—surface [E to NCS]		Ceylon, surface [E]
C. antarcticus		Central, 5,300 m	Antarctic, 3,600 m [SAI]
C. quadricuspis		North, surface [NCS]	
C. japonicus			
Elaphococcus furcatus	Tropical, surface [E]		
E. dichotimus	Arctic, surface		
E. umbellatus		South-east, surface [SCS]	
Spongiomma radiatum		Central, surface [E]	
S. asteroides	South, surface [SCS]		
Spongopila dichotoma	Tropical, surface [E]		
S. verticillata		Tropical, surface [E]	
Centrocubus octostylus		Central, surface [E]	
Octodendron verticillatum		South, surface [SCS]	
Spongosphaera polyacantha	Med., (Nice), surface [E]		
Spongodrymus elaphococcus	Tropical, surface [E]		
S. quadricuspis		Central, surface [E]	
Cannartus violina		Central, 5,300 m	
Cyphonium coscinoides		North, surface [NCS]	
C. ethmarium	Equatorial, surface [E]		
C. diattus			
Panartus tetraplus	Equatorial, 4,410 m		
P. tetracolus	Cosmopolitan, various depths in all oceans		Western, Zanzibar, 4,000 m
P. tetrathalamus		Central, 4,300–5,300 m	
Porodiscus orbiculatus	Cosmopolitan, surface		

Table 3-7. continued

Species	Atlantic Ocean	Location Pacific Ocean	Indian Ocean
Spongodiscus mediterraneus	Med.		
S. radiatus	North (Greenland) surface [SAS]	Central, 5,300 m	
S. favus			Between Ceylon & Socotra, Surface [E]
Tholospira nautiloides			
T. spinosa		South, surface [SCS]	
Phorticium pylonium	Cosmopolitan, common in all oceans, surface and various depths		
Soreuma irregulare		North, 5,300 m	
S. acervulina		South, 2,700 m	
S. setosum		Central, 4,400 m	
NASSELLARIA			
Plagonium sphaerozoum	Equatorial, surface [E]	North, Surface [NCS]	
P. lampoxanthium			
P. arborescens		Central, 5,300 m	Madagascar, Surface [E]
P. trigeminum		South, Surface [SCS]	
P. distractis			
Cortina tripus	Cosmopolitan, common in all oceans, surface, and various depths		
C. typus	Tropical, 4,500 m	Central, 4,300-5,300 m	
C. dendroides		Central, 4,900-5,300 m	
Semantis biforis			
S. distephanus	Tropical, surface [E]		

Species			
Dorcadospuris dentata		Central, 4,300–5,300	
Carpocanium diadema	Cosmopolitan in many locations in all oceans, surface		
C. laeve	Med., tropical, 4,470 m		
C. cylindricum	South, 4,000 m [AAB]	Central, 4,400 m	
Pterocanium gravidum		Central, 5,000 m	Zanzibar, 4,000 m [AAB]
P. pyramis		Central, 4,400–5,400 m [AAB]	
P. depressum	Med. (Messina), surface		
P. trilobum			
Podocyrtis tripodiscus	Cosmopolitan, tropical Atlantic and Pacific, 4,300–5,300 m [AAB]		
P. conica			Madagascar [E]
P. tridactyla	Med., Surface [E]		
P. ovata	Tropical, 4,600 m		
P. argulus			
Theocorys turgidula		Tropical, surface [E]	
T. veneris	Cosmopolitan, abundant, surface		
T. obliqua		Central, 5,300 m	
T. martis		South, 2,700 m	
Lithochytris cortina		Central, 5,000 m	
L. pyriformis	Tropical, 4,470 m		
L. lucerna		South, 3,200 m	
Callimitra carolotae		Central, 5,300 m	
C. emmae		Central, 5,000 m	
Cyrtocapsa tetrapera		Western, Tropical, 8,200 m	
C. cornuta		Central, 4,700 m	
C. diploconus	Tropical, 4,100 m		
C. fusulus		South, 2,700 m	

Table 3-7. continued

Species	Atlantic Ocean	Location Pacific Ocean	Indian Ocean
PHAEODARIA			
Cannoraphis spinulosa			
C. lamphoxanthium			
C. spathillata		North, Surface	Cocos Islands, surface [E]
		South, 4,700 m	
Aulacantha scolymantha	Cosmospolitan, common, surface and at various depths		
A. cannulata		South, surface	
A. clavata	South, 3,723 m		
Aulographis pandora	Cosmopolitan, in all oceans, surface to various depths		
A. bovicornis	South, surface [SCS]		
A. triangulum			
A. stellata		South, 4,700 m	Madagascar, surface
Aulosphaera trigonopa	Cosmopolitan, all oceans, surface		
A. flexuosa	North (Faerōe Channel), surface [TS to SAS]		
A. verticillata		South, surface [SCS]	
A. dendrophora		Central, 4,400 m	
Challengeria naresii	Cosmopolitan, 1,800–5,500 [AAB]		
C. sigmodon		North, 4,100 m	
C. pyramidalis	South, 3,700 m		
C. elephas			
C. murrayi		Northwestern, 4,100 m	Cocos Islands, surface [E]
Tuscarora tubulosa		North, 3,700–5,600 m [AAB]	
T. tetrahedra	Tropical, 4,470 m		

T. belknapii	South, 3,700 m
Coelodendrum ramosissimum	Cosmopolitan, surface to various depths
C. lappaceum	South, 2,700–4,700 m [AAB]
C. digitatum	Madagascar, surface [E]
Coelotholus octonus	Southeastern, 2,500 m
Coelographis regina	southeastern, 3,200 m
C. hexastyla	North, 4,100 m
C. gracillima	Med., Surface [E]

[a] Based on data from Haeckel (1887). Abbreviations used: Med., Mediterranean; Isl., Island. Depth of station is given in meters (m). Symbols in parentheses are codes for Biogeographical Zones (Casey, 1971a): AAB, Antarctic Bottom Faunal Zone; E, Equatorial, NCS, North Central Shallow Zone; SAI, Subarctic Intermediate Faunal Zone; SAS, Subarctic Shallow Faunal Zone; and SCS, South Central Shallow Faunal Zone. These are assigned based on the best estimate from Haeckel's information. No assignment is made when the information is too limited to do so.

and in the South Atlantic. The distribution of *Ceratospyris hyperborea* is characteristic of these bisubpolar species. *C. hyperborea* is confined to the east side of the Reykjanes Ridge in the North Atlantic and ranges as far south as latitude 45°N. In the South Atlantic it occurs abundantly north of the Antarctic Convergence. The Benguela Current undoubtedly carries it northward, and low numbers of *C. hyperborea* are found in sediments as a narrow tongue adjacent to the coast of Africa to latitude 24°N. *Pterocanium trilobum* has a similar distribution. Some radiolaria in the Atlantic Ocean are restricted to the southern equatorial regions. Among these are *Tholospyris scaphipes* and *Ceratospyris hyperborea*. *Tholospyris scaphipes*, however, is also abundant north of the Antarctic Convergence and is absent north of latitude 15°N. A small number of species represented by *Octopyle stenozona* are present in sediments underlying the Equatorial, Gulf Stream, and Canary Currents. Many species are found in sediments underlying both the Equatorial Current and the Gulf Stream. *Pterocanium praetextum* is representative of this group which is broadly distributed in sediments across the equatorial Atlantic and the Caribbean Sea. Their range encompasses a narrow recurved tongue adjacent to the coast of North America to latitude 45°N and longitude 45°W. *Tholospyris devexa* is most abundant in samples from the Caribbean Sea, and its abundance in the sediments declines toward the east. It occurs in a narrow band across the equatorial Atlantic below the water mass boundary between the North Atlantic Central Water and the South Atlantic Central Water.

A survey of polycystine radiolarian abundance in surface water of the North Atlantic during March and April of 1962 was made by Cifelli and Sachs (1966). They obtained samples along a cruise track extended along longitude 65°W, from the edge of the Nova Scotia shelf at latitude 41°N to the Caribbean Sea as far south as 13°N. This traverse included all of the water bodies commonly recognized in the North Atlantic which include: slope waters, Gulf Stream, northern Sargasso Sea, southern Sargasso Sea, Antilles Current, and Caribbean Sea. The maximum abundance of radiolaria was 27,000/1000 m³ seawater in the Caribbean Sea, latitude 17°N, and the minimum (600 specimens/1000 m³) was recorded in the Slope Waters, latitude 41°N where interestingly, planktonic foraminifera were most abundant. On the whole, radiolaria were most abundant in the Caribbean Sea, with densities ranging from 14,000 to 27,000/1000 m³. In the southern Sargasso Sea between 26 and 22°N, they were moderately abundant ranging from 2,500 to 4,500 specimens/

Fig. 3-8. Abundance and preservation of radiolaria in surface sediments of the North Atlantic Ocean. (Goll and Bjørklund, 1971)

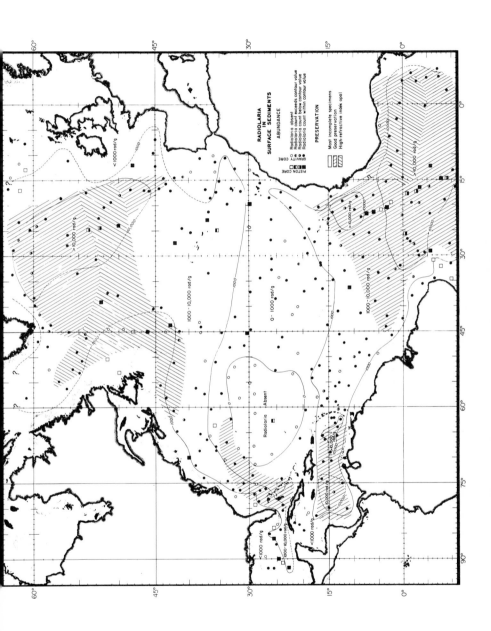

1000 m³. Lower concentrations of radiolaria occurred at the stations northward from latitude 30°N, including the northern Sargasso Sea, Gulf Stream, and Slope Waters. The numerical values recorded at these stations ranged from 600 to 13,500 radiolaria/1000 m³. In general, the pattern shows a gradual increase in abundance of radiolaria with increasing water temperature, thus further substantiating the principle that radiolaria are most abundant in warm tropical waters. Cifelli and Sachs do not report abundance measurements of individual species.

Takahashi and Honjo (1981) have presented some interesting taxon-quantitative data on radiolarian abundance and vertical flux of radiolarian skeletons in the western, tropical Atlantic Ocean using samples obtained from sediment traps. The sediment traps were deployed for 98 days from November 1977 through February 1978 at a station located at latitude 13° 32.2'N, longitude 54°1'W. Some of the most abundant species collected at the four sampling depths are presented in Table 3-8. The total number of skeletons accumulated per square meter per day are reported in this condensed table. A complete list of specimens collected is given in the paper by Takahashi and Honjo. Their data include polycystine and phaeodarian species and, therefore, it is one of the few modern research studies to provide estimates of phaeodarian abundance in the tropical Atlantic Ocean. Among the samples collected at the four depths, Spumellaria comprised approximately 20–30% of the total sample; Nassellaria, 60–70%, and Phaeodaria, 6–8%. The samples also yielded one of the highest coefficients of diversity ever reported in the water column, i.e., 3.3 to 3.6 (natural bels unit) based on the diversity index of Pielou (1969). This is probably due to the efficiency of the traps that catch substantial amounts of all particulates settling in the water column. The number of species encountered in all samples in each group was 89 Spumellaria, 84 Nassellaria, 34 Phaeodaria, and 1 Sticholonchidae. The diversity index of Nassellaria increased substantially from 389 to 988 m. This probably represents the contribution of deep-water species (Reshetnyak, 1955; Casey et al., 1979a,b) and tropical submergent species (Casey, 1971a; Casey and McMillen, 1977; Casey et al., 1979a,b). Among the species reported by Takahashi and Honjo (1981) that conform to Casey's deep water forms are *Cyrtopera languncula*, *Peripyramis circumtexta*, *Litharachnium tentorium*, and *Cornutella profunda*. It is interesting to note that several species of Phaeodaria occur appreciably in the upper water layer above 389 m. These include *Challengeron willemoesii*, *Protocystis xiphodon*, *Euphysetta pusilla*, *Borgertella caudata*, and *Conchidium caudatum*. It is difficult to make estimates about deep-dwelling forms, as the skeletons of Phaeodaria are more fragile and may be degraded during settling. As mentioned by Takahashi and Honjo, this also makes it difficult to make precise estimates of their vertical flux, as some of the shells occur only as pieces in the deep traps. The total radiolarian flux ranged from 16.0 x 10^3 to 23.7 × 10^3 shells/m²/day. An earlier estimate by Honjo (1978) obtained in the Sargasso during winter was 14.0 x 10^3 shells/m²/day, which further corroborates a lower productivity in this region when the water mass is colder.

In a study of living plankton assemblages in the South Atlantic, Boltovskoy and Riedel (1979) examined 59 plankton samples from an area Southwest of Uruguay defined by 36°10.5'S, 39°45.9'S, 54°59.8'W, and 49°27.4'W (Fig. 3-9). They identified 98 species of polycystine radiolaria collected in this area during August and September of 1975. Several types of collecting gear were used, including nets with different mesh sizes and filtering pumps. Most of the samples contained polycystine species, although very few had abundant numbers and some were totally barren of radiolaria. The species diversity in the area was quite low. A majority of the species collected were recorded in only a few samples, less than five, and only a few were common to 40% or more of the samples. Although the species diversity was low, the region was qualitatively rich in species, considering nearly 100 species were identified. The dominant species in the area belong to the Spumellaria and include the following, listed in order of decreasing importance: *Spongodiscus resurgens, Stylodicta multispina, Dictyocoryne profunda, Actinomma antarcticum, Acanthosphaera actinota, Porodiscus micromma, P. sp. aff. P. micromma,* and *Collosphaera huxleyi*. This is in contrast to the results of other researchers (e.g., Petrushevskaya, 1966; Renz, 1976; Riedel, 1958; Nigrini, 1967; and Haeckel, 1887), who have found a larger proportion of Nassellaria than Spumellaria in diverse locations of the world oceans. With the exception of some Plagioniidae, which were present in large numbers in some samples, but not identified to species level, the remaining abundant nassellarians were few. These included *Theocorythium trachelium, Triceraspyris antarctica,* and *Lipmanella dictyoceras*. Few of the species obtained in these living assemblages matched those observed in the sediments by Goll and Bjørklund (1974), who also examined the sedimentary record in this area. Boltovskoy and Riedel attribute this to a rapidly changing fauna in the region. This was documented by the observation that at one sampling site in early August only 5 species were found, whereas later in September, 26 species were obtained at the site. Some of these variations may be attributed to the hydrological conditions in the area. Several different types of water occur within its limits; i.e., purely subantarctic waters of the Malvinas (Falkland) Current, subtropical–subantarctic mixed waters in the Convergence Zone, and isolated spots of purely subtropical waters from the warm Brazil Current (Boltovskoy, 1970; Boltovskoy, 1975). Hence both cold-water and warm-water species occur here. Boltovskoy and Riedel's paper is nicely illustrated and contains concise descriptions of the species identified in the area.

Table 3-8. Some abundant solitary radiolaria collected at varying depths in sediment traps in the southwestern Atlantic Ocean[a]

Taxon	Collection depth (m)			
	389	988	3,755	5,068
SPUMELLARIA				
Family Liosphaeridae Haeckel				
Cenosphaera huxleyi Müller	5	46	107	38
Styptosphaera sp.	74	74	162	143
Plegomosphaera lepticali Renz	21	0	15	18
Total LIOSPHAERIDAE	100	120	284	199
Family ACTINOMMIDAE Haeckel				
Theocosphaera inermis (?) (Haeckel)	251	58	409	194
Actinomma arcadophorum Haeckel	56	40	56	65
Trilobatum (?) acuferum Popofsky	238	58	673	437
Stylosphaera melpomene Haeckel	132	64	52	110
Ommatartus tetrathalamus (Haeckel) subsp. A	104	161	156	149
Total ACTINOMMIDAE	1130	695	2077	1610
Family PHACODISCIDAE Haeckel				
Spongodiscus resugens Ehrenberg Spongodiscus sp. B	808	929	1337	1498
Spongotrochus glacialis (Popofsky) Total PHACODISCIDAE	920	1027	1497	1631
Family PORODISCIDAE Haeckel				
Euchitonia furcata Ehrenberg	106	32	128	46
Hymeniastrum euclidis Haeckel	81	72	126	102
Porodiscus micromma (Harting)	76	160	257	114
Stylochlamydium asteriscus Haeckel	121	72	55	91
Total PORODISCIDAE	410	375	677	429
Family LITHELIIDAE Haeckel				
Tholospira cervicornis Haeckel group	191	365	176	371
Pylonena armata Haeckel group	163	85	186	111
Tetrapyle octacantha Müller	513	444	947	458
Octopyle stenozona Haeckel	53	56	72	37
Total LITHELLIDAE	954	956	1402	1013
NASSELLARIA				
Family PLAGONIIDAE Haeckel				
Cladoscenium anacoratum Haeckel	123	320	267	419
Obeliscus pseudocuboidea Popofsky	83	131	167	113
Phormacantha hystrix (Jørgensen)	150	141	330	260
Peridium spinipes Haeckel	1251	2184	3114	2173
Lophophaena cylindrica (Cleve) Peromelissa phalacra Haeckel }	2445	1283	1925	1570
Acanthocorys cf. variabilis Popofsky	594	538	982	146
Helotholus histricosa Jørgensen	313	251	238	120
Total PLAGONIIDAE	5112	5334	7464	5134
Family ACANTHODESMIIDAE Haeckel				
Zygocircus capulosus Popofsky	248	596	795	285
Zygocircus productus (Hertwig)	453	436	634	431
Acanthodesmia vinculata (Müller)	108	91	69	54
Dictyospyris sp. A and B	109	54	36	9
Amphyspyris costata Haeckel	128	175	56	56
Total ACANTHODESMIIDAE	1197	1570	1768	1068

Table 3-8. continued

Taxon	Collection depth (m)			
	389	988	3,755	5,068
Family THEOPERIDAE Haeckel				
Cornutella profunda Ehrenberg	64	141	367	160
Pterocanium praetextum (Ehrenberg)	207	158	102	88
Eucyrtidium hexastichum (Haeckel)	120	45	133	40
Theocalyptra davisiana cornutoides (Petrushevskaya)	82	354	784	207
Theocalyptra davisiana davisiana (Ehrenberg)	83	122	197	75
Total THEOPERIDAE	707	1244	2467	1137
Family PTEROCORYTHIDAE Haeckel				
Pterocorys zancleus (Müller) ⎫ *Pterocorys campanula* (Haekel) ⎭	2966	2381	2698	1654
Total PTEROCORYTHIDAE	2975	2418	2820	1733
Family ARTOSTROBIIDAE Riedel				
Spirocyrtis scalaris Haeckel	91	79	67	101
Spirocyrtis sp. aff. *S. seriata/S. subscalaris*	1496	629	653	660
Carpocanarium papillosum (Ehrenberg)	0	7	2	56
Total ARTOSTROBIIDAE	1587	748	874	860
Family CANNOBOTRYIDAE Haeckel				
Acrobotrys sp. A, B, and C	123	76	126	68
Botryocyrtis scutum (Harting)	226	167	212	111
Total CANNOBOTRYIDAE	349	243	338	178
PHAEODARIA (Haeckel)				
Family CHALLENGERIIDAE Murray				
Challengeron willemoesii Haeckel	128	110	62	36
Protocystis ziphodon (Haeckel)	135	64	36	40
Total CHALLENGERIDAE	291	199	137	109
Family MEDUSETTIDAE Haeckel				
Euphysetta pusilla Cleve	407	229	188	21
Medusetta ansata Borgert	73	15	59	33
Total MEDUSETTIDAE	488	330	345	147
Family LIRELLIDAE Ehrenberg				
Borgertella caudata (Wallich)	252	62	149	180
Lirella bullata (Stadum and Ling)	2	384	603	705
Total LIRELLIDAE	254	453	1132	1084
Family CONCHARIIDAE Haeckel				
†*Conchidium caudatum* (Haeckel)	103	2	25	30
†*Conchopsis compressa* Haeckel	0	3	0	0
†Total CONCHARIIDAE	130	6	35	32

† Each value is counted as a half (0.5) shell.
ᵃ Takahashi and Honjo (1981).

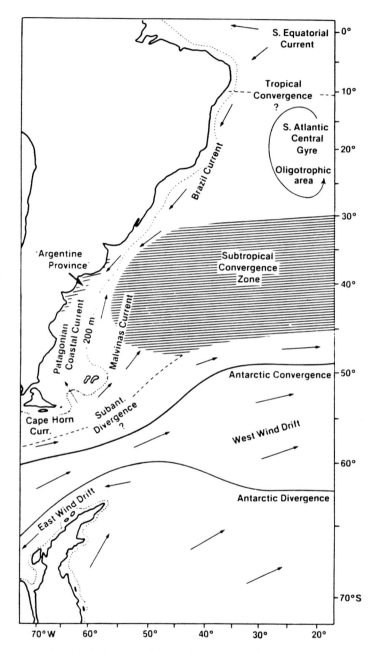

Fig. 3-9. Surface hydrology of the southwestern Atlantic Ocean. (Boltovskoy, 1979)

A complete understanding of the complex factors correlated with radiolarian abundance and geographical patterns of distribution in the South Atlantic will undoubtedly require a more thorough analysis of hydrological and bioecological variables that may influence radiolarian survival and distribution. Some of the potentially significant hydrological factors for zooplankton ecology in the South Atlantic have been summarized by Boltovskoy (1979). The major surface currents found in the South Atlantic are presented, beginning with the southern most ones and progressing northward (Fig. 3-9). The East Wind Drift Current and West Wind Drift Current form a counter flowing current pattern demarked by the Antarctic Divergence (ca. latitude 60° to 65°S). The Antarctic Convergence marks the southern limit to the Cape Horn Current. The Malvinas (Falkland) Current proceeds northward, lateral to the Patagonian Coastal Current. The Brazil Current flowing from the north is separated from the northern-most South Equatorial Current, possibly by the Tropical Convergence.

Four major biogeographical zones are established in this region, and in some, there are distinct major biological groupings as summarized in the following (Boltovskoy, 1979):

1. Antarctic Complex (between Antarctica and the Antarctic Convergence). This region is characterized by a high primary productivity (up to 2,700 mg C/m²/day) dominated by diatoms among the phytoplankton and euphausiids among the zooplankton. It has the highest degree of endemic forms and the lowest diversity of species.

2. Subantarctic Complex (between the Antarctic Convergence and the Subtropical Convergence). Primary productivity is low (usually less than 10 mg C/m²/day). Diatoms and copepods are the major organisms. Two minor associations are also identified in this zone: (a) a neritic area over the large Patagonian shelf (ca. 100,000 km²) with slightly higher productivity and plankton composed largely of larvae and brackish-water organisms, and (b) transgressive areas with cyclic or occasional influence of warm waters (e.g., (1) the area south of Cape Horn, Argentine province, influenced 3 months each year by the Brazil Current; and (2) the Subtropical Convergence zone).

3. Subtropical Complex (between the Subtropical Convergence and the Tropical Convergence): The northern boundary has not been determined exactly. Primary productivity is extremely low (less than 1 mg C/m²/day). Diatoms, dinoflagellates, and Coccolithophora dominate the phytoplankton. The zooplankton is dominated by Copepoda but has an important share of Chaetognatha, Hydromedusae, and Appendicularia.

4. Tropical Complex (between the Tropical Convergence and the Equator): Primary productivity can be high in some neritic waters, but does not exceed 40–50 mg C/m²/day in the pelagic domain. Dinoflagellates, Coccolithophora and cyanophytes are the major phytoplankton, whereas copepods are the major zooplankton.

The complex comingling of currents in the southwestern region near the coast including the Patagonian Coastal Current, Malvinas Current, and Brazil Current may create fluctuating hydrological conditions resulting in varying ecosystems, thus contributing to rapid changes in faunal composition over a period of weeks or months.

Several research studies on radiolarian distributions have also attempted to explain variations in abundance as a result of environmental seasonal factors including cyclical upwelling events and variations in oceanic currents. Among these is an interesting study by Goll and Bjørklund (1974). Goll and Bjørklund examined the distribution, abundance, and diversity of some polycystine radiolaria in recent sediments of the South Atlantic in a study complementary to their earlier research on the faunal distribution in the North Atlantic (Goll and Bjørklund, 1971). They analyzed 456 new samples obtained from the South Atlantic Ocean in addition to 55 samples described in their previous research. They chose seven representative species to illustrate the general distribution patterns in the sediments of the South Atlantic: (1) *Triospyris antarctica*, (2) *Tholospyris scaphipes*, (3) *Corocalyptra craspedota*, (4) *Tristylospiris palmipes*, (5) *Tholospyris procera*, (6) *Panartus tetrathalmus*, and (7) *Spongaster tetras*. *Tholospyris scaphipes* presented a discontinuous and enigmatic distribution which was difficult to interpret; therefore, it will not be described here. A brief synposis of the distribution and interpretation of factors contributing to these patterns as described by Goll and Bjørklund is presented.

The first species *Triospyris antarctica* (*Triceraspyris antarctica*) (Riedel, 1958) is distributed in a pattern that roughly matches the zone of maximum radiolarian abundance in southern high-latitude sediments (a latitudinal band located approximately at latitude 50°S. and extending from the eastern most sampling sites longitude 30°E. to about 45°W.). This region is partially coincident with the Antarctic Convergence (Fig. 3-9). For *T. antarctica*, occurrences of more than 1,000 specimens per gram underlie the Antarctic Convergence in the Argentine Basin. Similar high abundance was found in the radiolarian sediments on the crest of the Atlantic–Indian Ridge. An expanding latitudinal band of high occurrence stretches east of this ridge. Lesser concentrations are found to the south, where all radiolaria are less abundant due to ice-rafted clastics. Based on the sediment sample distributions, Goll and Bjørklund conclude that *T. antarctica* occupies surface or subsurface waters north of the Antarctic Convergence on the east side of the Mid-Atlantic

Ridge and possibly on the west as well. They hypothesize, moreover, that the distribution of T. *antarctica* is more closely regulated by the circumpolar current system than by surface temperature and salinity characteristics. The distribution of *Corocalyptra craspedota* is representative of subantarctic species. It is not present south of the Antarctic Convergence, and rare instances of occurrence were found north of the Subtropical Convergence. In general, it is rare in the sediments and reaches its maximum occurrence of 390 specimens per gram of carbonate-free sediment east of the Cape Basin. Goll and Bjørklund hypothesize that it is probably too low in abundance to represent a stable reproductive population; and assuming that shell dissolution is not a major factor in its scarcity, then an alternate explanation may be that it is a hybrid, thus accounting for its occasional occurrence amid a more stable interbreeding population. Two species, *Tristylospiris palmipes* and *Tholospyris procera*, exemplify central or subtropical species. The northern limit of their distribution is approximately at latitude 7°S, and both species are seldom present south of the Subtropical Convergence. It appears that both T. *palmipes* and T. *procera* are transported into the Atlantic from the equatorial Indian Ocean by the Agulhas Current. They reach as far as the Cape of Good Hope, but appear to avoid the Benguela Current and are absent in the sediments of the West African continental margin.

Panartus tetrathalamus is representative of equatorial species. It is distributed north of the Subtropical Convergence and, in this respect, resembles its distribution in the Indian Ocean where it occurs in all sediments north of latitude 35°S (Nigrini, 1967). It is scarce in subtropical regions, and has its highest occurrence in the South Atlantic in sediments underlying the Benguela and South Equatorial Currents. The South Atlantic and Indian Ocean populations may be joined by transport in the Agulhas and Benguela Currents. *Spongaster tetras* is a more typical equatorial species occupying a broad subtropical zone south of Cape Town. To the north in the subtropical zone, only shelf and slope sediments beneath the Benguela Current contain specimens of S. *tetras* and similarly distributed equatorial species. Other species with equatorial distribution are: *Clathrocircus stapedius, Corocalyptra elizabethae, Giraffospyris laterispina, Tholocubus tesseralis, Semantis distephanus, Triopyle hexagona, Dendrospyris stabilis,* and *Botryocyrtis scutum*. With the exception of S. *tetras*, the remaining species had low occurrences throughout their range. Goll and Bjørklund suggest that this thin abundance may represent the annual accumulation of specimens that occur in high densities only sporadically during the year. Thus at certain seasons, their densities may be very high in the water column; however, at other seasons they may be more sparse. If plankton samples are taken during peak periods of occurrence, then the sediments may appear to be under-represented relative to this faunal "bloom." Goll and Bjørklund hypothesize that some of these infrequent species may be introduced into the equatorial Atlantic Ocean from the Indo-Pacific by way of the Agulhas and Benguela Currents, only on a seasonal basis. Thus, they may have sufficient numbers to reach a steady reproductive population level only during restricted seasons of the year. This is an interesting biogeographical interpretation that is offered in lieu of the more commonplace assumption that the shells are dissolved in the sediments, thus contributing to their sparsity. However, such interpretations based on an assumption of sexual reproduction in radiolaria must be considered cautiously in view of our limited knowledge about the frequency of sexual reproduction, if indeed it occurs at all, among radiolaria.

Pacific Ocean   Several comprehensive studies have appeared recently on the ecology and zoogeography of radiolaria in the Pacific Ocean and California coastal waters (e.g., Kling, 1966, 1976; Casey, 1966, 1971b; Petrushevskya, 1971a; Nigrini, 1968; Renz, 1976). These studies vary in their emphasis on living plankton assemblages and interpretation of extant faunal assemblages versus paleoecological considerations. The main findings relevant to modern assemblages of radiolaria are summarized here.

Petrushevskaya (1971a) examined polycystine assemblages at 10 stations located in the central Pacific. Seven of the stations were located in an area near the Equator between 20°N and 20°S, while the remaining three were located more northwest between 20 and 40°N. The largest quantity of radiolaria was found near the Equator confirming again that the warm equatorial waters are particularly favorable habitats for many polycystine radiolaria. The quantity of total live radiolarian species in the water column varied from 25–50 to 10,000–16,000/m³. Although radiolaria occurred throughout the water column from the surface down to 5,000 m, most of them were in the upper horizons at depths in a range from 0–25 to 100–150 m. The density of radiolaria at these depths was 5,000–15,000 specimens/m³, and the temperature was 23–28°C with a salinity of 34–36%. At deeper horizons, the number of living radiolaria is less than that in the upper horizons. With increasing depths from 100–150 to 300–500 m, the density of polycystine radiolaria was about 1,000/m³. At depths greater than 500 m, the number was no higher than 100–500 organisms/m³.

Two major assemblages of polycystines were found. One group occurs in the surface water (0–100 m) at a temperature of 20–28°C (optimum 23–25°C), and at a salinity of 33.9–35.9% (optimum ca. 25%). This includes *Psuedocubus obeliscus, Zygocircus archicircus, Psilomelissa calvata, Lithomelissa* sp., *Botryocyrtis scutum,* and others. They reach densities of about 1,000 individuals/m³. A second group lives at a depth of 75–100 m and deeper at salinities of 34.2–36% and a temperature of 7–19°C (optimum 9–13°C). They were absent at the surface and occur most abundantly at a depth of 300 m. As a whole, they are less abundant than many of the surface dwelling species. The greatest density was 100 individuals/m³ (*Litharachnium tentorium*) at a depth of 300 m. In other cases, there were only about 25–50 individuals of a given species/m³. Other species included in this assemblage are *Cornutella verrucosa*, *Dictyophimus clevei*, *Phormacantha hystrix*, and *Peridium longispinum*. In general, Petrushevskaya observed that there is a good correspondence between occurrence of species in the water column and that in underlying sediments; however, the numbers of specimens of any given species may vary considerably in the sediments compared to the water column. The problem of shell dissolution in the sedimentary record has been cited earlier in discussions of the Atlantic Ocean sedimentary assemblages. Species with thick walls and robust spongy skeletons have greater preservation potential than thinner-walled species. On the whole, it appears some regions of the Pacific provide greater preservation than the Atlantic; however, the interpretation of general patterns of faunal assemblages in sediments is complicated due to the differential dissolution effects that may occur among species. Petrushevskaya presents tabular data on preservation potential of various species occurring in the Pacific. This is a useful reference in interpreting research studies on Pacific sedimentary records.

Renz (1976) examined the distribution and abundance of radiolaria in plankton samples and sedimentary records in a transect in the central Pacific Ocean (28°N–25°S along longitude 155°W). This was a multiobjective study intended to elucidate the distribution and ecology of some living species of polycystine radiolaria and to examine the problem of correspondence between plankton assemblages and their representation in the underlying sedimentary record. Thirteen stations and 157 plankton samples were used in the study. Samples were taken during September 1968 and August–October 1969. Renz acknowledges that such a limited sampling program makes it difficult to develop strong inferences about correspondence between sedimentary assemblages and plankton sample composition; however, some insights are generated through statistical analyses. She concludes that the numbers of organisms collected in the plankton exceeded those in the sediment. There was a much greater similarity between plankton assemblages in adjacent plankton samples than between sedimentary assemblages. It appears that factors controlling the distribution of groups in the plankton are different from those affecting the sediment distribution.

Renz's study yields interesting distributional data on living species throughout the water column from the surface to 200 m. Of the 137 species collected in the plankton, only 26 were restricted to a particular depth. Among the latter group, 18 species occurred only below 75 m and reached maximum abundance at two stations near the northern equatorial boundary (11°N, 154°58′W, and 5°3.8′N, 155°W). There was a strong and relatively shallow thermocline that may be responsible for the zonation of these species in this region. They generally occurred at temperatures less than 25°C. Among the species with restricted distributions, only a *Spongodiscus* sp. (*S.* sp. aff. *S. resurgens*) occurred in the 0- to 25-m horizon. Among the remaining species, some representatives occurring at each major boundary are as follows: (1) *Anthocyrtidium zanguebaricum*, *Lamprospyris hookerii*, and *Stichopilium bicorne* occurred at 25 m and deeper. (2) *Cornutella profunda*, *Dictyocodon palladius*, *Lychnocarnium* sp., and *Tripocyrtis pteides* Renz occurred at 50 m and deeper. (3) Only *Cubotholus regularis* was found in the 75- to 150-m zone. (4) Fourteen species were found 75 m and deeper, including *Artopilium undulatum*, *Cyrtopera languncula*, *Lithomitra lineata*, *Spongocore diplocylindrica*, and *Verticillata hexacantha*. (5) *Eucyrtidium acuminatum*, *Lithostrobus cornutus*, and *Porodiscus microporus* occurred 150 m and deeper. In the northcentral Pacific, 5 species were found below 200 m: *Nathopyramis* sp., *Calocyclas monumentum*, *Saturnalis circularis*, *Sethophormis rotula*, and *Stichocampe bironec. S. circularis* was most abundant with a density of 3 individuals/m³ water.

Three planktonic areas or domains were found based on mappings of the plankton assemblages: A Northern Tropical Domain (NTD) delimited on its southern boundary approximately by the North Equatorial Current (between 28°5′N and 11°N). An Equatorial Domain (ED) occurring south of the NTD and delimited on its southern boundary by the South Equatorial Current (between 6°58′S and ca. 11°S). A Southern Tropical Domain (STD) extending to the south of ED. More than half of the species (67) occurred in all three domains. Four species were restricted to the Southern Tropical Domain: *Conarachnium umachrian*, *Cornutella profunda*, *Eucyrtidium anomalum*, and *Spongobrachium* sp. aff. *S. ellipticum*. Renz did not report extensive seasonal or annual variability owing to the small number of samples.

The relative abundance and distribution of some radiolarian species in Pacific sediments was investigated by Casey (1971b) and correlated with the biogeographical zones and water masses overlying the sediments (Table 3-9). *Cornutella profunda* and *Cyrtopera languncula* appear to be good indicators of North Pacific Intermediate water mass. They occur as part of the dominant community north of the Polar (Subarctic) Convergence. They also occur in the California Current at depths between 400 or 500 and 1,000 m and apparently are part of the continuous water mass that dives at the polar convergence and descends deeply under surface layers in middle and low latitude regions. The major faunal groups associated with other biogeographical zones are listed in the left hand column of the table. Within these major zones, Casey describes some interesting subunits containing either species not generally found in the major area or species more abundant than in the major area. Some of these structured subareas can be correlated with known oceanographically stable areas such as large eddy systems. Three of these areas were analyzed. One area was from the northeastern section of the Transition Zone, where the abundant species were (?) *Hexadoridium streptacanthum* and *Spongaster tetras* (rectangular form). *Theoconus zancleus*, *Pterocanium trilobum*, and *Litharachnium tentorium* were also present, but absent or rare in the rest of the Transition Zone. These species were sparse or absent in the western transitional part of the zone. However, the species *Peripyramis circumtexta*, *Siphocampe erucosa*, and *Spongopyle osculosa* were present in this western region and are part of the central water mass that forms and submerges at this region. Another zone with subunits occurs in the equatorial region. A zone south of Baja California is delimited by abundant *Euceryphalus* sp., *Clathrocanium ornatum*, and *Lithomelissa monoceras*. A second equatorial subunit occurs near Central America (Nigrini, 1968) delimited by abundant *Acrosphaera murrayana*, *Botryocyrtis* sp., *Larcospira quadrangula*, and *Lamprocyclas maritalis*. Two species appear to have bipolar distributions: *Sethophormis rotula* and *?Hexadoridium streptacanthum*. These occur at high latitudes of the Pacific in Subarctic and Transition water masses of the Northern Hemisphere and Antarctic and Subantarctic water masses of the Southern Hemisphere.

Further studies in the central North Pacific (Kling, 1979) contribute insights into the vertical distribution of polycystine radiolaria at two stations in the northern part of the anticyclonic gyre, which dominates the circulation of the subtropical North Pacific. In general, the central gyre of the North Pacific is characterized by sluggish circulation and oligotrophic environments. The abundance of polycystines vertically forms a stratified pattern with high values at the surface (100 or more individuals/m^3), a sharp decrease to a minimum at about 50 m, a maximum exceeding surface levels at about 100 m, and a decrease to very low values (ca. an order of magnitude lower than surface values) at 300 m. The decrease in radiolarian abundance persists at further depths in the water column. Faunal profiles yield four vertical zones that correlate with water mass characteristics: (1) A Surface Zone that is characterized by decreasing abundance with depth corresponding to the temperature decrease in the thermocline. Species occurring in this zone are *Heliodiscus asteriscus*, *Lithopera bacca*, *Ommatartus tetrathalamus*, *Pterocanium praetextum eucolpum*, *Cypassis irregularis*, *Tetrapyle octacantha*, and *Spongaster tetras irregularis*. (2) An upper Subsurface Maximum Zone characterized by species that are absent or scare at the surface, but reach a pronounced maximum at about 50–100 m. Major hydrographic features include a shallow salinity minimum and oxygen maximum, with a chlorophyll maximum at about the same depth as the radiolarian maximum. Species are *Phormospyris stabilis*, *Styloch-*

Table 3-9. Relative abundance of radiolarians in Pacific sediments[a]

Sediment zones	Subarctic			Transition				Central (north)				Equatorial						Calif. current								Central (south)		Subantarctic		Antarctic
Station coordinates	164°43′E, 51°30′N	162°30′E, 51°15′N	176°15′W, 53°20′N	164°50′W, 46°57′N	173°02′W, 44°45′N	158°38′W, 36°56′N	139°16′W, 53°02′N	126°38′W, 28°25′N	135°53′W, 28°25′N	154°55′W, 20°27′N	157°26′W, 19°12′N	160°30′W, 5°21′N	125°30′W, 14°20′N	125°26′W, 10°21′N	131°31′W, 3°12′N	116°13′W, 1°27′N	105°09′W, 10°53′N	125°39′W, 42°05′N	128°12′W, 40°08′N	126°24′W, 38°03′N	123°19′W, 35°08′N	126°02′W, 34°29′N	118°30′W, 33°10′N	118°42′W, 28°35′N	109°31′W, 17°52′N	154°45′W, 24°41′S	164°08′W, 40°37′S	178°57′W, 52°37′S	169°12′E, 57°43′S	165°56′W, 64°11′S
Biogeographic Faunal zones																														
Subarctic Transition Fauna																														
Pterocanium sp.			C	R			A									R		C	C		R	R				R	R	R		
Sethophormis rotula																			A		R		C	R		C	C	C	R	C
Spongotrochus glacialis	A	A	A	A	A	RC	C									R		C	RC	A	R				R	R	R	C	A	A
Transition Fauna																														
Pterocorys hirundo	R	C	A	A	A	A	A	R	R	C		R	R	R		R		C	C	C	C	RC	R	R	R	RC		R		R
?Hexadoridium streptacanthum		CA	C	C	A	A	C	R	R			R	R	R	R	R	R	C		RC	C	RC	R	R	R	R				R
Subarctic Intermediate Fauna																														
Cornutella profunda	R	RC	A	C	C	C	A	R		RC	RC	R	R	R	RC	R	R	C	A	C	C	A	C	R	C	R	A	C		A
Cyrtopera langungula		R	A	A	CA	R	R	C	C	RC	C	R	R	R	R	C	C	A	A	RC	R	R	C	R	C	R	RC	RC	C	A
Transition Central Fauna																														
Peripyramis circumtexta	R	R	CA	CA	C	C	A	R	R	R	C	R	RC	C	R	R	R	R	C	RC	R	A	C	C	R	RC	RC	R	R	C
Siphocampe erucosa	A		C	A	A	A	A	A	C	RC	C	RC	C	RC	C	C		C	A	A	C	C	C	C	R	RC	RC	C	R	C
Spongopyle osculosa	C		RC	RC	CA	C	C	CA	R	RC	A	R	C	RC	C	C	CA	RC	A	C	C	A	C	C	R	R	R	C	A	C
Central 'Shallow' Fauna																														
Calocyclas amicae								A	A	A	A	R	CA	RC	C	C		RC			A		C	C	RC	R	CA			
Euchitonia furcata								A	A	CA	A	CA	RC	C	C		C						C	C	C	A	R			
Eucyrtidium hertwigii					R			A	CA	A	R	RC			CA		CA						C	C						

Equatorial Fauna
- Acrosphaera murrayana
- Acrobotrissa cribrosa
- Amphispyris costata-thorax
- Anthocyrtidium cineraria
- ?Clathrocorys murrayi
- Dictyoceras vichowii
- Eucecryphalus sp.
- Lithomelissa monoceras
- Peridium spinipes
- Tristylospyris scaphipes

Equatorial Central 'Shallow' Fauna
- Botrycyrtis sp.
- Carpocanium sp.
- Clathrocanium ornatum
- Dictyocoryne profunda
- Eucyrtidium hexagonatum
- Lamprocyclas maritalis
- Larcospira quadrangula
- Lithamphora furcaspiculata
- Litharachnium tentorium
- Pterocanium praetextum
- Pterocanium trilobum
- Theoconus zancleus
- Theophormis callipilium
- Spongaster tetras
- Spongocore puella

[a] From Casey (1971b); A = abundant; CA = common to abundant; C = common; RC = rare to common; R = rare

lamidium venustum, Lophocorys polyacantha, Anthocyrtidium ophi-rense, Stichopilium bicorne, Theocorys veneris, Phormostichoartus corbula, Neosemanis distephanus, Pterocorys clausus, Phormospyris scaphipes, Lithaarachnium tentorium, Eucecryphalus craspedota, and *Spongodiscus resurgens.* (3) A Lower Subsurface Maximum Zone containing species that reach a peak here but are poorly represented in the upper zone and vice versa. It is located about 100 m below the upper zone (number 2) and occurs below the shallow salinity minimum and within the depth of the subadjacent salinity maximum. Species occurring here are *Lamprocyclas maritalis polypora, Theocalyptra bicornis, Theocalyptra davisiana cornutoides, Spongocore puella, Lithocampe* sp., *Hymeniastrum euclidis, Dictyophimus infabricatus,* and *Psuedodictyophimus gracilipes.* (4) A Deep Zone characterized by a sparse population restricted to deep water and consistently absent from the upper few hundred meters. This cold dysphotic to aphotic zone is inhabited by the following species: *Botryostrobus aquilonaris, Lampocyrtis nigriniae, L.* (?) *hannai, Cornutella profunda, Stichopera pectinata,* and *Peripyramis circumtexta.* These, moreover, are often well represented in sediment fossil samples.

Much work has been done on the ecology of polycystine radiolaria; however, the Phaeodaria have been relatively neglected. Although they are usually poorly represented in the sedimentary record compared to polycystine radiolaria, their remarkable diversity of distribution in the water column including some surface dwelling species and others that occur only at great depth suggests they are of potentially significant ecological interest. Kling (1976) examined the horizontal and vertical distribution of some phaeodarian species largely in the Family Castanellidae and two in the Family Circoporidae. Samples were obtained in the eastern North Pacific at stations west to southwest of the continental margin, roughly within the region bounded by the Equator and latitude 55°N, and longitudes 90 to 160°W. Most samples were taken with stratified opening/closing net tows. Among the species examined, most exhbitied well defined vertical and horizontal biogeographical zones. Although no two species had identical distributions, the patterns of some species were closely related to others so that groups of biogeographically related species could be defined. Three major kinds of distribution patterns were noted and related to subsurface hydrography: (1) Species occur in shallow and deep northern waters but extend into and are restricted to deeper waters to the south. (2) A second group occurs at moderate depths south of the first group. At greater depths they overlap with the deep-dwelling species of the first group, but occur deeper to the south. (3) A third group is concentrated at the surface in the region of the equator at the southern end of the sampling zone. This pattern of species distribution is consistent with Casey's general observations (1971a,b, 1977) that cold water species dwelling in the north form a continuous distribution with deep water species in more south-ern latitudes (Subantarctic Faunal Zones, Fig. 3-7). The third group corresponds to Casey's equatorial faunal assembly dwelling in the sur-face or near surface of the warmer waters at the equator. Kling (1976) has carefully analyzed the distribution of several phaeodarian species in relation to the quality of the subsurface water mass in these regions. He identified three principal hydrographic features in a transect from north to south: (1) The Subarctic Water mass (of Sverdrup) and the Intermediate Water characterized by cool, low salinity surface water north of about 44°N with a basal halocline, from which the Intermediate Water extends at depth to the south, centered on a salinity minimum. (2) A more southern zone (ca. between 35 and 44°N) with rather con-torted isohalines marked by a subsurface salinity maximum. (3) A surface zone of maximum salinities (Central water mass) and maximum temperatures (approximately south of latitude 35°N). This zone is sep-arated from the first two zones by a northward shoaling halocline and thermocline.

Among the species identified, two (*Castanidium apsteini* and *Cas-tanidium variable*) are typical of the northern and intermediate water fauna. *C. apsteini* occurred in the subarctic region below 25- to 50-m, with greatest abundance between about 100 and 200 m. This distri-bution continued to the south (at longitude 155°W) at gradually greater depths to where the upper limit was 300 m. It also occurs to a lesser degree off the west coast of North America in the southern part of the Alaska Current. The pattern of distribution is clearly limited to the cold, low-salinity waters of the subarctic region and the low salinity intermediate water mass that submerges southward in midocean. *C. variable* has a similar distribution, but ranges farther south along the west coast of North America coincident with the subarctic water mass carried south in the California Current. The geographic pattern of dis-tribution is characteristic of "Subarctic" zooplankton distributions. Ad-ditional species of radiolaria occupying a similar pattern of distribution but at greater depths are *Haeckeliana irregularis* and *Haeckeliana por-cellana*. A second group of Phaeodaria were found in the salinity max-imum zone (second water mass type). A *Castanissa* sp. identified as sp. 2 was most abundant in the central region and represents a group of species occupying a narrow, midlatitude zone (ca. between 30 and 44°N, generally at depths greater than 100 m). This zone extends east-ward into the southern Alaska Current and the California Current. Faunal assemblages in this region are absent or infrequent in the subarctic and equatorial regions. Some variability in distribution was observed among the species. *Castanissa circumvallata* occurred in the central region similar to *Castanissa* sp. 2; however, it was most abundant in the Cal-ifornia Current and at locations about 100 miles offshore in a narrow zone between about 100 and 200 m. *Castanea amphora, Castanella*

wyvillei, Castanella thomsoni, Castanea henseni, and *Castanea globosa*
were similarly distributed with abundances apportioned about equally
between the central region and the California Current. They occurred
largely in the northern central region, in a zone with temperatures and
salinities transitional between the high values of the surface layer above
and to the south and the low temperatures and minimum salinities at
the core of the Intermediate Water below. Most species occurred largely
well above the core of the Intermediate Water mass and near the subsurface salinity maximum that gives this zone hydrographic identity.
Few or none occur in the warm, high-salinity surface water. A sharp
northern boundary occurs in a zone of rapid temperature changes, which
seems to be determined largely by eastward transport of water by the
North Pacific Drift.

The Central Water Mass near the Equator includes a faunal assembly
that occurs largely in the upper 200 m of the water column. This borders
on the north of Casey's biogeographical zone known as the Equatorial
Faunal Zone, which he describes as a broad surface band of water
gradually tapering as a wedge down to depths of about 350 m in the
equatorial Pacific. The species occurring in this zone include: (1) *Castanidium longispinum* a shallow dwelling form (0-to 20-m) that also
descended to depths as great as 500 m, but consistently was found
above 50 m; and (2) *Castanidium elegans* most abundant in the upper
200 m of the central region at approximately latitude 28°N. It did not
occur in any of the equatorial samples collected by Kling (1966, 1976).

In general, the distribution of these Phaeodaria follows predictable
patterns within broad hydrographic and geographical parameters. The
subarctic species dwelling in the cold, near-surface water of the north
follow the submergence of the cold water into greater depths at mid-
latitudes. More southerly faunal assemblages are located in regions of
rather distinctive water masses including the region marked by a subsurface salinity maximum and the southerly zone of high surface salinities and temperature. Phaeodarian species distributions occur,
therefore, in three faunal zones located in three strata of water that are
segregated vertically by density differences and maintained horizontally
by closed circulation within the strata.

Nigrini (1968) has examined recent sediments in the eastern tropical
Pacific with the aim of correlating species assemblages and hydrographic features in that area. The samplng site was in the region of the
Peru Current, the South and North Equatorial Currents, and the Equatorial Countercurrent. This is in an area from latitudes 23°N to 13°S
and westward to longitude 156°W, approximately due west from the
coast of Central America and northern Venezuela. Samples were taken
only from the core tops with the intent of representing very recent
sedimentary assemblages. The major hydrographic features dominating
the region include the North and South Equatorial Currents, which

appear to correlate with a faunal pattern extending westward from the Central and South American coasts. These currents are divided by a tongue of water corresponding to the Equatorial Counter-current. Several species appear to have distributional patterns correlated with the major hydrographic features. *Polysolenia murrayana*, *Theoconus minythorax*, and *Lamprocyclas maritalis ventricosa* are relatively common in the region of the Peru Current and along the Central American coast. They are carried westward by the North and South Equatorial Currents and decrease in abundance to the west. They are rare or absent in the Equatorial Countercurrent tongue. A restricted, but similar pattern of distribution, was found for a *Carpocanium* sp. occurring north of the Equator. *Pterocanium gandiporus* and *Dictyophimus infabricatus* appear to inhabit coastal regions and are not carried oceanward in significant numbers. By contrast, some tropical species occur in a complementary pattern to *P. murrayana* and associated species. *Spongaster tetras tetras* is abundant in the Equatorial Countercurrent tongue, but diminishes in number in the North and South Equatorial Currents and along the coasts of Central and South America. *Panartus tetrathalamus* decreases in abundance in the vicinity of the Peru and North Equatorial Currents, but is quite abundant in the regions of the South Equatorial Current and the Equatorial Countercurrent. *Euchitonia* spp. are least abundant near the coasts of Central and South America, but increase in abundance toward the open ocean.

Indian Ocean and Adjacent Antarctic   Our knowledge of radiolarian distribution in the Indian Ocean is limited largely to sediment studies (e.g., Petrushevskaya, 1971b; Nigrini, 1967; and Riedel, 1951a, b). Relevant reports of plankton sample studies (Cleve, 1900, 1901; Petrushevskaya, 1971b) indicate that some species inhabit only warm surface waters. These include *Euchitona elegans*, *Botryocyrtis scutum*, *Pterocanium praetextum*, *Phorticium pylonium*, and *Eucyrtidium acuminatum*. In total, about 11 warm surface water species are identified by Petrushevskaya (1971b). She also lists 24 species occurring at subsurface depths of 50–400 m. These include *Hymeniastrum profundum*, *Cornutella verrucosa*, *Lamprocyclas maritalis*, *Heliodiscus asteriscus*, and *Axoprunum stauraxonium*. Also included in more cold water forms are 11 of these deep dwelling species, such as *Theocalyptra bicornis*, *Antarctissa* spp., *Spongotrochus glacialis*, *Spongodiscus favus*, and *Lithelius nautiloides*.

Nigrini (1967) examined the distribution of radiolaria in recent pelagic sediments from middle (30–45°S) and low (10°N–30°S) latitudes. Two areas of radiolarian abundance were identified: (1) A low-latitude belt (10°N to approximately 20°S) composed of 12 species. Among the species occurring here are *Panartus tetrathalamus*, *Pterocanium praetextum praetextum*, *P. trilobum*, *Otosphaera auriculata*, and *Botry-*

ocyrtis scutum. (2) A middle latitude belt (35–45°S) encompassed 7 of the described species, including *Eucyrtidium acuminatum, Lithocampe* sp., *Cornutella profunda, Dictyophimus crisiae,* and *Actinomma medianum.* These two faunal belts are separated by an area corresponding to the subtropical anticyclonic gyre, which was essentially barren of radiolaria. On the whole, however, the tropical equatorial zone is rich in radiolaria (Riedel, 1951a, b) and probably represents the enhancing effect of warm water on radiolarian productivity as found in the Atlantic and Pacific Oceans.

In a pair of companion papers, Johnson and Nigrini (1980, 1982) analyzed the distribution of radiolaria in recent sediments of the western and eastern Indian Oceans and correlated radiolarian assemblages with hydrographic features in the two halves of the ocean. A marked east–west asymmetry in faunal distribution patterns was found across the Indian Ocean and was partially explained in terms of major hydrographic features characterizing the two regions. Only a brief summary of their substantial papers can be presented here with a focus on ecologically relevant findings. There are several oceanographic factors that may contribute to producing a significant east–west asymmetry in the major flow patterns of the currents in the Indian Ocean. They are cited as a background for the discussion of the major radiolarian assemblages reported by Johnson and Nigrini. A strong upwelling occurs off the coasts of Somalia and the Arabian peninsula in the western half of the Indian Ocean during the strong southwest monsoon of summer in the Northern Hemisphere. There are no comparable regions of upwelling along the eastern margin of the Indian Ocean. There are strikingly different water source regions in the eastern and western sides of the northern Indian Ocean. Outflows from the Persian Gulf and the Red Sea enter the Arabian Sea increasing the subsurface salinity. In the Bay of Bengal, by contrast, an enormous fresh water inflow via the Ganges and Brahmaputra rivers provides a low-salinity cover extending far to the south. A major influx of Pacific surface waters enters into the eastern Indian Ocean via the straits and passages through the Indonesian archipelago. Moreover, owing to the rotation of the earth, there is a narrowing and intensification of current flow along the western edge of major current gyres. Hence, the southern Indian Ocean exhibits a well-defined boundary current flowing southward near Madagascar; however, the corresponding northward flow along western Australia is imperceptible. Topographic features may obstruct current flow causing deflections in the course of the currents, thus influencing radiolarian patterns of distribution. For example, latitudinal fluctuations in the mean position of the Subtropical Convergence and the Antarctic Convergence are in many instances a consequence of this effect. In regions where the flow of water is sufficiently thick to reach the sea floor, there may be perturbations in the microfossil distribution reflecting the current pattern. Deep currents in general may displace microfossils over considerable distances and must also be considered in interpreting the role of hydrographic factors in radiolarian distributional patterns.

Nine radiolarian assemblages characterizing the entire Indian Ocean north of the Antarctic Convergence (near 50° S) were identified by Johnson and Nigrini. These assemblages were composed of subgroups of radiolaria (six groups based on a statistical recurrent analysis technique of Fager (1957) that were correlated with major biogeographical features). Recurrent group analysis is a frequency analysis of the occurrence of radiolaria in pairs within a segment of the core. A recurrent group represents the largest possible separate unit, wherein all pairs of species show an affinity for each other. Thus, an affinity of 0.5 means that pairs of species are found together in at least 50% of their recorded occurrences (e.g., Nigrini, 1970). A synopsis of the nine assemblages based on recurrent group analyses is presented:

1. Arabian Upwelling Assemblage: This group of species was well defined in three samples nearest the Arabian margin and is characterized by the exclusive occurrence of three species (*Collosphaera* sp. cf., *C. huxleyi*, *Lamprocyclas nigriniae*, and *Lamprocyclas maritalis ventricosa*). This radiolarian assemblage augments criteria previously applied by Prell and Curry (1981), who employed the changes in $\delta^{18}O$ composition of shells from surface-dwelling planktonic foraminifera (*Globigerina bulloides*) as indicators of deep water upwelling in this region.

2. Northern Indian Ocean Assemblage: This is characterized by three assemblages including the one near the Arabian Coast. The southern limit of this assemblage, near the Equator, identifies the southern boundary for the assemblage. Also included here is a group of 46 species that extend into the tropical latitudes (e.g., *Acrosphaera* sp., *Collosphaera* sp., *Cornutella profunda*, *Dictyocoryne* sp., *Heliodiscus* sp., *Pterocanium* sp., and *Spongaster tetras tetras*).

3. A Tropical Assemblage: This extends over a broad region from near the Equator to ca. 16°S. It is defined largely by the 46 species cited in the foregoing assemblage. It is further marked by the absence of the species found in the Arabian upwelling assemblage and those of the subtropical latitudes cited in the next group.

4. The South Equatorial Current Assemblage: This occurs south of ca. 16°S and is composed of a distinctive ensemble of species, which closely follows the position of the South Equatorial Current. It extends to ca. 20°S over the eastern three-quarters of the Indian Ocean, and to near 25°S in the vicinity of Madagascar. The assemblage consists of the 46 species in the Tropical Latitude group and 6 species of the Subtropical Latitude Assemblage; e.g., Am-

phiropalum cf., *Tessarastrum straussii*, *Collosphaera huxleyi*, and *Eucyrtidium acuminatum*.

5. A Central Assemblage: This occurs in the eastern three-quarters of the Indian Ocean between ca. 20 and 35°S, and is marked by relatively poor preservation of radiolarian fossils. The Subtropical Latitude Assemblage occurs in a rather coherent pattern in this region. This assemblage corresponds closely in its distribution pattern with the core of the subtropical gyre. The east–west asymmetry of its distribution is consistent with the relatively weak boundary current off western Australia compared to the more intense southward limb of the gyre near Madagascar.

6. A Temperate Assemblage: This corresponds with the region of strong eastward circumpolar flow north of the Subtropical Convergence.

7. Transitional Assemblage: This is a group of species occurring south of the Subtropical Convergence with a peculiar asymmetry in distribution. The characteristic species have identical northern limits west of ca. 70°E; however, there is apparently a difference of ca. 4° between their northern limits in the eastern region. Hence they are named Transitional Assemblage. The assemblage is based on only four samples and therefore Johnson and Nigrini recommend further analyses to determine its validity.

8. The Subpolar Assemblage: This occurs in a narrow zone extending from ca. 48°S to the northern limit of the Temperate Latitude species.

9. A Polar Front Assemblage: This is not well marked, and is characterized by the absence of the Temperate Latitude species (*Pterocanium praetextum eucolpum*, *Theocorythium trachelium dianae*, and *Trigonastrum* sp.) and the presence of five species in a bimodal distribution including *Actinomma medianum*, *Botryostrobus aquilonaria*, and *Saturnalis circularis*. These occurred in two zones, one west of Sumatra, extending southward to approximately latitude 15°S, and a second group in a zone immediately south of latitude 40°S. In general, radiolarian abundance decreased markedly as the Antarctic Convergence (Polar Front) was approached. The reason for the bimodal distribution of the five species is uncertain. However, Johnson and Nigrini hypothesize that inflow of Pacific water into the subtropical gyre may cause extensive regional alternations in the quality of the ocean water that is particularly unfavorable to these species, thus eliminating them from a broad zone between approximately latitudes 15 and 40°S. They further suggest that the ancient environment may have been more favorable for a cosmopolitan distribution of the species, owing to a possible land bridge between Australia and Indonesia that reduced considerably or closed off the westward near-surface flow from the Pacific.

Regional Abundance and Seasonality   Little is known about the impact of environmental variables on radiolarian mortality or seasonal patterns of abundance owing largely to the lack of long term data on plankton composition and environmental variation in any given locale. Casey et al. (1970), however, have made some clever deductions about radiolarian survival and seasonality in coastal waters over the Santa Catalina Basin near California. By comparing faunal densities in the plankton with accumulation rates of skeletons of each species in a varved sediment (Santa Barbara Basin), they were able to calculate radiolarian mortality rates and seasonal abundances. In this region of the Pacific, a heterogeneous group of radiolaria is swept into near coastal water by ocean currents. Thus, a mixture of species from a variety of biogeographical zones comingle in this area, providing an excellent opportunity to determine the effects of the environment on radiolarian survival and seasonal fluctuations in abundance. Striking differences in mortality rates were found depending on the biogeographical zone from which the radiolaria come. Species dwelling in equatorial–central shallow water (e.g., *Theoconus zancleus*) survived longest (mean life span = 37.0 days), as might be expected. The water mass over the Santa Catalina Basin is comparable in quality to their preferred habitat. Radiolaria with markedly different preferred habitats yielded shorter life spans in these alien waters. *Lithomelissa monoceras* and some collosphaerids associated with equatorial shallow water survived on the average of 8.7 days. *Spongotrochus glacialis* normally associated with subarctic transition shallow water masses survived on the average of 11.35 days. *Spongopyle osculosa* and *Siphocampe erucosa* (transition central zone species) yielded mean survival of 9.7 days. The subartcic intermediate species, *Cornutella profunda*, survived on the average for only 0.44 days. From the data, it is clear that within the error of computation, there are certain faunal differences in mean survival times that are probably related to the biological sensitivity of the species in addition to the degree of alienation relative to their preferred environmental zone. For example, even though *Spongotrochus glacialis* originates in cold shallow water in the northern-most region, its mean survival time is about 11 to 12 days compared to a survival time of about 4 to 7 days for species originating in shallow water of the transition zone (*Pterocorys hirundo* and *Hexadoridium streptacanthum*). Presumably, these differences can be attributed to differences in the hardiness or biological adaptability of the different species.

Seasonal variations in abundance were noted in the upper 200 m off the coast of California, where variations in water temperature and quality follow seasonal cycles. At greater depths, the populations were more uniform throughout the year due to the constancy of the environment. In the upper 200 m *Pterocorys hirundo*, *Spongotrochus glacialis*, and *Hexadoridium (?) streptocanthum* occurred most abundantly in winter and reflect the transport of water into the area from higher latitudes. *Lithomelissa monoceras* and some collosphaerids appeared most frequently in summer and reflect the transport of water from lower latitudes. Some shallow water species (e.g., *Theoconus zancleus* and *Eucyrtidium hexagonatum*) occur throughout the year and are believed to be within their normal range and reproductive area.

In further research by Casey, species diversity and seasonal abundance of some polycystine radiolaria were investigated using plankton samples obtained at three water depths (0–25, 25–50, and 50 + m) during winter, spring, and fall seasons in four transects taken across the south Texas shelf. These data elucidate seasonal cycles in the region and contribute further evidence of radiolarian species as indicators of oceanic water masses (Casey, 1971a,b, 1977). These data have been correlated with results from plankton analyses of radiolarian assemblages in the Gulf of Mexico and Caribbean Seas (McMillen and Casey, 1978), southern Sargasso Sea (Spaw, 1979), and western South Atlantic (Boltovskoy and Riedel, 1979). The physical oceanographic setting in the area as summarized by Casey *et al.* (1981) is characterized in winter by a mixing of shelf water to a depth of about 70 m, during turbulence generated by southerly directed winds accompanying passage of cold fronts. The general circulation pattern is wind-driven from north to south along the east coast of southern Texas. At the time these studies were in progress, an upwelling occurred at the midshelf and appears to have continued into the spring. Spring is generally dominated by a lens of brackish water that flows through the area from north to south and is generated by local land runoff or contributions from the Mississippi. Summer is characterized by the development and thickening of the thermocline. Water masses have longer residence times in summer (nearly a month) compared to winter (days or weeks) due to the variable north–south winds that occur in summer. In conjunction with these hydrographic features, Casey *et al.* conclude that relatively high radiolarian densities and diversities in the water overlying the south Texas shelf appear to represent incursions of open ocean waters. Regions with densities of about 100 individuals/m^3 probably represent mid-Gulf of Mexico waters, whereas significantly higher densities are probably from invasions of Loop Current waters by anticyclonic gyres that are known to detach from the Loop Current and move across the Gulf of Mexico (Molinari *et al.*, 1978). The abundance of several water mass indicator species of radiolaria during winter, spring, and fall is summarized in Table 3-10. Comparisons are made with other published data on distributions. The pattern of radiolarian densities in the region is presented as Fig. 3-10. In general, the pattern of abundance conforms to a reasonable cyclical peak of deep-dwelling species occurring near the surface in winter, probably occasioned by the upwelling and sustained by colder surface water conditions. Open Gulf Surface or shallow-dwelling

Table 3-10. Some peak radiolarian densities and depth of occurrence over the south Texas continental shelf[a]

STOCS radiolarian indicator group	Winter	Spring	Fall	McM/C	SPAW	B/R
Winter/deep/upwelling						
Helotholus spp.	0.6 (25–50)			—	I	S
Lithelius minor	0.2 (0–25)			S-U	UN	—
Spongopyle osculosa	0.1 (0–50)			D	I	S-"U"
Spongotrochus glacialis	1.8–2.0 (0–50)		0.2–0.3 (0–25, 50+)	—	I	S-"U"
Summer above thermocline						
Euchitonia elegans	0.3 (25–50)		0.9 (0–25)	S	S	—
Eucyrtidium acuminatum	0.1 (25–50)		0.3 (25–50)	—	S	—
Lamprocylas maritalis			0.1 (25–50)	—	I	S
Ommatartus tetrathalmus	0.3 (25–50)		0.4 (0–25)	S	S	S
Summer below thermocline						
Anthocyrtidium ophirense			0.7 (50+)	—	—	—
Lamprocyclas nupitalis			0.2 (0–25)	S-U	S-U	—
Pterocorys zancleus	0.1 (25–50)		0.4 (0–25)	S	S	—
Warm and cold morphotypes						
Spongaster tetras tetras (warm)	0.2 (0–25)		0.3 (25–50)	S	S-I	S
Spongaster tetras irregularis (cold)	0.1 (0–25)			—	—	—
Pterocanium praetextum praetextum (warm)	0.1 (0–25)	0.1 (0–50)	0.2 (0–25, 50+)	S	S	—
Pterocanium praetextum eucolpum (cold)	0.1 (0–25)	0.1 (0–25)		S	—	—
Open gulf surface or shallow						
Disolenia zanquebarica		17.0 (25–50)	0.4 (50+)	S	—	S
Polysolenia lappacea		8.4 (25–50)		S	—	—
Mesotrophic and shelf circulation						
Dictyocoryne profunda-truncatum group	0.6 (0–25)		1.7–3.1 (0–50)	S-U	I	S-"U"
Subtropical underwater						
Amphirhopalum ypsilon		0.1 (50+)		U	—	S-"U"
Year-round form						
Spongosphaera streptacantha	1.7 (0–25)	2.5 (0–25)	0.3 (25–50)	S-U	S-I	—

[a] Casey et al. (1981).

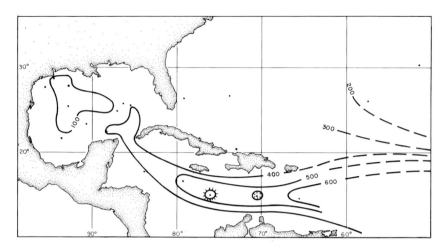

Fig. 3-10. Contours of radiolarian densities in numbers of radiolarians per cubic meter for the upper 50 m of the water column for the western North Atlantic, Caribbean, and Gulf of Mexico. (Casey *et al.*, 1981)

species occur in near surface waters most abundantly in spring at a depth of 25–50 m. Species dwelling in the stratum labeled "Summer Above Thermocline" were most abundant in fall in surface waters, but showed a slight persistence in winter at surface and subsurface depths (25–50 m). Assays of radiolarian richness and diversity during the winter months show a pond of high diversity water at the surface and at shelf-intermediate depths along a transect, latitude 27°N at the outer and midshelf stations (ca. between 96.5 and 97°W). During spring, higher diversities were generally confined to the outer shelf, except for an encroachment of higher diversities at shelf-intermediate depths. High radiolarian diversities at all depths on the shelf accompanied by encroachment of Gulf waters occurs during autumn.

Additional research on species in warm water locales has been pursued in the equatorial Pacific and East China Sea. A quantitative investigation of the distribution of *Sticholonche* sp. in the Equatorial Pacific (Takahashi and Ling, 1980) at depths of 0 to 800 m revealed that these fragile, motile, radiolarian-like organisms constitute a substantial part of the total radiolarian standing stock. The highest standing stock occurred between 55 and 110 m, at a depth coincident with the pycnocline and thermocline. Maximum densities as high as 25 *Sticholonche*/liter of water were found in this horizon. The number of *Sticholonche* organisms in the surface waters varies from 0.8 to 10.4/ liter. This corresponds favorably with values obtained in the Atlantic Ocean (Emery and Honjo, 1979) and in the East China Sea (Tan *et al.*, 1978). The latter region is characterized by a maximum density occurring in surface waters (0–20 m). This difference probably reflects variations in biophysicochemical parameters between the Pacific and East China Sea. Although *Sticholonche* was collected at depths as great as 3,000 to 4,949 m in the Equatorial Pacific, their abundance appears to decline sharply below 800 m. Takahashi and Ling conclude that the *Sticholonche* recovered from deeper water, though few in number were alive, since their delicate skeletons dissolve rapidly after death and would not likely descend to such great depth. Large populations of *Sticholonche zanclea* are not likely to occur at great depth since these organisms prefer warm-water habitats. Tan et al. (1978) found that S. *zanclea* occurs most abundantly in the western part of the East China Sea between longitudes 122 and 123°E during the summer months. A direct relationship was found between seasonal abundance and temperature. Its numbers are highest in regions of high phosphate and phytoplankton concentrations. Land run-off appears to enhance its abundance through enrichment of the ocean water.

Further research by Tan and Tchang (1976) on the radiolarian composition of plankton in the East China Sea revealed that Spumellaria were among the most abundant radiolaria comprising 55 species in a total of 98 examined. Nassellaria accounted for 36 species, Phaeodaria 6 species, and Sticholonchea 1 species. The relatively large numbers of Spumellaria compared to Nassellaria are of interest in comparison to pelagic waters of the major oceans where Nassellaria appear to be most abundant. This is certainly true for eutrophic waters in the California current of the Pacific (Casey et al., 1979a) and apparently in the pelagic waters of the Atlantic (Takahashi and Honjo, 1981). Casey et al. (1979a), however, report substantially larger numbers of Spumellaria than Nassellaria in plankton from the open ocean Gulf of Mexico, and the Gulf Stream near the coast of Florida and eastward in the Sargasso Sea. And, as previously reported, Spumellaria are often more abundant than Nassellaria in some regions of the South Atlantic. Possible hydrographic and biological factors contributing to these differences need to be explored.

Relatively little is known about seasonality of radiolaria in northern waters, particularly in geographically isolated zones. Among the earliest studies in the North Sea and northeastern Atlantic, Mielck (1913) reported vertical zonation of radiolaria, and stated that during spring and summer radiolaria occurred exclusively at depths greater than 100 m, whereas in autumn and winter they also were present in the upper 100 m. Other researchers have focused on taxonomic descriptions of localized species in the open sea near Spitzbergen (Cleve, 1899), at various locations near the Norwegian coast (Gran, 1902) and in the Norwegian fjords (Cleve, 1899; Jörgensen, 1900, 1905).

One of the most recent detailed studies of seasonal abundance of the more commonly occurring radiolaria in Norwegian fjords (Bjørklund, 1974) examined the taxonomy and vertical zonation of radiolaria in

Korsfjorden, Western Norway. Bjørklund established a biological sampling station at latitude 60°11′30″N and longitude 5°13′00″E where there is a depth of 690 m. Material was collected from November 1969 until May 1971. The general hydrographic characteristics in this region are fairly well established. During summer and autumn, there is a fairly stable water mass bearing a native fauna of radiolaria; however, this is augmented each late winter and early spring by an influx of Atlantic water. During the period of Bjørklund's investigation, there was a shallow water mass of different composition from the deep water demarked by the 34% isohaline located fairly steadily between 75 and 100 m. The general pattern of annual deep water movements include a static water mass approximately June to November, followed by flushing in December and January leading to incursions by oceanic surface water (January–February) and vertical mixing through spring until about the end of May (Matthews and Sands, 1973). Sixty-eight species were found in the fjord, but the five most common were studied in more detail, i.e., *Echinomma leptodermum, Cromyechinus borealis, Lithomelissa setosa, Protocystis tridens,* and *Challengeron diodon.*

There was a substantial fluctuation in total radiolarian abundance throughout the year. A peak abundance of approximately 27,000 estimated in the water column above an area of 0.07 m^2 occurred in June through September, but was considerably lower (ca. 2,000–6,000) during the remaining part of the year. A mean density of about 6,000 radiolarians in a column above 0.07 m^2 was obtained for the whole sampling period. Species diversity tended to be high during January–March probably due to the influx of open-ocean water augmenting local fauna with Atlantic Ocean species. These may compete with the indigenous fauna, thus explaining a generally lower total abundance during this period, but thereafter when the fauna brought in with the ocean water gradually dies off, the local fauna increase in numbers and reach a maximum density at about August. Bjørklund did a thorough analysis of abundance of each of the major five species with depth in the water column. The occurrence in the 0- to 100-m zone and 100- to 200-m zone was presented in detail. The pattern in deeper zones paralleled the 100- to 200-m zones, though there was a tendency toward higher densities for *C. diodon.*

The 0- to 100-m zone was dominated in general by *L. setosa, E. leptodermum,* and *C. borealis.* During the period investigated by Bjørklund, there was an apparent seasonal trend of dominance with *L. setosa* most abundant in the winter season; and *E. leptodermum* and *C. borealis,* dominant in spring, summer, and autumn. The 100- to 200-m zone was characterized by an abundance of the phaeodarians, *P. tridens* and *C. diodon,* and a proportional decrease of *E. leptodermum* and *C. borealis,* and to a lesser extent of *L. setosa.*

There is good evidence for a periodicity in radiolarian numbers between the 0- to 100- and 100- to 200-m zones. A distinct peak of radiolarian numbers occurs in the uppermost 100 m from December to February, and from November to January, which coincides with the first influx of open ocean water and vertical mixing. In the remaining months of the year, the major densities occur in the 100- to 200-m zone.

Four faunal zones were identified based on the average population depth. Faunal Zone 0 was characterized among the 45 species by a dominance of *L. setosa*, *E. leptodermum*, and *C. borealis* (0–100 m). This is a peculiar zone defined by the hydrographic conditions (34% isohaline occurring at 75–100 m throughout) and faunal assemblage. Faunal Zone 1 is mainly occupied by *E. leptodemum*, *C. borealis*, *L. setosa*, and *P. tridens* (188–292 m). Faunal Zone 2 occurs at 307-360 m and only *C. diodon* is dominant here. Faunal Zone 2 is dominated by *E. leptodermum*, *C. borealis*, and *L. setosa* with an average population depth of 483–493 m. To further characterize the annual cycle of abundance, Bjørklund analyzed the population densities in each faunal zone during three time periods: (1) November 1970–March 1970, (2) April 1970–September 1970, and (3) October 1970–May 1971. In Faunal Zone 0, *L. setosa* is most abundant in periods 1 and 3, whereas *E. leptodermum* and *C. borealis* occurred predominantly in period 2. In Faunal Zone 1, a similar pattern occurred; *L. setosa* is dominant during periods 1 and 3, and the remaining species occur abundantly during period 2. *C. diodon* in Faunal Zone 2 occurs abundantly throughout the three periods. Faunal Zone 3 species have a peak abundance during period 2. It is interesting to note that during period 2 when the major peak in abundance of all radiolaria occurs, ciliates are the main other zooplanktons in abundance which may be a significant source of food for radiolaria in the fjord. Our knowledge of the various physical and biological factors correlated with radiolarian abundances in these northern regions is quite limited and highlights the need for additional ecological research in these localized environments.

In further studies of northern water masses, Petrushevskaya and Bjørklund (1974) examined the ecology of some major species of radiolaria commonly occurring in Holocene sediments of the Norwegian–Greenland Seas. They found that the cold water cosmopolitan species *Diplocyclas davisiana* and *Lithometra lineata* as well as *Artostrobus annulatus* occurred commonly in deep water sediments and were rare in shallow water sediments. A second group of species are also widely distributed in cold water. But, they were not entirely restricted to deep water; they also inhabit the surface or subsurface layers even close to the shore in regions of cold water. These species include: *Pseudodictyophimus gracilipes*, *Spongotrochus glacialis*, *Spongodiscus resurgens*, and *Cromyechinus borealis*. Although they have a broad range it appears they prefer water temperatures no higher than 10°C. *Phorticium clevei* and *Amphimelissa setosa* occur in sediments underlying the cold Labrador and East Greenland Currents, which suggests they are strongly associated with cold-water habitats. A third biogeographical zone occurs in the region of the Labrador and East Greenland Currents, and the Irminger Sea. *Botryostrobus tumidulus* is well represented here as is *Lithomitra arachnea*, but in a more restricted area; it is not common where there is a downflow of current. *B. tumidulus* is not only present in cold currents, but also occurs in the region of the North Atlantic Drift, probably at greater depth, based on the distribution pattern in the underlying sediments. A fourth group comprised of *Artobotrys borealis* and *Eucercyphalus craspedota* is present in shallow water sediments, but is rare in cold water regions. Petrushevskaya and Bjørklund suggest that these species may inhabit the waters of the North Atlantic Drift and the Norwegian Atlantic Current. Of all species studied in the Norwegian–Greenland Sea area, the data for *E. craspedota* in surface sediments are the most comprehensive in providing a reliable perspective on the distribution of the species. Its southern boundary is more or less coincident with the northern border of some tropical species such as *Toctopyle stenzona* and *Pterocanium praetextum* (Goll and Bjørklund, 1971; Petrushevskaya, 1971b). Its northern or northwestern boundary is not well defined; however, the northern border of its main area of distribution coincides with the northern boundary of the boreal zone, as defined by Geptner (1936) or of the subarctic zone, as defined by McIntyre and Bé (1967). These boundaries roughly follow the contour of the coast line of Labrador and continue north-eastward paralleling the southeast coast of Greenland, but veer more to the east at about latitude 70°N. *L. arachnea* is the only species among those examined that is restricted to Arctic water. Its skeletons are common only in the region of the Labrador current. Petrushevskaya and Bjørklund present rather detailed maps of the distributions of the species and discuss their faunal assemblages in relation to other cold-water species as indicators of water masses and ecological variables.

Colonial Radiolaria

There have been few modern exploration programs examining the abundance and geographical distribution of colonial radiolaria. The Russian investigators, Pavshtiks and Pan'Kova (1966), Khmeleva (1967), and Strelkov and Reshetnyak (1971) have reported data on colonial radiolarian morphology and abundance in diverse geographical locations including the Atlantic, Pacific, and Indian Oceans, the Antarctic, the Davis Straits, and the Gulf of Aden. A substantial study of colonial radiolaria collected in situ in glass jars by SCUBA divers was reported by Swanberg (1979). His data are very valuable in comparison to conventional sampling data using nets or other mechanical devices, since some delicate colonial species are lost by damage when collected in these devices. The merit of Swanberg's SCUBA sampling approach is clearly indicated by the fact that he identified some new, delicate colonial radiolaria that undoubtedly would have been damaged beyond recognition if collected by conventional mechanical means. Some pertinent findings from these studies are summarized.

Assessment of colonial radiolarian abundance is difficult due to their patchiness and tendency to congregate near the surface of the ocean when the water is calm. Thus, during extended periods of calm weather, most of the colonial radiolaria in the water column are located in a narrow stratum near the surface. Estimates of abundance based on observations in this concentrated zone clearly yield inflated values, unless the data are expressed as surface layer densities and not used to extrapolate abundance throughout the water column. When the ocean surface is rough or turbulent, the radiolaria are dispersed deeper into the water, and it is possible to make more valid estimates of abundance per unit volume of water. Under the best of conditions, however, abundance estimates are prone to considerable error and probably are valid only within approximately an order of magnitude. Traditional methods of using a net and flowmeter to determine the density per unit volume water sampled are prone to error due to destruction of the sample in the net. Swanberg (1974, 1979) has developed an efficient method of visually estimating abundance by use of a plastic hoop enclosing a 0.5 m² area that is carried into the water by a SCUBA diver. The hoop is passed along a line marked into 1-m segments. The number of colonial radiolaria passing through the hoop as it is moved through a meter distance is used to calculate density per unit volume. This method is particularly valuable in making estimates of radiolaria at varying depths and in relation to particular communities or patches of growth. The diver is able to determine more accurately the location of the sample and the general abundance of the radiolarian swarm while the sample is taken. An alternate approach to the grid method of Swanberg is to estimate the densities based on drift rate calculations described by Harbison et al. (1978).

Swanberg (1979) estimated abundances of colonial radiolaria at 16 stations (Table 3-11) in the Sargasso Sea (Stations 420–425) and near the geographic equator in the Indian Ocean (466–481) between the northwest and the southwest monsoons. The other stations were in the North Atlantic central gyre and the Sargasso Sea. Colonies were most abundant in summer in the central gyres in calm weather and least abundant in relatively eutrophic coastal stations at all times of the year. Further evidence of radiolarian distribution and general abundance was presented by Swanberg (1979) for 452 stations. Colonial radiolaria were found at 89% of the stations (402 of the 452), and were rated as abundant or very abundant (>1/100 m³) at 114 of the stations.

The geographical distribution and abundance of 42 species and subspecies of colonial radiolaria were examined by Strelkov and Reshetnyak (1971). The distribution of some representative species is presented in Table 3-12. These observations on samples collected from Arctic,

Table 3-11. Abundance of colonial radiolaria at some geographical locations in the Atlantic Ocean[a]

Station	Coordinates	Total radiolaria counted[b]	Radiolaria/ m³	Observation collection depth (m)
420	33.65°N, 71.05°W	20	1.10	15
423	33.82°N, 71.90°W	6	0.20	—
425	34.02°N, 71.88°W	4	0.40	15
425	34.02°N, 71.88°W	6	0.70	8
415	34.02°N, 71.88°W	4	0.40	5
466	3.00°S, 55.83°E	5	0.20	10
467	2.50°S, 55.50°E	4	0.20	10
468	1.00°S, 55.50°E	1	0.04	—
469	0.50°N, 55.50°E	1	0.04	10
469	0.50°N, 55.50°E	62	540.00	Surface
470	2.05°N, 55.50°E	29	0.60	10
470	2.05°N, 55.50°E	27	0.50	2–3
471	1.25°N, 55.50°E	7	0.10	10
471	1.25°N, 55.50°E	12	0.20	1
471	1.25°N, 55.50°E	51	113.00	0.10
472	3.00°S, 55.50°E	3	.06	10
472	3.00°S, 55.50°E	3	.07	1
472	3.00°S, 55.50°E	33	220.00	Surface
473	1.50°S, 55.50°E	21	.22	10
474	3.00°S, 55.50°E	118	2.00	10
474	3.00°S, 55.50°E	46	0.92	1
478	0.50°N, 55.50°E	156	14.00	—
481	0.00°, 55.50°E	40	150.00	Surface
578	34.62°N, 60.08°W	61	0.07	0–10
592	37.13°N, 66.75°W	4	3.00	0–20

[a] From Swanberg (1979).
[b] The reliability of the radiolarian abundance estimates must be judged in relation to the number of radiolaria counted.

subtropical, and tropical water confirms a well-established opinion, based on analyses of sediment samples, that most colonial radiolaria inhabit warm waters largely in the tropical areas of the ocean. They found that the numbers of colonial radiolaria decline sharply in the temperate zone (boreal as well as notal). Based on the total cumulated data, Strelkov and Reshetnyak conclude that among colonial radiolaria there are those species that inhabit a very restricted horizon in the water column (stenobathic species) and those that are not confined to a narrow horizon (eurybathic species). A further analysis based on geographical distribution yields three basic groups: (1) widely distributed eurybathic species, (2) circumtropically distributed stenobathic species comprising the majority of colonial radiolaria, and (3) endemics restricted to certain geographical regions.

Six species are included in the first group, i.e., *Collosphaera huxleyi*, *Acrosphaera spinosa*, *Collosphaera tuberosa*, *Siphonosphaera cyathina*, *Collozoum inerme*, and *Sphaerozoum punctatum*. These species were found in the boreal zone and even the Arctic and Antarctic in addition to tropical zones. Four species, *Collosphaera huxleyi*, *Acrosphaera spinosa*, *Collozoum inerme*, and *Sphaerozoum punctatum*, were found in the boreal zones of the Atlantic, and three species, *Acrosphaera spinosa*, *Collosphaera tuberosa*, and *Siphonosphaera cyathina*, were discovered in plankton samples from the Antarctic. *Collozoum inerme* is a very widely distributed species and has been reported to occur in the Arctic plankton of the Atlantic (Cleve, 1900). Their widespread occurrence undoubtedly contributes to their being carried even into Arctic waters by ocean currents. Pavshtiks (1956) documents great masses of *Collozoum inerme* occasionally being carried in surface water of the Spitzbergen current northward up to 80°N. It is interesting to note at this point that even if the prey normally consumed by *C. inerme* were no longer available in these northernmost regions, it is possible that the continued survival of *C. inerme* may be explained by its ability to use symbionts as a source of nourishment (Anderson, 1976b) in addition to its potential to adapt to environments varying greatly in temperature and perhaps salinity.

The second group of colonial radiolaria examined by Streklov and Reshetnyak was found everywhere in the tropical zone at a water temperature of 20–25°C and a salinity of 25%. They encompassed 29 forms as follows: *Collosphaera macropora*, *Acrosphaera lappacea*, *Siphonosphaera socialis*, *S. compacta*, *Solenosphaera zanguebarica typica*, *S. zanguebarica pyriformis*, *S. chierchiae*, *Rhaphidozoum acuferum*, *Siphonosphaera tubulosa*, *Collosphaera polygona*, *Acrosphaera circumtexta*, *Buccinosphaera invaginata*, *Solenosphaera polysolenia*, *Collozoum amoeboides*, *Solenosphaera pandora*, *Sphaerozoum verticillatum*, *Siphonosphaera macropora*, *S. martensi*, *S. tenera*, *Solenosphaera collina*, *S. tenuissima*, *Collozoum ovatum*, *C. contortum*, *C. pelagicum*, *C. serpentinum*, *Collosphaera armata*, *Rhaphidozoum neapolitanum*, *Acrosphaera murrayana*, and *A. cyrtodon*. The first 9 species were encountered in all three oceans, whereas the remaining 20 species were noted only in the tropical zone of two oceans; 9 species (*Soophonosphaera martensi*, *S. tenera*, *Solenosphaera collina*, *S. tenuissima*, *Collozoum ovatum*, *C. contortum*, *C. pelagicum*, *C. serpentium* and *C. amoeboides*) were found in the Atlantic and Pacific; 8 species (*Collosphaera polygona*, *Acrosphaera cicumtexta*, *Buccinosphaera invaginata*, *Solenosphaera polysolenia*, *S. pandora*, *Collozoum amoeboides*, *Sphaerozoum verticillatum*, and *Siphonosphaera macropora*) occurred in the Indian and Pacific Oceans, whereas 3 species (*Rhaphidozoum neapolitanum*, *Acrosphaera murrayana*, and *A. cyrtodon*) came from the Indian and Atlantic Oceans. The third group of

Table 3-12. Distribution of some colonial radiolaria[a]

Species	Atlantic Ocean	Location Pacific Ocean	Indian Ocean
Collozoum			
C. inerme	Southeast part of Gulf of Mexico, Caribbean Sea, Strait of Florida, Bahama Channel, Mediterranean Sea	Tropical Zone and south China Sea	Present
C. ovatum	Southwest part Gulf of Mexico to west of Cuba	Tropical zone and south China Sea	—
C. contortum	Tropical zone (Haeckel, 1887)	Tropical zone	—
C. amoeboides	—	Tropical zone	Tropical zone
C. pelagicum	Tropical zone, southeast part of Gulf of Mexico and Florida Strait; Mediterranean Sea	Tropical zone	—
C. serpentinum	Tropical zone, Bahama Channel near Canary Islands	Tropical zone	—
Sphaerozoum			
S. punctatum	Med. Sea, southeast Gulf of Mexico to west of Cuba; Caribbean Sea, Florida Strait, and Bahama Channel	Tropical zone and south China Sea	Gulf of Aden
S. verticillatum	—	Tropical zone	Surface water (Haeckel, 1887)

Species			
Raphidozoum			
R. acuferum	Tropical zone, Mediterranean Sea	Tropical zone	Tropical zone
R. neapolitanum	Southeast part Gulf of Mexico and Mediterranean Sea	—	Present
Collosphaera			
C. huxleyi	Boreal and tropical zones	Tropical zone and Antarctic	Tropical zone
C. tuberosa	Tropical zone	Tropical zone and Antarctic	Tropical zone
C. armata	Tropical zone	—	Tropical zone
C. macropora	Tropical zone	Tropical zone	Tropical zone
C. polygona	Tropical zone	Tropical zone	Tropical zone
Acrosphaera			
A. spinosa	Tropical zone	Antarctic and Tropical zone	Tropical zone
A. lappacea	Area between 35 and 45°N	Central part	Central and southern parts
A. cyrtodon	Tropical zone	—	Tropical zone
A. circumtexta	—	Tropical zone	Tropical zone
A. murrayana	Tropical zone	—	Tropical zone
Tribonosphaera			
T. centripetalis	—	Central part; endemic	—
Siphonosphaera			
S. tubulosa	Tropical zone	Tropical zone	Tropical zone
S. cyathina	Tropical zone	Tropical zone	Tropical zone
S. socialis	Tropical zone	Tropical zone	Tropical zone
S. compacta	Tropical zone, Sargasso Sea, and north and south Equatorial Currents	Tropical zone	Tropical zone

Table 3-12. continued

Species	Location		
	Atlantic Ocean	Pacific Ocean	Indian Ocean
S. martensi	Tropical zone	Tropical zone	—
S. tenera	Sargasso Sea and Mediterranean	Tropical zone	—
S. macropora	—	Tropical zone	
Solenosphaera			
S. zanguebarica	Tropical zone	Tropical zone	Tropical zone
S. pandora	—	Tropical zone	Tropical zone
S. inflata	Endemic (northern part)	—	—
S. chierchiae	Tropical zone	Tropical zone	Tropical zone
S. tenuissima	Tropical zone, Sargasso Sea	Tropical zone	—
S. polysolenia	—	Sparse	Sparse
Buccinosphaera			
B. invaginata	—	Tropical zone	Tropical zone

[a] Based on data from Strelkov and Reshetnyak (1971).

endemics includes 5 tropical species (roughly located between latitudes 30°N and 30°S). Four are found in the tropical zone of the Pacific: *Tribonosphaera centripetalis, Siphonosphaera socialis tubuliloba, S. socialis mazosphaeroides,* and *Siphonosphaera* sp.; and 1 species, *Solenosphaera inflata* occurs in the tropical zone of the Atlantic. Two species, as follows, were endemics of the boreal zone of the Pacific Ocean, extending from the boundary of the Arctic zone (ca. 60°N) to 45°N: *Sphaerozoum bathybium* was encountered only in the Okhotsk, and *S. hamutalum* was found in the Bering Seas. Further information on the distribution of colonial species examined by Strelkov and Reshetnyak is presented in Table 3-12.

Abundance data were obtained only for samples collected near Cuba, owing to their excellent state of preservation. Strelkov and Reshetnyak report that no more than 10 species were found in this area of the Atlantic, which exemplifies the paucity of colonial radiolaria occurring in this region. No representatives of the genera *Siphonosphaera* or *Buccinosphaera* were found near Cuba. There were, however, over 2,000 colonies taken at 56 stations and of these the key species for that region, *Sphaerozoum punctatum,* reached 983 or about 50% of the colonies encountered. *Collosphaera huxleyii,* represented by 424 colonies (or 21% of the total taken), and *Collozoum inerme,* represented by 203 colonies (or 10% of the total colonies), were encountered more rarely. Two species occurred moderately abundant in about equal numbers: *Collosphaera tuberosa* (149 colonies or 7% of the total) and *Collozoum ovatum* (182 colonies or 2% of the total). The following 5 species were among the rarest in occurrence: *Collozoum serpentinum* (2 colonies), *Solenosphaera zanguebarica, Acrosphaera spinosa,* and *Collozoum pelagicum* with 5 colonies each (or 0.25%) and the species *Rhaphidozoum neapolitanum* (45 colonies or 2%).

In the Pacific and Indian Oceans, Strelkov and Reshetnyak report that a greater variety of colonial radiolaria and less quantities of any one species are found as compared to the paucity of species with a few dominant forms found near Cuba. Furthermore, analyses of samples from the western part of the Pacific Ocean reveals a marked increase in the variety of species of colonial radiolaria in the warm waters of the Kuroshio Current, thus highlighting once again the salutary effect of warm-water masses for growth and diversity of colonial radiolaria. Thirty-two species of colonial radiolaria were found at 61 stations out of the 107 stations that were explored. This is a particularly rich diversity, since only about 50 species of colonial radiolaria are known to exist in total. Furthermore, restricted data on species abundance have been published by Pavshtiks and Pan'Kova (1966), who reported finding 3,000–4,000 colonies of *Collozoum*/m^3 in the top 50 m in the Davis Straits during September 1964. An even greater number of *Collozoum* was reported by Khmeleva (1967), who found 16,000–20,000 colonies/

m³ in the Gulf of Aden. Such large densities of these rather large colonial protists illustrate further the likelihood that the colonies are capable of a fairly independent trophic existence, due to their nutritional augmentation by the abundant algal symbionts typically associated with the colonies.

On the whole, it is clear that we need much additional information on the trophic autonomy of all colonial radiolaria and particularly the more abundant forms. The diversity and abundance of prey in various regions of the ocean, and the physical and chemical characteristics of the water masses in regions of the ocean exhibiting marked differences in colonial radiolarian abundance, need to be analyzed using multivariate statistical designs. This analysis may indicate what factors in the ocean environment are most likely to account for the variance in diversity and abundance of colonial radiolarian species. This is equally true for solitary radiolarian species; however, with these usually small forms, the task can become more complex, due to the wide diversity of species and the labor intensive task of thoroughly sorting samples from many locations and many time intervals. Research of this kind with living species is undoubtedly very valuable, as it may supply essential insight into modern ecosystems as models for interpreting paleoecological systems.

General Ecological Perspectives

There are few well-documented generalizations that can be made about the ecology of a given radiolarian species, owing to the broad scope of species studied, the varied and sometimes occasional locales examined, and often the lack of simultaneous physical and chemical monitoring of environmental variables during faunal sampling. Some reasonable perspectives on radiolarian patterns of interaction with the biological and physical environment emerge, however, from the sum total of research on radiolarian physiology, patterns of geographical distribution, and environmental correlates with faunal assemblages.

There is clear evidence from much of the research literature that many species of radiolaria inhabit identifiable masses of ocean water (e.g., Casey, 1971a, 1977; Casey et al., 1979a; Kling, 1966, 1978; Petrushevskaya, 1971a; Boltovskoy and Riedel, 1979) and occupy faunal niches or biographical zones comparable with other zooplankton (McGowan, 1971; Beers and Stewart, 1969a, b, 1971; Renz, 1976; Casey et al., 1979a). In some cases, the major controlling factor is the temperature–salinity profile of the water and perhaps other physicochemical correlates. The potential food organisms or detrital particles and light intensity, depending on depth, are also important factors as energy sources for radiolarian productivity.

In the major oceans of the world, there appears to be an abundant growth and rich diversity of radiolarian species in the warm waters of the tropical equatorial zone. Many solitary and colonial species occur in these highly productive regions. In the Atlantic and Pacific Oceans, there is good evidence that total radiolarian productivity and diversity decline with increasing latitude, though clearly there are species endemic to, or characteristic of, more northern water masses. Within this broad perspective, there are finer ecological discriminations based on localized or broader hydrographic patterns. Current patterns and localized annual cycles of water mass exchange (e.g., lateral influxes and upwellings) provide additional structure to large and small regions of the ocean, thus compartmentalizing it into finer ecological domains. The complex faunal assemblages occurring in the southwest Atlantic Ocean occasioned by the complex comingling of cold- and warm-water currents (Boltovskoy and Riedel, 1979) exemplifies the role of ocean currents in creating finer oceanic domains within larger provinces. Likewise, the pattern of currents near Labrador and Greenland in the North Atlantic and the complex intersection of the North and South Equatorial currents off the Coast of Central America and Venezuela in the Pacific (Petrushevskaya and Bjørklund, 1974; Nigrini, 1968) further illustrate the salient effects of water mass movements on radiolarian diversity and abundance. The annual influx of open ocean water into deep water layers of Korsfjorden (Western Norway) and subsequent vertical mixing have been shown to correlate well with seasonal cycles of radiolarian abundance (Bjørklund, 1974). The distribution of radiolaria, moreover, can be traced to transport in currents that can carry them far from regions of localized abundance into alien ocean environments, or from one ocean to another where they may have variable reproductive success (Goll and Bjørklund, 1971, 1974; Casey et al., 1979a). In some cases, as occurs in the California Coastal Current (Casey et al., 1970), the movement of water masses often subjects the fauna to changing environments that result in variations in abundance or diversity of the species in the dislocated communities. The variations in species survival under the conditions of these changing environments yield insight into the adaptability and environmental specificity of many species. By contrast, the submergence of cold northern water under warmer masses of surface water in the middle and low latitudes provides a continuum of a relatively homogeneous subarctic environment, where a characteristic faunal assemblage exists. As the cold water descends deeper in the water column at lower latitudes, the associated faunal assemblages likewise are more deeply located. On the whole, the succession of faunal zones occurring horizontally in the northern geographical regions become "stacked" vertically in the lower latitudes as submergent strata of cold water become layered under the warmer surface waters. In many oceanic locations, the abundance of fauna declines from the near surface to deep water; however, regional conditions such as subsurface haloclines or presence of a thermocline and associated chlorophyll maxima can cause perturbations in the typical decline in abundance with depth. A subsurface radiolarian density maximum may occur in the water column coincident with the thermocline; however, at deeper levels it gradually or abruptly declines as is typical with increasing depth. Examples of subsurface radiolarian population maxima have been found associated with chlorophyll maxima in the Gulf of Mexico and associated Seas (Casey et al., 1979a) and for isohaline maxima in the Pacific (Kling, 1976).

Associated with these physical and chemical variations in water mass, there are also variations in phytoplankton and zooplankton productivity. The kind and abundance of these potential food oragnisms no doubt influence the productivity of radiolaria. Physiological research on radiolarian trophic activity clearly shows that many of the larger spinose and nonspinose Spumellaria as well as colonial radiolaria are omnivorous (Anderson, 1976a, 1978a, 1980; Swanberg, 1979). However, the details of their trophic activity remain to be investigated. The small Nassellaria and some small-sized Spumellaria may be microherbivores, bacterial feeders, and perhaps organic detrital feeders. Fine structure evidence from ultrathin sections of small Nassellaria (Anderson, 1976b) shows digestive vacuoles in the extracapsular rhizopodial network, containing partially digested bacteria and other large particles of matter, thus indicating they are particle feeders and not entirely dependent on their symbionts for nutrition nor on dissolved organics as sources of nutrition.

Considerable differentiation in trophic patterns may exist among radiolarian species, and therefore they could occupy diverse ecological niches. This may partially explain the large diversity and abundance of species occupying the same region. They may not be competing as heavily for the same food resources as previously assumed. Other factors contributing to stable multispecies communities in highly productive regions of the ocean could be that phytoplankton and zooplankton productivity exceeds grazing pressures by the radiolaria, thus further reducing interspecific trophic competition.

The presence of algal symbionts associated with many polycystine radiolaria, including many of the small, very abundant Nassellaria in pelagic environments, contributes further complexity and structure to radiolarian niches. There is some physiological evidence to show that symbiont primary productivity contributes to the radiolarian host carbon metabolism, and may provide additional sources of nourishment besides predation in oligotrophic environments (Anderson, 1978b, 1980).The symbiont contribution to radiolarian nutrition may be sufficient to provide a fair degree of trophic autonomy for many species of radiolaria, thus reducing their dependence on prey as a source of nutrition. We do not know what part of their total nutritional requirement can be satisfied by symbiont contributions; however, research with *Thalassicolla* sp. (Gamble, 1909; Anderson, 1980) and *Collozoum inerme* (Anderson, 1976b) shows that they can subsist for many weeks without substantial sources of exogenous food when illuminated. Symbiont contributions to radiolarian nutrition may also partially alleviate interspecies trophic competition within the photic zone (Casey et al., 1979a). Generalizations about niche limitations derived from studies on protozoa lacking symbionts (Hardin, 1960) may not apply to the more complex and trophically diversified algal symbiotic radiolarian species.

The Phaeodaria, without algal symbionts, have received less research attention than the polycystine radiolaria, hence it is difficult to generalize about their ecology. The very deep-dwelling forms occupying horizons near the ocean floor arouse intriguing questions about their metabolism, longevity, and trophic interactions at these great depths. They may be detrital feeders snaring flocculent organic matter descending in the water column (Silver and Alldredge, 1981). However, we cannot dismiss offhand the possibility that they are predators. Other species of Phaeodaria occur at many levels in the water column including the near surface horizons; however, they have not been carefully examined to determine food vacuole content nor their subcellular metabolic or structural specializations.

Casey et al. (1979a) have made some interesting inferences about phaeodarian and polycystine niches based on cumulative plankton data from studies in the Gulf of Mexico and adjacent seas. This interesting and intellectually provocative paper provides some plausible hypotheses to guide future research. They have assembled data regarding radiolarian abundance from three oceanic zones varying in productivity: (1) the eutrophic Californian Current, (2) a mesotrophic region in the Gulf Stream, and (3) an oligotrophic region in the open ocean Gulf of Mexico. Comparative data are also used from a eutrophic neritic environment above the Galveston shelf. Polycystine radiolaria appeared to have their greatest density (400 organisms/m^3) and diversity in the eutrophic waters of the California Current with decreasing densities of 122 and 19 organisms/m^3, respectively, in the Gulf Stream and open ocean water of the Gulf of Mexico. Very low populations of polycystines were observed in the eutrophic neritic water, and these were usually forms possessing symbiotic algae. This observation may imply that the eutrophic environment provided nutrients that enhanced symbiont productivity and therefore led to increased productivity of the radiolaria. Since only limited information is available on the circumstances associated with the observation, it is not possible to conclude if this was a sporadic occurrence unrelated to eutrophy or a causally connected phenomenon. Clearly additional research is warranted on the effects of varying the kind and amount of nutrients on radiolarian symbiont abundance, productivity, and contribution to host nutrition. These studies should also include environmental variables simulating various depths within the photic zone.

The density of polycystine and phaeodarian species occurring at various depths in the water column was plotted by Casey et al. for samples from the Sargasso Sea and Gulf Stream. Phaeodarians peaked at a depth equivalent to one-half the photic zone (ca. 50 m) in both geographical regions, and increased in abundance again at depths below 150 m in the Sargasso Sea. Spumellaria exhibited two peaks in the Sargasso Sea, at about one-half the photic zone depth and at about 150 m depth. Nassellaria were similarly distributed, but occurred with maximum densities at higher levels in the photic zone. They occurred largely in surface water in the Gulf Stream within the photic zone. Spumellaria were distributed in increasing numbers with depth below about 50 m. Acantharians occurred abundantly in the photic zone down to depths of 200 m. Polycystine radiolaria were excluded from the depths where Acantharia were abundant. Four basic broad niches are suggested for polycystine radiolaria: (1) nannoherbivore, (2) bacterivore, (3) detritivore, and (4) association with symbiotic algae. To these there must be added an omnivore niche (Anderson, 1980) to represent fully our current understanding of radiolarian trophic activity. It is also clear that the niches are complex and that radiolarian species may vary in their niche association across seasons, with varying depth, and depending on the availability of prey. Many of the smaller polycystines may be infusorial and nannophytoplankton feeders, based on the contents of food vacuoles in Nassellaria and light microscopic observations of protozoan and algal prey snared in the extracapsulum of Spumellaria (Anderson, 1980). These smaller species could compete with juvenile

---

Fig. 3-11. Current knowledge of trophic activity among major groups of radiolaria: solitary large Spumellaria (S1) with diameters in the millimeter range, small Spumellaria (S2) of microscopic dimensions, colonial radiolaria (C), exhibiting central capsules (CC) and symbionts (Sy), Nassellaria (N), and Phaeodaria (P). Each column represents a trophic regime: Bacterivore (bacterial-consuming species and perhaps detrital feeders), Herbivore (phytoplankton consumers preying, for example, on diatoms, flagellates, and other microphytoplankton), Omnivore (preying upon phytoplankton and zooplankton), and Symbiotroph (symbiont-bearing species that obtain part or all of their nutrition from the Symbiont). Pictures spanning two columns indicate trophic activity in both categories. An asterisk indicates laboratory data in addition to field-based observations. A question mark indicates uncertain category assignment. (Based on data from Anderson, 1976b, 1978a,b, 1980)

Bacterivore	Herbivore	Omnivore	Symbiotroph

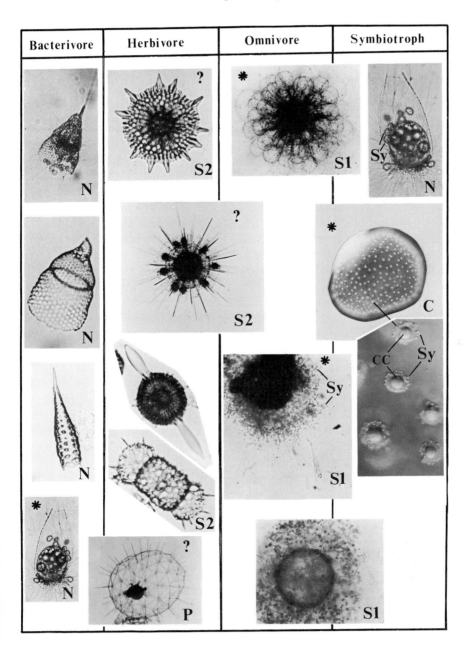

planktonic foraminifera that are too small to snare larger microzooplankton prey. Indeed electron microscopic observations of food vacuoles in some juvenile planktonic foraminifera show microalgae and
what appears to be digested remains of unicellular invertebrates, perhaps protozoa. Nonetheless, polycystine radiolaria and planktonic foraminifera overlap considerably in their distribution according to the
data of Casey et al. (1979a). Many of these planktonic foraminifera are
also algal symbiont-bearing organisms (Anderson and Bé, 1976; Bé et
al., 1977) and may be able to utilize their symbiont-derived nutrition
in the photic zone to ameliorate partially trophic competition among
themselves and with the radiolaria. The various depth zones inhabited
by radiolaria undoubtedly influences their ability to exploit niches.
Below the photic zone, there is little contribution of photosynthetically
derived symbiont nutrition; however, we do not know to what extent
the dinoflagellate symbionts are able to use heterotrophic (perhaps os-
motrophic) nutrition to help sustain the algal–host association.

Our current knowledge of radiolarian trophic activity derived from
field-based observations and laboratory experiments (asterisk) is summarized in Fig. 3-11. Some of the microscopic-sized Nassellaria are
bacterial feeders and may consume other small particles. They also
possess algal symbionts and therefore are included in both the Bacterivore and Symbiotroph categories. The term "symbiotroph" is used for
organisms deriving part or all of their nutrition from their symbionts.
The trophic category of the Phaeodaria and some small Spumellaria is
uncertain as indicated by a question mark. Those organisms occupying
two adjacent trophic categories are displayed in pictures spanning the
two category columns. For example, colonial radiolaria are consumers
of zooplankton and algae, but also derive some of their nutrition from
their algal symbionts; therefore the picture (C) is placed in the Symbiotroph column and partially extended into the omnivore column.
Likewise some large Spumellaria (S1) are omnivores, but also bear algal
symbionts and therefore are placed largely in the omnivore column,
but also span into the symbiotroph column.

Although the pelagic environment is relatively unstructured compared to terrestrial environments (McGowan, 1971), it is clear that the
diverse regions inhabited by radiolaria and the complexity of their
physiology and trophic activity impose an infrastructure on this otherwise fairly uniform environment. When local variations in hydrographic features where radiolaria occur are also considered, the
remarkable complexity of their ecology becomes more clearly apparent
and the challenge of our task assumes exciting proportions.

Chapter Four
Paleoecology and Evolution

Fossil radiolaria are significant sources of information in the earth sciences for micropaleontological research on stratigraphy, paleoecology, and cognate disciplines. Consequently, a substantial body of literature has appeared on the morphology of radiolarian skeletons, their relation to the history of ancient and modern environments, and the events that shaped the evolution of the marine environment over vast periods of time. As with any historical research, the evidence used in reconstructing past environments is of necessity inferential. Much of the evidence must be considered post facto, since it is not possible to recreate or to find in the present environment those exact or even very similar circumstances that surrounded these early events. In some cases, the conditions are sufficiently uniform over time that good extrapolations can be made from correlations with modern events. It is clearly impossible, however, to recreate experimentally vast ecosystems simulating early Earth environments. Thus the fields of paleoecology and evolution are epistemologically grounded in a type of historical research paradigm that excludes the precision of controlled experimental studies in elucidating causal relations. Nonetheless, with careful attention to collection of evidence from as wide a base as possible, and taking care to cross correlate the validity of inferences using evidence from the fossil record internally and in relation to modern events, substantial progress has been made in clarifying the paleoenvironment and in deducing some of the ecological settings that accompanied biological evolution. Moreover, through increasing efforts to apply experimental research methods and careful observational studies to the analysis of extant ecosystems and the effects of environmental variables on radiolarian physiology, survival, and reproduction, a substantial body of knowledge can be accumulated to refine inferential conclusions made from fossil evidence. The study of radiolarian paleoecology and evolution is enhanced by the superb fossil record deposited by many species (e.g., Riedel, 1951a; Sancetta, 1978, 1979a,b), particularly among the Polycystina. The lineages of many forms can be partially or substantially deduced from ocean sediment samples obtained from successively deeper horizons in the ocean floor. Riedel and Sanfilippo (1981) and earlier researchers (e.g., Haeckel, 1887; Popofsky, 1913; Deflandre, 1953; Hollande and Enjumet, 1960) have been highly productive in applying imaginative, but well-grounded theories to the interpretation of radiolarian evolution. The extensive body of literature in this interdisciplinary field of paleohistory is so large and ranges so broadly from clearly physical geological themes to more biologically organized themes, that some careful selection must be made in reviewing the field. Within the thematic framework of this book, only those research studies most pertinent to the biological history of radiolaria are considered. When possible, however, cross-disciplinary references are made to link the biologically relevant research to themes in cognate disciplines of the earth sciences. In the process of pruning and merging research data, some significant reports may be excluded largely due to thematic criteria rather than to criteria of quality or importance. This is clearly a biologist's perspective on the literature in radiolarian paleocology and evolution and, while hopefully not too limited by this disciplinary viewpoint, it may exclude areas of research in topics on basic micropaleontology or physical geology that some readers would prefer to see included.

In selecting the research papers to be reviewed in this brief account, I have chosen largely those papers that (1) pertain clearly to radiolarian paleoecology and evolution, (2) address major generalities or principles in the field rather than more isolated, factual findings, and (3) provide critical or heuristic perspectives on the questions and problems inherent in this field of inquiry. The chapter is organized on a continuum extending from research fundamentally focused on paleoecology toward more phylogenetic research. Thus the conditions that may have existed in earlier environments where radiolaria have lived and evolved are considered first, followed by current views on the evolutionary history of selected groups of radiolaria. Where appropriate, cross references between paleoecology and radiolarian evolution will be made during the course of the exposition.

Ideally, in oceanic paleoecological research, we would like to obtain as much information as possible on critical environmental variables that are known to influence life. These include temperature, salinity, available dissolved and particulate nutrients, light quality and intensity at depths in the ocean, and the prevailing currents. Clearly, given the inferential nature of the evidence, it is seldom possible to reconstruct all of these parameters. The amount of present information is also limited by the fact that the most intensive analysis of sedimentary data has occurred only in recent decades. This has been occasioned by the substantial exploratory work on ocean sediments including piston coring and Deep Sea Drilling Project (Initial Reports of the Deep Sea Drilling Project, National Science Foundation; Washington, D.C.). Much of the research effort has been directed toward biostratigraphic zonation using radiolarian and other microfossil species in various sedimentary strata. Some of this research has been reviewed critically by Berger and Roth (1975) or included in general surveys on the micropaleontology of radiolaria (e.g., Goll and Merinfeld, 1979; Kling, 1978). The voluminous reports of the Deep Sea Drilling Project bear ample testimony to the vigor and scholarly productivity of the earth scientists who have analyzed the biostratigraphy of radiolaria and interpreted their distributions in light of prevailing theory. From this research and previous terrestrial geological investigations, a fair understanding of radiolarian distribution over geological time has been accumulated.

Radiolaria in Geological Time

Radiolaria appear to have evolved first during late Precambrian or early Paleozoic time, and their skeletons have remained a significant part of the sedimentary record into modern times. We know little about the origins of the skeletonless species, as they leave no trace in the fossil record. It is difficult to assess, moreover, the relative abundance of skeleton-bearing radiolaria over geological time, due to differential dissolution of some species and their loss from the fossil record. However, some general estimates have been made of the diversity of radiolarian species based on available fossil skeletons in the stratigraphic record (e.g., Tappan and Loeblich, 1973). These data must be interpreted in light of possible biases introduced by differential preservation of fossil skeletons among the different species and the uncertainties in taxonomic identification of some species. Within these limitations, however, some broad estimates of relative patterns of diversity can be made. As an aid in conceptualizing the variations in diversity of radiolaria in geological time, a chart (Fig. 4-1) is presented showing comparative data for radiolaria, other microplankton including possible food sources (diatoms and dinoflagellates), and other microzooplankton. The radiolarian diversity is plotted from 500 million years B.P. to the present. During the Paleozoic, Spumellaria dominated the early fauna, but by the Carboniferous, Nassellaria began to diversify markedly. Following the heavy extinctions and lag in diversity during the Permo-triassic, a substantial elaboration occurred and continued to the present. The Spumellaria, however, never achieved the relative dominant diversity that they exhibited in the early fauna. Nassellaria continue to diversify markedly with the exception of the Tertiary. The magnitude of the attrition in the Tertiary is uncertain at present, as some of the disap-

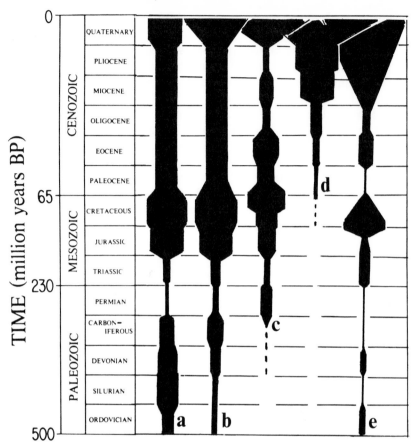

Fig. 4-1. Relative diversity of Spumellaria (a), Nassellaria (b), dinoflagellates (c), diatoms (d), and tintinnids (e) during the geological history. Bar widths indicate species diversity. Major expansion of diversity has occurred among most groups in recent time, although the species diversity of Spumellaria may be less than that of Nassellaria. (Based on data from Tappan and Loeblich, 1973)

pearance of species during this interval may be attributed to poor preservation rather than disappearance of species. There is, however, general consensus that some attrition in diversity occurred during this period based on several lines of fossil evidence. The diversity of Spumellaria and Nassellaria in recent time is apparently expanding. If only the number of different species is considered, the Nassellaria appear to be diversifying more than the Spumellaria. If, however, the abundance of each species is considered in addition to the number of different species, the Spumellaria show evidence of being equally or more diversified compared to the Nassellaria (Lombari and Boden, 1982).

The radiolarian patterns of diversity vary concomitantly with diversity of the phytoplankton. In general, modern zooplankton diversity follows phytoplankton diversity, and this appears to be the case also in geological time. There is a fair correspondence between radiolarian diversity and dinoflagellate diversity, which may indicate a trophic relationship or symbiotic interdependence (Anderson, 1980). This correlation is particularly striking in the Jurassic and Cretaceous periods, when both the dinoflagellates and the radiolaria were expanding in diversity. These conclusions, however, must remain hypothetical until more exact information is available on the abundance and diversity of various species of dinoflagellates in paleoenvironments and their biological interdependence with radiolaria. Ultimately these major patterns of diversity and the complementary observations on general radiolarian abundance should be explainable in terms of the changes in paleoenvironmental variables throughout geological time. With increasing evidence from sedimentary records becoming available, some insights into the qualities of the ancient environment are beginning to emerge and to form the environmental data base for paleoecological reconstructions. Some of this research has been aided by comparison of fossil fauna to known existing faunal–environmental relationships (e.g., Ericson et al., 1964; Ericson and Wollin, 1968; Casey, 1971a; Casey, et al., 1982a,b).

The Paleoenvironment

The correlation of fossil radiolarian assemblages with modern fauna and their environments has been used to interpret the paleoenvironment, particularly with respect to temperature and biotic factors. Prior to 1965, much of the limited research on polycystine radiolaria addressed the question of whether these forms dwelled in shallow- or deep-water habitats. Some of this research has been reviewed by Goll and Merinfeld (1979) and more extensively by Campbell and Holm (1957). However, the question has not been fully resolved. More recent research has focused on paleoclimatic variables, particularly temperature profiles in the ancient environment. Much of the work has been based on fossil data from high latitudes. The polar regions exhibit greater faunal fluctuations over geologic time spans than environments at lower latitudes and therefore have attracted a major, although not exclusive, part of research attention. This research thrust is illustrated by the work of Hays and Donahue (1972), who reconstructed temperature profiles for the Quaternary Antarctic during the last 3 million years. During this time, five species of radiolaria found in Antarctic deep-sea sediments became extinct. Three of these (Clathrocyclas bi-

cornis, Helotholus vema, and *Desmospyris spongiosa*) appear to be endemic to high, southern latitudes. The other species *Eucyrtidium calvertense* and *Stylatractus universus* are cosmopolitan occurring also in low and high northern latitudes (Hays, 1965; Hays and Opdyke, 1967; Hays et al., 1969; Hays, 1970). The stratigraphy of nine diatom species was also examined in conjunction with the five radiolarian species. The combined fossil record indicated a general cooling trend during the last 2.5 million years. A pronounced cooling occurred at approximately 700,000 years which was followed by about seven strong temperature oscillations with a stronger warm fluctuation interspersed at about 400,000 to 500,000 years ago (Fig. 4-2). Several lines of evidence indicate that the early Pleistocene (2–1 million years B.P.) marine environment in the Antarctic was warmer than that in the late Pleistocene. The radiolaria *Pterocanium trilobum* and *Saturnalis planetes,* although common in Upper Pleistocene sediments of lower latitudes (warmer waters), are rare or absent in Upper Pleistocene Antarctic sediments. They are common, however, in the Lower Pleistocene sediments, thus indicating a warmer water mass at this time. Diatom extinctions and sediment composition also support these conclusions. The first appearance in abundance of endemic Antarctic diatoms, usually inhabiting sea ice, occurred about 0.8 to 0.7 million years B.P., indicating a cooling at this time. Although the exact paleoenvironmental variables associated with these changes are not known, Hays and Donahue hypothesized that with extensive cooling, the freezing of pack ice around Antarctica was increased substantially by the formation of large ice shelves. This event could initiate strong vertical circulations around Antarctica (Gordon, 1971), thus increasing surface sources of nutrient-

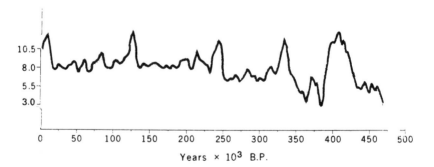

Fig. 4-2. Oscillations in climatic temperature (°C, vertical axis) of the Antarctic during the past 500,000 years based on [18]O isotopic data. (Reproduced from Hays et. al., 1976b; copyright 1976 by the American Association for the Advancement of Science.)

rich water and stimulating primary productivity. The sediments reflect this change by an upward transition from nearly barren clay to a highly biogenic ooze.

Sources of Error in Paleoenvironmental Research

The foregoing study illustrates the kind of inferential paleoecological data that can be assembled from analysis of carefully selected fossil samples. As a general critical perspective on paleobiological research, some sources of potential error in this kind of research are discussed. Inherent in any inferential analysis is the problem of error introduced by human miscoding of information—as, for example, misidentifying taxonomic categories of fossil specimens, and experimental error introduced by natural variations in the phenomena under analysis. These latter variations may be biological in origin including natural genetic variability in the initial population, bioturbation resulting from burrowing action of animals that upset the natural layering of the sediments, and phylogenetic variability over time. In the latter case, species that normally inhabit a niche at one point in history may change by natural selection and inhabit different niches. If a model assumes constancy of species niche, an error will be introduced in the conclusions.

Further sources of error can arise from inconsistencies between assumptions in the theoretical model and sample parameters. For example, if the model includes a mathematical correlation technique relating sedimentary samples with extant surface samples, certain requirements for homogeneity of variance in the data set, continuity of the sample variable, or normality of the distribution may not be met. Hence great care must be taken to ensure correspondence between the formal logical rules of the model and the data set under analysis.

Statistical Techniques

With these kinds of criteria as a guide, several sophisticated statistical methods have been developed to correlate faunal distributions with various paleoenvironmental variables.

For example, a rather sophisticated factor analytic approach to paleoecological analysis was developed by Imbrie and Kipp (1971). The fundamental procedures in the Imbrie–Kipp analysis comprise five sequential steps: (1) Paleontological core-top data are filtered by mathematical analysis to eliminate samples unrepresentative of Holocene conditions and rare (less than 2%) species. (2) These data are resolved into several assemblages by factor analysis techniques. This involves statistical resolution of the core samples into groups that have characteristics in common. (3) A least-squares technique is used to write a set of paleoecological equations relating the factor description of these assemblages to observed oceanographic parameters. For example, sample groups (assemblages) may be found to correspond to ocean regions

4. Paleoecology and Evolution

with characteristic temperature and/or current patterns. (4) The paleontological data for samples taken downcore are expressed in terms
of the core-top assemblages. (5) The paleoecological equations are used
to estimate paleoenvironmental parameters. These equations, as illustrated at a later point, are regression-type equations that express paleoecological variables (e.g., surface water temperature) as a function
of the factor-derived assemblages and appropriate coefficients. The
assumptions of this and other modern quantitative techniques in paleoecology have been critically discussed by Malmgren and Haq (1982).

Hays and co-workers (e.g., Hays et al., 1976b; Lozano and Hays, 1976)
used the Imbrie–Kipp technique to estimate surface water temperatures
in the paleoenvironment by statistical analysis of sediment core samples. Sediment samples were taken from cores obtained in an area of
the ocean floor in the southern parts of the western Indian Ocean and
the Atlantic Ocean from latitude 35 to 60°S and longitude 80°E and
55°W. Seventy-two samples were analyzed and yielded three factors
accounting for cumulative variances of 59.9, 95.0, and 97.8%. These
percentages represent the amount of variation in the environmental
variable that can be predicted by each factor. These factors were found
to correspond to three assemblages of radiolaria: (A) Subtropical Assemblage, (B) Antarctic Assemblage, and (C) Subantarctic Assemblage.
Each of these groups (A through C) have a factor analysis coefficient
(factor loading) assigned to them as part of the factor analysis procedure.
Using these data, it is possible to derive an equation using multivariate
statistical techniques to predict the paleotemperature for a given month
of the year based on the fossil evidence. The multivariate equations
representing surface water temperature in the southern hemisphere for
August (T_w) and for February (T_s) in degrees Celsius as derived by
Lozano and Hays are:

$$T_w = 15.266A + 1.542B + 9.984C - 1.984$$
$$T_s = 13.966A - 2.904B + 10.545C + 5.103,$$

where A, B, and C represent the factor loadings for the three respective
assemblages obtained using the factor analysis technique. The error of
estimate for both equations is less than 10% of the observed temperature
range; i.e., 17.7°C (-1.7 to 16.0°C) for August and 20.5°C (0.8 to 21.3°C)
for February. Additional data on radiolarian abundance showed a strong
correlation between oceanographic properties and underlying sediment
types. For example, the geographic line bisecting the zone with highest
values of radiolaria per gram of $CaCO_3$-free sediment nearly coincided
with the Antarctic Polar Front between longitude 15°W and 20°E. The
corresponding zone in the Indian Ocean sector lies parallel to and just
north of the Antarctic Polar Front. Further data on the depth of the

lysocline and the calcium carbonate compensation depth were presented and discussed in relation to sediment composition.

With the foregoing conceptual discussion as a guide, some research on the Arctic, Atlantic, and Pacific paleoenvironments is presented.

Paleoenvironmental Analyses

Radiolaria have a long and varied evolutionary history spanning from the Paleozoic (e.g., Holdsworth, 1977) through the Mesozoic (Pessagno, 1977) into modern time (Riedel and Sanfilippo, 1977, 1981). Although Paleozoic species are often distinctly different from post-Paleozoic forms (Holdsworth, 1977); and considerable diversity of species and variations in abundance occur throughout geological history (e.g., Chen, 1975; Swain, 1977; Kennett, 1979), radiolarian fossils have proved to be an excellent tool in the analysis of paleoenvironments. Much current research has focused on the Pleistocene, particularly with reference to a period about 18,000 years B.P., when there was a glacial maximum within an interglacial period. Some of this research has been interpreted and conveniently assembled in a series of papers produced by the CLIMAP research team (e.g., Cline and Hays, 1976). They have employed diverse fossil biota including Coccolithophorida and foraminifera (McIntyre et al., 1976; Kellogg, 1976; Bé et al., 1976; and Ruddiman and McIntyre, 1976); and radiolaria (Pisias, 1976; Hays et al., 1976b) obtained from the Antarctic, Atlantic, and Pacific Basins to reconstruct paleoenvironments. Some of the major findings of the CLIMAP team and other researchers are presented, particularly in relation to data derived from fossil radiolaria.

Climatic Cycles and Glacial Events   Biostratigraphic analyses of faunal assemblages in the Pacific (Kennett, 1970; Nigrini, 1970; Moore, 1973; Sachs, 1973; Johnson and Knoll, 1974; and Pisias, 1978) and Atlantic (Gardner and Hays, 1976; Ruddiman et al., 1976) consistently indicate that several climatic cycles of alternating warm and cold periods have occurred in recent geological history as previously deduced by Hays and Donahue (1972). Sachs (1973) concludes that in the North Pacific (longitude 44°22'N, latitude 163°33'W) waters in the mixed layer were as warm as those in the Recent, only three times during the past 800,000 years: 100,000–120,000; 380,000–430,000 and 740,000 years B.P. Likewise, Johnson and Knoll (1974), using two cores from the eastern equatorial Pacific Ocean, found paleotemperature curves closely resembling those reported by Sachs (1973). A warm interglacial interval at approximately 110,000 years B.P. correlates well with the warm interval reported by Sachs at 100,000-120,000 years B.P. Further research on early Pliocene to middle Miocene radiolarian assemblages in the eastern Pacific from northern to Baja California (Weaver et al., 1981) indicates a basically temperate association of species during the period 15 to 14

million years B.P. These species including *Antarctissa* and *Botryocyrtis*
sp. are abruptly replaced at ca. 14 million years B.P. by an assemblage
including *theocorys edondoensis*, indicating a dramatic change in world
climate. This event appears to correlate well with the major buildup
of the Antarctic ice-sheet occurring between 15.8 and 14.0 million years
B.P. A continued trend toward a cool-temperate assemblage or transitional fauna emerges during the interval of 14–10 million years B.P.
A distinct cold-water fauna develops during the period of 1–6 million
years B.P. Known deep-dwelling radiolaria such as *Cyrtopera languncula, Peripyramis circumtexta,* and *Cornutella profunda* increase markedly. In general, Ruddiman and McIntyre (1976), working with
calcareous fossil plankton, concluded that during the past 600,000 years,
there may have been as many as seven complete climatic cycles, which
is consonant with the seven intervals of more intense ice rafting observed in the North Pacific (Kent et al., 1971).

The paleoenvironment during the interval of the major glacial event
at ca. 18,000 years B.P. has received substantial research attention (e.g.,
McIntyre et al., 1976; Morley, 1977, 1979; Morley and Hays, 1979).
Estimates of surface temperatures for the South Atlantic at the glacial
maximum using the Imbrie–Kipp technique indicated that the largest
changes in that time period compared to today occurred in the subantarctic region between the Antarctic Polar Front and the Subtropical
Convergence (Morley and Hays, 1979). Temperatures in this region
were 2 to 5°C cooler during the last glacial maximum. Cooler sea-surface
terperatures during this period were also recorded in sediment cores
in the eastern equatorial Atlantic and off the southwest coast of Africa.
Major oceanographic features also exhibited a northward shift, including a substantial northward displacement of up to 5° of latitude of the
Antarctic Polar Front with a lesser displacement of the Subtropical
Convergence and Equatorial Divergence. These displacements toward
the Equator of major convergences in the southern hemispehre are consistent with the findings of Moore (1978) in the North Pacific at the
same period of time. Based on a Q-mode factor analysis, Moore identified seven factorial groups, which correlate well with surface water
masses and/or major ocean currents. Comparisons of the inferred oceanographic conditions at 18,000 years B.P. were made with the present.
The Tropical Group which is restricted primarily to the eastern half of
the ocean in modern times, ranged across the entire ocean at 18,000
years B.P. and extended into the western boundary currents. A Subtropical Group, now found mainly in the western subarctic, expanded
to the east and south toward the Equator at 18,000 years B.P. and strongly
resembled an assemblage found in the subantarctic area. The expansion
of these two assemblages was coincident with a decline of the Western
Pacific Temperate and Transitional Groups. Moore hypothesizes that
these differences in the 18,000 years B.P. distributions may be caused

by an increase in the influence of Arctic air masses in the North Pacific and by a general increase in the wind-driven zonal flow in the Pacific, a general "spin-up" in the circulation of the central oceanic gyres, and a steepening of oceanographic gradients in the subpolar regions.

Paleoceanography The merit of paleohydrographic knowledge derived from radiolarian faunal analysis can be enhanced, if it is linked with additional information on the topographical features in the ancient oceans that have influenced current patterns and distribution of radiolaria. Examples of this kind of research are Boltovskoy and Riedel (1979), Johnson and Nigrini (1979, 1982), and others cited in the biographic section of Chapter 3. In a recent example pertinent to this chapter, Casey *et al.* (1979c) examined extant and quaternary radiolarian faunas in the Gulf of Mexico. These populations may have become isolated in the Gulf of Mexico, when the uplift of the Panamanian block isolated the equatorial and temperate Atlantic waters and thus blocked Atlantic radiolarian faunas from entering the Pacific. Pan tropical species that were hypothetically isolated in the Gulf of Mexico include *Spongaster pentas, Ommatartus penultimus,* and perhaps *Ommatartus avitus.* The radiolarian species *Collosphaera tuberosa, Buccinosphaera invaginata,* and *Amorphophalum ypsilon* are believed to have evolved in the Pacific since the blockage of Panama. They may have invaded the Atlantic by rounding Africa in off-shoots of the Agulhas Current. If these events could be assigned a date, it would indicate where "warmer water invaded the Atlantic." The association with symbiotic algae in many of these radiolaria may enhance their ability to invade as an expatriate, and to survive as an expatriated breeding population, in a new environment. The supportive trophic role of symbiotic algae for host nutrition shown by cytochemical and isotopic labeling experiments (Anderson, 1976b, 1978a, 1980) could aid their survival, while entering or traversing oligotrophic regions. If the symbionts also recycle host waste products as a source of essential nutrients to sustain primary productivity, the host–algal system would have a remarkable degree of stability in oligotrophic environments.

Toward More Comprehensive Paleoecological Models

Although temperature profiles, current patterns, and topographical features are significant factors in understanding the paleoecology of radiolaria, a much more comprehensive picture of the dynamics of the ecosystem requires better resolution of overall nutrient levels, relative abundance of significant major nutrients such as phosphorous, nitrogen, and silica, and the amount of particulate matter in the ocean. These factors are at present difficult to reconstruct based on fossil evidence.

Some general suppositions about surface nutrient levels can be made based on the richness and composition of diatom assemblages associated with the radiolaria as, for example, reported by Bramlette (1965), Newell (1966), and Weaver et al. (1981). However, these inferences are of necessity rather broad generalizations about relative amounts of upwelling and possible sources of nutrient replenishment from advection and current distribution. Correlations of core top samples with extant surface concentrations of individual nutrients seldom yield statistically reliable predictions for samples down core.

Radiolaria and Plankton Productivity

A substantial effort has been mounted to understand the geochemical processes that control productivity in extant oceans (e.g., Bramlette, 1965; Tappan, 1968; and Broecker, 1971a,b); however, at present our knowledge of the geological controls on oceanic fertility is limited. Berger and Roth (1975), however, have proposed an interesting rather straightforward, if not highly simplified, model of oceanic fertility that offers some insight into possible sources of fossil data that may aid better resolution of surface nutrient conditions in the paleoenvironment. It is assumed on good physical evidence that upwelling is a major source of surface nutrients and that this supply is dependent on adequate contact betwen the upwelling water and the bottom sediments where large stores of nutrient occur. Part of the energy driving this process comes from the constant supply of cold, ocean-bottom water generated in the polar regions and from saline outflows of shelf areas. Continuous production of bottom water, as at present, stimulates backflow of recycled nutrients into the surface water. Conversely, appropriately spaced strong pulses would hinder it by providing a heavy (cold or saline) stagnating bottom layer. There is good evidence that the precipitation and the rate of remineralization and recycling of biogenic phosphorous compounds are crucial in controlling fertility of the ocean on a geologic time scale. It is generally assumed, however, that the limiting factor at any point in time for productivity is available nitrogen ("bionitrogen") rather than phosphate. The possible role of silica as a limiting nutrient for radiolaria needs additional research to assess properly its importance in relation to nitrogen and phosphorous. Research on nutrient cycling in the oceanic biosphere shows that in the euphotic zone, nitrate essentially disappears before the phosphate does (Desrosieres, 1969). The nitrogen cycle in the biosphere is characterized by (1) fixation on land and in the ocean, (2) internal recycling within ocean ecosystems, and (3) eventual removal by deposition and denitrification. The removal is especially enhanced under anaerobic conditions. The nitrate and phosphate cycles must be coupled, as both compounds reach limiting levels in surface water almost simultaneously and in general are highly correlated. However, the nature of this coupling is not fully understood. Berger and Roth suggest that phosphate may regulate the coupling of the cycles by determining how much new nitrogen can be fixed in oligotrophic waters by blue green algae and other organisms (Stewart, 1971; Mague et al., 1974). The nitrate source will fluctuate with the available phosphate in the water, thus closely linking the two cycles. The nitrogen to phosphorous ratio in the nitrogen fixers is an important determinant of total availability of these nutrients in the biosphere; however, the problem of which of these two factors leads in the nutrient cycle is a complex one that has been critically discussed by Broecker (1974). The relative proportions of available nitrogen and phosphorous in the ocean biosphere is further complicated by differential rates of removal through sedimentation in a variety of biogenic solids that vary in their solubility. The turnover of biophosphate is in part determined by the solubility of the skeletons, scales, and other biogenous solids that trap the phosphate and carry into deep water. The silica cycle is also of clear importance to radiolarian paleoecology, both as a source of skeletal building matter and, more indirectly, as an indispensable element in the metabolism of diatoms that at least partially contribute to the food chain of which radiolaria are a part. The well known fact that silica reaches near limiting values, substantially before nitrogen and phosphorous usually become limiting, indicates that silica is a significant regulating factor in the abundance and productivity of diatoms and radiolaria. Hence a parcel of replenished nutrient-rich water will first lose its silica through diatom "luxury" assimilation, then its bionitrogen through primary productivity of other algae including calcareous secreting algae such as coccolithophorids, and then its phosphate by nitrogen-fixing primary producers. These basic principles are synthesized by Berger and Roth into a general model of sedimentary evidence for paleoenvironmental nutrient conditions. In this model, bionitrogen and phosphate are closely linked and control silica and carbonate precipitation. If the rate of supply of either silica or carbonate exceeds the rate of assimilation, an excess could build up until the oceans become saturated and precipitation commences. Cycling of silica and carbonate would show appreciably before this point is reached, however, as the biogenic solids will exhibit increasingly slower dissolution with depth as the waters become closer to saturation. Eventually, a limiting point short of inorganic precipitation will be reached where the fossil skeletons achieve maximum preservation, due to near chemical saturation of the overlying water. These differential effects of varying nutrient levels and concomitant variations in solubility of silica and calcium carbonate fixed in biogenous solids should produce differences in the sedimentary records as summarized in Table 4-1. It may be difficult to make quantitative estimates of the parameters cited in the table particularly with reference to variations in shell morphology and diagenesis. The model, however,

Table 4-1. Fossil record and deposits in high- and low-fertility state[a]

Type of deposit	High fertility	Low fertility
All biogenous and related chemical deposits	Tendency toward disequilibrium; basin–basin fractionation (fertility gradients controlling: vertical mixing patterns)	Tendency toward equilibrium, basin–shelf fractionation, humid–arid fractionation (physical gradient controlling: depth, oxygenation)
Calcareous fossils and associated deposits, authigenic carbonates	Absent or poorly preserved over wide areas, but abundant, well preserved, and rather pure in others; strongly developed stable CCD.[b] Aragonitic fossils generally rare in deep ocean; heavy-shelled planktonic forams common; authigenic carbonates rare	Interrelated with other types over wide areas; in shallow-water deposits abundant fine-grained muds and ooid types; CCD unstable; aragonitic fossils relatively common (may be diagenetically altered); thin-shelled, small planktonic forams common; authigenic carbonate common
Siliceous fossils, biogenous chert, and authigenic silicates	Restricted sites of preservation, below high-fertility areas and in topographic depressions. Relatively delicate frustules supplied to sea floor but strongly selected for preservation of heavy shells. Oceanic diatom oozes rare, only most resistant forms preserved (e.g., *Ethmodiscus*)	Siliceous remains, chert, and authigenic silicates (palygorskite, sepiolite, zeolite, glauconite) widespread; intercalated with calcareous and/or phosphatic deposits. Accumulation especially in broad belts of regionally increased fertility. Supply of wide range of delicate and heavy frustules, delicate ones being destroyed during diagenesis

Phosphatic fossils	Chert and authigenic silicates restricted to areas of initial preservation of biogenic silica; restricted to areas of upwelling; rarely associated with carbonate	If bionitrogen strongly limiting, fossil widespread. Phosphate no longer limiting nutrient. Association with carbonates and replacement of carbonate by apatite common
Organic fossils	Restricted to areas of upwelling in a well-oxygenated ocean; relatively low diversity	In an anaerobic ocean (reduced overturn, bionitrogen strongly limiting) organic fossils are widespread, well preserved; high diversity

[a] Berger and Roth (1975).
[b] CCD, calcium carbonate compensation depth.

points the direction to some potentially significant fossil lines of evidence that may be fruitfully explored toward refining inferences about paleoenvironmental nutrient levels.

A Biological Oceanographic Model

Based on data from Holocene sediments of the world ocean, Casey et al. (1982a,b) have constructed a model of radiolarian distribution in relation to major oceanographic parameters such as currents, divergences, convergences, water masses of characteristic temperature, and oligotrophic to eutrophic conditions. This model is an extension of earlier biogeographical zone models (Casey, 1971a, 1977; McGowan, 1971, and others), and involves a detailed analysis of major correlations of radiolarian assemblages with hydrographic factors mapped onto a hypothetical global ocean model. The major surface hydrographic features of the hypothetical world ocean are presented in Fig. 4-3. Casey and co-workers also present subsurface hydrographic diagrams and distributional maps that provide additional detail beyond the summary given here. The hypothetical model contains wind-driven surface and some subsurface currents as labeled in the diagram.

The basic pattern of shallow water circulation is described as large subtropical anticyclonic gyres driven by the easterly trades and westerlies, and subpolar cyclonic gyres driven by the westerlies and polar easterlies. These currents either impel water into (anticyclonic) or out of (cyclonic) the gyres because of the coriolis effect, resulting in either polar divergences (PD) or tropical convergences (TC). Convergences and divergences in general are generated by the impact of the coriolis effects and water density effects on specific currents of the gyres and the equatorial countercurrent. Differences in density of water, occasioned by poleward transport, cooling, and evaporation resulting in increased salinity, produce submergence, as these water masses are once agin thrust toward the Equator and collide with warmer, lower-latitude water masses. The dense water submerges beneath the less dense warmer water piled into the subtropical anticyclonic gyre. Casey designates this as a subtropical convergence (STC). Similarly a polar convergence (PC) occurs in the northern-most waters, where cold polar air masses and polar ice caps cool the poleward arms of the subpolar cyclonic gyres, increasing the water density to the degree that these waters dive under the less dense waters of the southern limb. Water masses of characteristic warm or cold temperature are also designated in the hypothetical model as "upper warm water sphere" lying above the central waters, and "lower warm water sphere" coincident with the central water zone. The warm water sphere is separated from the underlying cold water sphere by a pycnocline (rapid change in density with distance). Further discriminations among cold water masses at varying depth are made, depending on whether the water is formed by

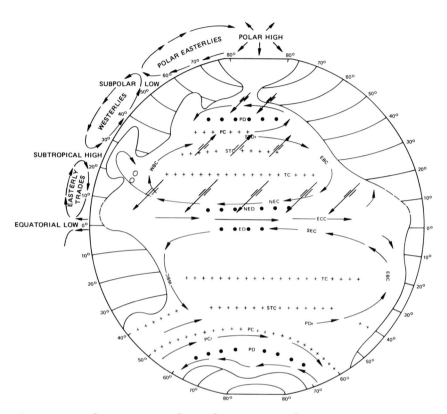

Fig. 4-3. Circulation patterns for surface waters and major convergences and divergences in a hypothetical ocean. PD, Polar Divergence; PC, Polar Convergence; STC, Subtropical Convergence; TC, Tropical Convergence; NED, North Equatorial Divergence; ED, Equatorial Divergence; arrow between bars, Winds at Surface of Ocean; arrow, Currents at Surface of Ocean; WBC, Western Boundary Current; EBC, Eastern Boundary Current; PDr, Polar Drift; PCr, Polar Current; NEC, North Equatorial Current; ECC, Equatorial Counter Current; SEC, South Equatorial Current; ●, Water Upwelling; +, Water Downwelling. (Casey et al., 1982)

surface cooling of a parcel that sinks (primary water mass) or mixing of primary waters to yield secondary water masses.

The dominant distributions of spumellarian and nassellarian groups (essentially radiolarian families) are mapped by Casey on the hypothetical ocean surface by matching core sample assemblages with their respective overlying water mass. The warm water sphere, surface, or subsurface forms include the beloids, collosphaerids, saturnulins, artiscins, phacodiscids, spongasterins, dictyocornis, spongurins, pylonids, lithellids, and stylodictins. The intermediate- and deep-water (cold-water sphere) forms include spongotrochins, spongopylins, tholonids, and orospherids. The nassellarian groups, lophophaenins, sethoperins, carpocanids, eucyrtidins, pterocoryids, cannobotryids, acanthodesmids, and some artostrobids, dominate in the warm water sphere and transitional-central water region. Plectopyramins, artostrobids (robust), sethoperins, and theocalyptrids appear to be excluded from the warm water sphere. Different species among these groups dominate in either the transitional-central or cold-water sphere, whereas others are restricted to waters greater than 2,000 m (Spaw, 1979). However, no spumellarians are known to be restricted to waters greater than 2,000 m. Antactiscins are the dominant form in the Antarctic. In general, there are major vertical discontinuities in radiolarian assemblages. Surface forms are separated from subsurface forms (dwelling from about 50 to 200 m). Although several biological factors may account for this abrupt discontinuity including bottom of the photic zone, phytoplankton compensation depth, or pigment layer, the most identifiable oceanographic break separating surface and subsurface appears to be the nutricline—a region of increasing nutrient concentration beneath the region of uniform nutrient-depleted water. Various groups of radiolaria mark the submergent water assemblages including a subtropical underwater population dwelling at about 200 m in many temperate and tropical seas, and the transitional-central water fauna that become submerged in tropical zones. Other factors contributing to variations in radiolarian abundance may include exclusion by competition for the same food source, or chemical inhibition by competitors that release toxic substances.

In overall view, the model predicts that warm water sphere radiolarians exhibit a major poleward boundary in their distributions at the subtropical convergences, with some forms extending to the polar convergences. Some warm water forms exhibit poleward extensions in the western boundary currents. Collosphaerids appear to be enhanced in sediments under the oligotrophic anticyclonic gyres and the eutrophic eastern tropical regions. These enhancements may be explained by the primary productivity of the symbionts and their contributions to host nutrition, as explained in the second chapter. The *Dictyocoryne, profunda-truncatum* group appears to be indicative of warm-water-sphere mesotrophic conditions, as occur in the equatorial regions and the poleward extensions of their gyre currents. Cold-water-sphere radiolaria are abundant poleward of the polar convergences (PC, Fig. 4-3), but also occur in abundance under the eastern boundary currents and equatorial divergences. Intermediate- and deep-water radiolarian species appear to be enhanced in Holocene sediments under the polar cyclonic gyres, eastern boundary currents, and the oceanic convergences and divergences.

A synthesis of ocean fertility paleoenvironmental models (e.g., Berger and Roth, 1975) with hydrographic radiolarian distribution models (Casey et al., 1982a,b) may yield substantial improvements in generating ecologically significant inferences about paleoenvironments and thus enhance our understanding of the environmental factors associated with radiolarian evolution. Additional references to research on radiolarian paleoecology, evolution, and pertinent biostratigraphic sources organized according to geographic locales are: Atlantic Ocean (Maurasse, 1976; Spaw, 1979), Pacific Ocean (Theyer and Hammond, 1974; Sachs, 1975; Kruglikova, 1976; Reynolds, 1979; Labracherie, 1980; Sloan, 1981), Indian Ocean (Keany, 1973; Dow, 1978), and Polar regions or general global (Hays, 1965; Fillon, 1973; Blank and Margolis, 1975; Keany, 1976, 1977; Blueford, 1978; Maurasse, 1979; Stanley, 1981).

Evolution

Considerable interest in plankton evolution and paleoenvironment has developed in recent years, as witnessed by the number and variety of papers on relevant topics, including the general abundance and diversity of plankton in geological time (e.g., Bolli et al., 1957; Banner et al., 1967; Cutbill and Funnell, 1967; Downie, 1967; Sarjeant, 1967; Lipps, 1970; and Tappan and Loeblich, 1972), and theory of plankton–environmental interactions during evolution (Tappan and Loeblich, 1973; Harper and Knoll, 1975). The problem of interpreting organismic–environmental interactions based on fossil evidence is accentuated by the fact that the fossilized specimens frequently retain only their hard parts, and therefore it is not possible to make conclusive statements about the organization of the living substance. We must assume, where appropriate, that the ancient species possessed living matter much as with modern forms. This assumption is clearly prone to considerable error, as we cannot be confident that certain adaptive structures observed in the living substance of modern forms were also present in phylogenetically earlier forms. If, for example, the fossil radiolaria possessed somewhat elaborate silica skeletons, we are probably justified in assuming that they possessed a cytokalymma or other living membrane-lined structure that mediated silica secretion and determined skeletal form. However, whether or not the rhizopodia were as elaborate as those in modern species, or the flotation mechanisms were similar to those of extant forms, etc., is a point we cannot resolve. It is probable that fundamental physiological processes of respiration, food ingestion and digestion, and composition of essential subcellular organelles (mitochondria, Golgi bodies, endoplasmic reticulum, nucleus, etc.) were much like those in modern forms. These attributes appear to be rather conservative features among all extant eucaryotic organisms (those with a membrane-enclosed nucleus) across a wide spectrum of species with varying degrees of phylogenetic development.

The first occurrence of symbiotic algae within radiolaria is not known, as they do not leave fossil impressions. It may be possible to detect the presence of symbionts, however, by examining the organic matrices of the fossil skeletons to determine if there is isotopic fractionation suggesting contribution by algae to the organic substance of the skeleton. There is substantial evidence that symbiont-bearing calcareous organisms (e.g., hermatypic corals and foraminifera) are enriched in the stable isotope ^{13}C compared to non symbiont bearing species due to depletion of ^{12}C by preferential uptake by the symbionts (Erez, 1978; Fairbanks and Dodge, 1979; Cummings and McCarty, 1982). If sufficient organic matter can be recovered from radiolarian skeletons, similar $^{13}C/^{12}C$ analyses may be used to indicate the presence of algal symbionts in fossil species. In this case, the organic products should be enriched in ^{12}C.

To understand more fully the kind of environmental pressures that were impressed upon radiolaria, it is clearly of paramount importance to know whether ancient species possessed algal symbionts and the duration of these associations. Their food sources, moreover, may or may not be the same as extant, apparently related, species based on comparative skeletal morphology. This is perhaps one of the most vexing problems, as there is no reason to believe a priori that a fossil specimen with a skeleton similar to a living form would have the same diet. Diet is clearly determined more by metabolic pathways of a predator than by its skeletal morphology. Once again, if we can refine our techniques of skeletal preparation and cleaning to permit high-resolution chemical analysis of the organic content of skeletons, there may be some correlation of proportions of organic constituents with the kind of prey consumed. This could be determined in part by examining a wide variety of living species, whose feeding habits are known, and correlating the skeletal organic composition with their trophic activity. Although our knowledge of feeding behavior in ancient forms is limited, there is a fair amount of evidence that they were coupled to oceanic productivity much as with modern forms. Their fossil diversity clearly correlates with presently known prey such as diatoms and dinoflagellates (Fig. 4-1), which are also a substantial part of the food chain in modern oceans. We may expect therefore that they reflect the general productivity of primary producers and the potential productivity of consumers.

Given the presence of a siliceous fossil skeleton, it is clear that the amount of available silica in the seawater was a significant selection factor in addition to primary nutrients. In general, it is known that silica secreting plankton (especially diatoms and also radiolaria) are most abundant in silica-enriched surface water as occurs for example in cold, nutrient-rich water in upwelling regions. Both modern radiolaria and diatoms possess remarkable capacities to absorb and concentrate silica within their skeletons against exceptionally severe concentration gradients. In diatoms, it is known that the proportion of silica to nonsiliceous matter decreases as the available seawater silicate declines; however, in culture vessels, the available silicate is almost completely depleted by diatom growth. We do not know the comparable "silicate depletion" capacity of radiolaria nor at what point they may succumb for lack of available silica. However, if earlier ancestral forms were equally competent in silica biomineralization, they undoubtedly could have survived in comparably relatively silicate-limited environments as exist today in large parts of the open ocean. In general, it is presumed that radiolarian productivity correlates fairly closely with available silicate. Consequently, the massive chert deposits in the Atlantic underlying calcium–carbonate-rich deposits in the modern sediments suggests a period of heavy biogenic silicate production and/or enhanced preservation of biogenic particulate silica in the paleoenvironment.

Paleoecology and Plankton Evolution

Some of the fundamental ecological variables pertinent to plankton evolution in environments of varying nutrient concentration, temperature, and biological productivity have been discussed in an essay by Tappan and Loeblich (1973). Their model may serve as an interesting heuristic guide for further research to determine the validity of the assumptions with a wide variety of plankton including species of solitary and colonial radiolaria. They hypothesize that fluctuations in light quality and intensity in the water column, variations in nutrient sources in planktonic environments, and temperature variations are among the significant factors driving evolutionary change. Diversity among plankton species has been correlated with both vertical and latitudinal temperature zonation (Cifelli, 1969; Lipps, 1970). Periods of evolutionary diversification have been correlated to temperature changes in the Cretaceous (Douglas and Sliter, 1966; Neagu, 1970) and Tertiary (Berggren, 1969; Cifelli, 1969). These physical environmental stresses favor the survival of organisms that are better able to regulate their buoyancy, compensate for current displacement into alien environments (e.g., through diversification of trophic activity, utilization of symbiont primary productivity or adaptive adjustments to changes in light intensity), and adapt to variations in temperature. Most cosmopolitan species are eurythermal or have adapted to varying depth habitats (e.g., submergent species) to match their stenothermal demands. Other cosmopolitan species are euryhaline. Smayda (1958) has noted that temperature effects are complex. The increased metabolic rate in warmer temperatures also increases nutrient requirements; thus, some Arctic species of plank-

ton can grow successfully in warm waters only at high nutrient levels,
as in polluted regions. Other pressures toward diversification include
changes in depth of water on continental shelves during times of actively
spreading rift systems, thus favoring survival of progressively holoplanktonic adaptive forms (Frerichs, 1970). Increased ecological niches,
associated with continental displacements forming extensive epicon-
tinental seas and complex and lengthened coastlines, favored adaptive
radiation into new environments (Valentine and Moores, 1970). During
periods of stress, as, for example, in highly competitive environments,
where numerous species are competing for the same niche, or in high
latitudes, where there are extremes of temperature or nutrient supply
including availability of light, the most competitive or better-adapted
organisms will tend to dominate. These conditions are characterized
by low total diversity and a high ratio of productivity to consumption.
The ratio of productivity to biomass may also be high. As environments
stabilize over many generations and/or biota become better adapted to
their habitat, "mature" plankton communities develop. These communities have longer food chains, and their components exhibit a greater
size range and greater total diversity. Individuals per species are fewer
and endemism is enhanced. Life spans may be longer, means of dispersal more restricted, and interactions among biota more delicately
adjusted. Intraspecific morphological variations are less, but genera
and species may possess more ornate skeletal structures and a greater
percentage of inert matter composing the skeleton (Margalef, 1968). In
large part, much of this theoretical literature concerns nonradiolarian
plankton, including substantial amounts on planktonic foraminifera
and phytoplankton. Additional research on the application of these
principles to radiolarian evolution is clearly warranted.

Interspecies Competition and Radiolarian Evolution

In a concise and interesting paper, Harper and Knoll (1975) have applied
the principles of interspecific competition for resources (in this case
soluble silica) and biogeochemical cycling of silica in the ocean to
explain selection pressures on radiolaria during the Cenozoic that could
account for modifications in the robustness of their skeletons. During
the Cenozoic, radiolaria exhibit a marked decrease in test thickness and
weight. Berger (1968) attributed this to selective dissolution of more
fragile tests in older sediments, hence biasing the fossil sample toward
heavy skeletons. Moore (1969), however, has presented evidence that
the differences in test composition cannot be attributed solely to selective dissolution effects, since the younger sediments lack the robust
forms found in the older Paleogene deposits. Moreover, the changes
in skeletal morphology include structural alterations in addition to
decreasing wall thickness. Pore size increases with a concomitant decrease in bar width, and there is a reduction or loss of test processes

accompanied by a more regular pore alignment. These events may represent evolutionary trends within individual lineages as documented by Goll (1968, 1972a), for the Trissocyclidae; Sanfilippo and Riedel (1970), for the Theoperidae; or Riedel (1959) and Riedel and Sanfilippo (1971), for the *Ommatartus* lineage. Harper and Knoll hypothesize that these changes can be explained by competition for available silica between the radiolaria and the diatoms that were becoming increasingly significant as silica fixing organisms. Diatoms entered the oceanic silica cycle in Jurrassic time, but now account for as much as 90% of the suspended silica in the oceans. The radiolaria originated much earlier and probably were well established by the time diatoms become strong competitors for soluble silica.

This competition during Cenozoic time could have affected the silica budget in at least three ways: (1) diatoms may *never* have depleted silica in the surface waters sufficiently to upset the radiolarian–silica balance; (2) there may have been a monotonic increase in the steady-state concentration of dissolved silica in the surface waters of the oceans throughout the Cenozoic Era; or (3) through increasing diatom competition for silica, less has been available to the radiolaria during Cenozoic time. The first two possibilities are rejected by Harper and Knoll, based on empirical evidence of the robust silica assimilating capacity of diatoms (Carley, 1969; Lisitzin, 1971), and paleogeological evidence of decreased sources of silica input to the oceans during the Cenozoic. The remaining third possibility becomes the probable one. Radiolaria have had to compensate for an ever-decreasing steady-state concentration of available silica. Two compensatory strategies are possible; either radiolaria become less abundant than they were in the past or they use less silica per test. There is a fair amount of evidence that no great decrease in radiolarian numbers occurred during the Cenozoic, as would be required to compensate for diatom competition. Indeed Moore (1978) has found that radiolarian abundance in the quaternary sediments from the equatorial Pacific Ocean is not significantly different from that of lower Tertiary sediments. There is good evidence to suggest that radiolarian test weight decreased as diatom species diversity increased. Hence, it appears evolution is operating through positive selection pressure favoring phenotypes that can use less silica and still conserve the functional integrity of the skeleton. Harper and Knoll acknowledge that other competing explanations for the observed changes need to be assessed, including the possibility that thinner skeletons aided buoyancy or that the changes are the result of orthogenic evolution (evolution through change due to intrinsic genetic control mechanisms). These, however, are dismissed as less probable, based on physiological and evolutionary sources of evidence. Hence, it appears that modification of the oceanic cycle by diatoms has exerted a primary selection pressure on radiolarian evolution. Radiolaria have adapted by secreting thinner skeletons rather than by improving their capacity to compete for silica. It is interesting to note in passing that the persistence of the siliceous skeleton in many radiolarian species, in the presence of the heavy selection pressure imposed by an ever decreasing supply of silica, indicates a significant adaptive value of the skeleton for the radiolaria.

The quality of the logical argument and the nature of the evidence presented by Harper and Knoll are illustrative of the kind of detailed analysis required to draw strong inferences in paleobiological research. In order to have a strong inference, one must show that the necessary conditions have been met to draw a conclusion, and that other possible competing explanations have been adequately dismissed, so as to strengthen the conclusion as a sufficient one. Weak inferences may suffer from a variety of omissions, including inadequate evidence to show that necessary conditions have been met to permit application of an evolutionary principle to a paleoenvironemtnal problem or to draw a conclusion based on the evidence. Even if the necessary conditions have been adequately documented, it is essential to consider critically other possible competing explanations and to marshal evidence to dismiss them, or otherwise to incorporate them in the conclusions. One cannot identify all possible alternative explanations, as this is epistemologically impossible; however, sufficient reasonable alternatives should be addressed to permit a strong conclusion. The precision and detail with which alternative explanations are generated and tested for adequacy is determined in part by the elegance and detail of prevailing theory as well as by the creative analytical skills of the researcher. In ex post-facto analyses of the type employed in evolutionary research, it becomes especially critical to examine competing explanations and to judge their merits relative to the hypothesized conclusion. This is especially necessary given the absence of controlled experimentation in these historical-based studies. In the Harper and Knoll study, the argument based on competition for available silica is strengthened by the fact that the geochemical cycles are probably sufficiently uniform over geologic time to permit extrapolations from the present to the past. Such extrapolations are not so easily made when the interspecific competition involves more subtle factors that cannot be readily inferred from the geologic record.

Interspecies competition during evolution is probably most blatant when the species are members of diverse groups, as with the diatom and radiolarian competition for silica. More subtle interactions occur when the competing organisms are closely related, occupy very similar niches, or represent recently divergent forms tending toward separate species. Kellogg (1976a) examined the possible role of exclusionary species interactions in explaining the selection pressure that enhances divergence of form in closely competing species during early stages of their divergence as species. This phenomenon of accentuated morphological, ecological or behavioral differences between two newly differentiated cognate species while in the same region is called character displacement. This phenomenon is known to occur during secondary geographical overlap following the allopatric (geographically separated) phase of speciation (Brown and Wilson, 1956), but may occur also in nonallopatric speciation. The accentuated differentiation of the evolving species may reduce competitive interactions, thus explaining the adaptive mechanisms driving the change.

Kellogg (1976a) examined the evidence supporting character displacement as a mechanism in the divergence of *Eucyrtidium matuyami* from *Eucyrtidium calvertense* (Hays, 1970). *Eucyrtidium matuyami* differs from *E. calvertense* in larger size, a more inflated fourth segment, and more prominent longitudinal furrows on the abdomen and fourth and fifth segments. The evolutionary history of these two species has been discussed by Hays (1970). Evidence presented by Kellogg for character displacement includes increasing divergence in size (width of fourth segment) with increasing depth in the core. *Eucyrtidium calvertense* exhibits nearly constant size until this species becomes common in the same core, whereupon it decreases markedly in size reaching a constant minimum shortly before the extinction of *E. matuyami* in the Brunhes period. There was an almost fourfold increase in the rate of evolution toward a smaller size by *E. calvertense* following the appearance of *E. matuyami*. When maximum divergence of shell morphology was reached shortly before the extinction of *E. matuyami*, both species achieved a fairly stable size. By contrast, the increase in size of *E. matuyami* coincident with the decrease in size of *E. calvertense* may not be a result of character displacement. Kellogg discussed several alternative explanations, including responses to environmental variables such as possible temperature changes, or complex patterns of gene flow and competition between sympatric populations.

Ancestral Forms and Phylogenetic Lineages

The problem of identifying ancestral forms in reconstructing phylogenetic sequences based on fossil evidence is a complex one, due in part to the inferential nature of the evidence and the uncertainty in the representativeness of the fossil record. Many of the factors contributing to bias in biostratigraphic research, including diagenesis, differential dissolution, bioturbation, and various forms of stratum displacement, also contribute to uncertainty in identifying ancestral–descendent relationships. The problem is perhaps more acute in reconstructing phylogenetic sequences, as it is tacitly assumed that the "exact" nearest ancestor for a descendent can be identified within the fossil record. The confusion generated by the profusion of coexistent species in a stratum representing a particularly productive era is troublesome enough without the additional sources of error contributed by sample perturbations. The epistemological question of whether it is indeed possible to determine accurately phylogenetic sequences based on ex post facto evidence has been discussed critically from the perspective of various disciplines, including taxonomy, biostratigraphy, paleontology, and statistical sampling theory. Viewpoints range from rejection of ancestor–descendent hypotheses, based on the postulate that they are not amenable to tests of truth or false value (e.g., Cracraft, 1974; Engleman and Wiley, 1977), to the affirmation of the merits of phylogenetic reconstructions (Harper, 1976; Szalay, 1977; Bretsky, 1979). Prothero and Lazarus (1980) discuss the semantic and sampling theory issues in addressing this question, and take the position that the problem is better restated as one of determining the adequacy of evidence for reconstructing phylogenies and assessing the likely validity of the fossil samples, rather than taking dialectical positions on the issue. Needless to say, their discussion is enlightened, however, by the knowledge that has been generated through the dialectics. They propose that it is possible to judge the adequacy of fossil data for reconstruction of lineages if two assumptions can be met (Prothero and Lazarus, 1980; p. 120):

1. The hypothetical ancestor must be older than its descendent.
2. No potentially ancestral population remains unsampled.

The first assumption is straightforward. The latter assumption is necessary, among other reasons, to ensure that a potentially low-abundant, direct ancestral form to a descendent is not overlooked either as a result of not detecting it or as being masked by more abundant forms that are not necessarily in the most "direct" line to a descendent. In most cases, it is not possible to satisfy rigorously the above assumptions. With these assumptions as a guide, however, it is possible to design adequate sampling programs and to obtain sufficiently representative replicate fossil samples to satisfy better the requirements stated in the assumptions. Moreover, the assumptions provide a conceptual guide for assessment of the validity of ancestor–descendent hypotheses and phylogenetic lineages based on fossil evidence. They are a useful part of the intellectual repertoire of a critical and reflective researcher. Prothero and Lazarus illustrate the merits of marine sedimentary microfossil samples in elucidating ancestor–descendent relationships. Based on the data of Hays (1970), they demonstrate how the use of multiple core samples from diverse locales and their cross correlation can be used to improve data quality and better satisfy the criteria of (1) temporal priority and (2) sample validity.

An ultimate question must also be addressed; namely, what do the reconstructed lineages actually represent? From a biological perspective, the lineages ideally would represent the continuity of genotypes or succession of gene pools characteristic of the populations throughout geologic time. It is difficult enough to make genetic maps of extant breeding populations, where we have access to empirical data about phenotype variability and can observe phenotypic changes in offspring within a reproducing population. The evidence remaining in a radiolarian fossil record, though superb relative to many other fossilized groups, is meager from a biological perspective. Much of the essential information about behavior, physiology, and basic reproductive patterns is obscure at best, and must be inferred from highly derived sources of evidence, i.e., whatever remains in the preserved skeleton. Hence, it is very difficult to declare what the genetic continuity is in a reconstructed lineage based on fossil evidence. By morphometric analyses of variability among fossil skeletons, putatively representing a species, within a stratigraphic section, it may be possible to make some informed guesses about genetic variability and the diversity of the gene pool throughout geologic time.

Inevitably, the discussion of lineages raises the question of what is meant by species in paleontological reconstruction of phylogenies. The general problem of defining species in paleontology and neontology has been critically discussed by Haldane (1956) among others, and by Shaw (1969) and Riedel (1978), from a more purely micropaleontological perspective. In a classical biological definition based on population criteria, a species is a group of organisms capable of interbreeding and coexisting in a common habitat. This definition preserves the concept of the gene pool as a fundamental species attribute; however, this is not readily applied in micropaleontological studies of radiolaria, due to limitations of the fossil evidence as discussed above. Consequently, a variety of species definitions have been used in micropaleontology depending on the context of the research. In the applied sciences, species definitions are often determined by pragmatic criteria of what attributes best predict a technological outcome. In most cases, morphological features of the skeleton are employed as species attributes; however, at present we have little knowledge of how much variability is attributed to intraspecific variation as opposed to interspecific differences. Consequently, it is difficult to make extrapolations from phenotypic definitions of species to genotypic definitions. Hence, critics such as Shaw (1969) suggest that the concept of species be discarded and other categorical names used that more accurately reflect the scientific context or technological perspective. Riedel (1978) argues for a more objective basis of describing radiolarian species by removing potentially confusing semantic descriptions, often laden with imprecise terminology, and replacing them with number-coded descriptors based on more exact morphometric attributes.

In sum total, it is clear that the best interests of scientific inquiry will be served, if a synthesis between micropaleontological paradigms and biological paradigms can be achieved. Thus, the two disciplines, which are of necessity epistemologically linked, will be better informed

by progress in the complementary cognate field. There is a clear need
for clarification of species definition in radiolarian research and the
meaning of ancestor–descendent relationships to provide the semantic
bridge necessary for crosscommunication between the disciplines. In
part, this could be realized if the term "species" were used in a classical
way to mean organismic attributes that are closely linked to inheritable,
genetically conservative characteristics. Other variations in fossil attributes that are employed for descriptive or diagnostic, applied purposes should not be used as species attributes; nor perhaps should
species names be given to forms that do not exhibit sufficiently distinctive characteristics to qualify as genetically stable, transmissable
"traits."

Until we gain additional knowledge about the natural variability of
skeletal form within living species, the extent to which environmental
factors influence skeletal form, and the roles of sexual reproduction (if
at all) and asexual reproduction in radiolarian biology, these issues will
be difficult to resolve on an empirical basis. The work of micropaleontologists in analyzing variablity within fossil groups and conceptualizing phylogenetic relationships is a welcome aid to biologists who
can profitably use this information to guide their search for interdisciplinary relevant research topics and pertinent experimental organisms. In overall view, the inferential data from micropaleontological
analyses of fossil lineages point strongly to sufficient genetic variability
between generations to provide the necessary grist for natural selection
to proceed efficiently and thus to enhance rates of evolutionary change.
Such rapid advances as well as evidence of possible hybridization in
fossil skeletons (Goll and Bjørklund, 1971, 1974; Kling, 1971; Lazarus
et al., 1982) strongly point toward sexual reproduction in radiolaria.

Phylogenetic Patterns

The extensive stratigraphic research during recent decades has spawned
considerable activity in construction of phylogenetic lineages and the
elucidation of species diversity and evolutionary patterns among fossil
and living radiolaria. Some of the findings have been summarized by
Riedel and Sanfilippo (1981) in an informative essay on current theory,
methods of quantification, and illustrative data on ranges and phyletic
pathways among some polycystine radiolaria. There are numerous reconstructions of species ranges and lineages in the current literature,
including those in a substantial number of reports in the many volumes
of the Initial Reports of the Deep Sea Drilling Project (e.g., Riedel and
Sanfilippo, 1970; Goll, 1972b; Kling, 1973; Chen, 1975; Sanfilippo et
al., 1981; Schaaf, 1981). There is little merit in reproducing many of
them in a chapter dedicated to conceptual issues and problems. Moreover, the constant revision of existing lineages as new data are acquired
often makes them out of date very soon. Hence, a summary of some

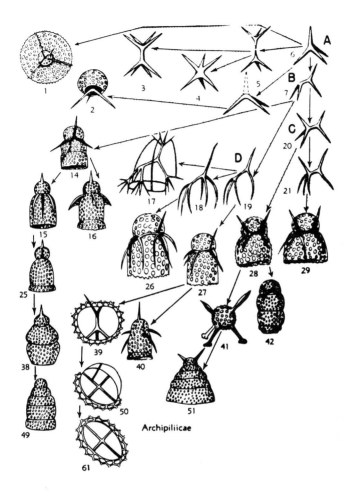

Fig. 4-4. Phylogenetic pathways of nassellarian skeletal development originating from simple spicules (A–D) and culminating in shells with a bell-like cephalis. (From Campbell, 1954, *Treatise on Invertebrate Paleontology*, courtesy of The Geological Society of America and University of Kansas)

current knowledge about radiolarian phylogenetic lineages and stratigraphic ranges is presented in the content of a historical perspective.

The earliest substantial consideration of radiolarian phylogeny was presented by Haeckel (1887). His lineages were largely constructed on comparative morphological principles; namely, examining species for common characteristics or where possible similar cytological features, and arranging them in phyletic trees proceeding from simple or less sophisticated forms to the more complex forms. Haeckel recognized the severe liability of not having extensive paleontological and onto-genetic data to aid his task, and conceded that his constructions were at best only the foundations of a natural system. Since some of the species and even some families that he erected were undoubtedly on-togenetic stages rather than natural phyletic groups, there is little value in reproducing his charts of lineages. He presents a rather extensive discussion of the rationale for his constructions (Haeckel, 1887, pp. ci–cxxvii) which provide some insight into the prevailing thought of the time. Popofsky (e.g, 1913, 1917) constructed geneological trees for major radiolarian groups, largely on the basis of common morphological fea-tures between hypothetical ancestor–descendent relationships. He was aided, however, by a growing body of literature in micropaleontology, particularly with regard to the excellent radiolarian fossil samples ob-tained from terrestrial strata on the island of Barbados. The substantial progress in ocean exploration by the German expeditions in the late 19th and early 20th centuries (e.g., Meteor Expedition and German South Polar Expedition) provided additional data from marine samples that were not available to Haeckel. Some examples of the kinds of constructions developed from these data and the logical relations in-ferred are illustrated in Fig. 4-4. Some progress made during the early 20th century in radiolarian fossil research has been presented by De-flandre (1953), and cytological contributions to the theory of radiolarian evolution have been discussed by Hollande and Enjumet (1960).

Recent advances in more sophisticated theories of evolution and more detailed biostratigraphic data have contributed to better resolution of phyletic sequences. A chronological range chart for some tropical and polar polycystine radiolaria is shown in Fig. 4-5 from Riedel and Sanfilippo (1981). A variety of criteria have been employed in delin-eating species ranges, including continuity of a morphotype from first appearance in the stratigraphic record to its disappearance, or a more sophisticated approach of distinguishing between a *morphotypical* and *evolutionary*, quantitatively defined, bottom (mB, eB) and top (mT, eT) of a taxon's range (Riedel and Sanfilippo, 1971). In the latter method, a distinction is made between the ranges of radiolaria with well-under-stood phylogenetic histories as compared to those that are less well understood. When there is incomplete knowledge of the phylogenetic relationships among the forms observed in the fossil record, then only the ranges of the morphotypes can be plotted. Thus the first occurrence and last occurrence of each morphotype are determined and can be plotted as a time line against geological periods as shown in Fig. 4-5. The result is that many of the morphotype ranges overlap, and we cannot immediately say which antecedent morphotype if any has given rise to later occurring forms. If evidence is available on the phyletic sequence of the morphotypes such that we can posit which ones may have given rise to subsequent ones, it is possible to redefine the ranges so that the

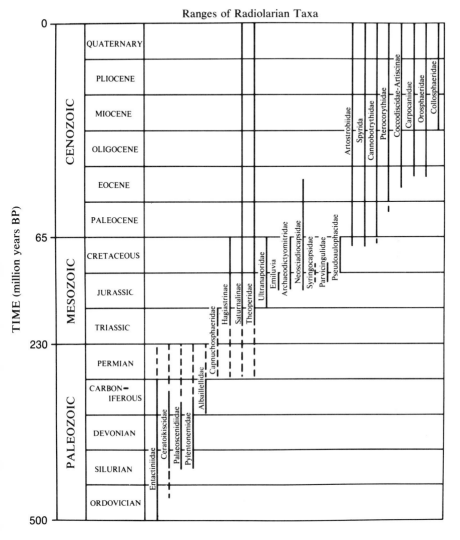

Fig. 4-5. Time ranges of radiolarian taxa (mostly families) which are believed to represent natural groupings, to show the pronounced changes near the Paleozoic/Mesozoic and Mesozoic/Cenozoic boundaries. Broken lines indicate doubtful limits. (Riedel and Sanfilippo, 1981)

ancestor range extends only to a point where the descendent range can be reasonably posited to begin. To make this discrimination of contiguous ranges, Riedel and Sanfilippo examined "populations" of specimens obtained from the fossil record. Within a stratigraphic sequence, population characteristics change gradually from ancestral to descendent characteristics, and therefore a statistical rule must be used to determine at what point a population has changed sufficiently to state that the descendent range should begin. The rule is applied that when 50% of a population exhibits ancestral or descendent characteristcs, a new range is begun. This convention makes it possible to reorder the overlapping "morphotypic" ranges so that they are colinear and contiguous in what is called an "evolutionary" time range. The beginning of an evolutionary range is denoted by eB and the top by eT. Clearly, the statistical convention of resolving range boundaries coincident with the 50% composition point of the populations is a convenient way of adlineating the evolutionary ranges; however, in the process, the information about the total range of the morphotypes is lost, unless these data are presented separately. The evolutionay ranges are, however, probably more useful and valid from a theoretical biological perspective than are the morphotypic ranges.

Through careful analysis of variations in morphology of groups of radiolarian fossil skeletons, sampled at succesive strata within the fossil record, it is also possible to infer the lineage relationships ("Phylogenetic trees") of the evolving radiolaria (e.g., Goll, 1976; Kling, 1978; Riedel and Sanfilippo, 1981; Lazarus *et al.*, 1982). The accuracy of these constructions depends in part on the completeness of the fossil record, the proper identification of intermediate forms within a lineage, and the separation of intraspecific steady-state variation from phyletic variation that accounts for "true" phylogenetic change.

Phylogenetic Lineages

A study of evolution in the radiolarian species—complex *Pterocanium* (Lazarus *et al.*, 1982)—illustrates this research paradigm. A global taxonomic and biometric study of five lineages of the radiolarian genus *Pterocanium* during the last 6 million years showed that phyletic change was gradual, but occurred at variable rates accompanied by a trend toward decreasing size in all lineages, but with concurrent reversals of

---

Fig. 4-6. Phylogeny of late Neogene *Petrocanium*. Solid bars show main lineages; shaded intervals show zones of morphologic intergradation between lineages. Figure is diagramatic, and vertical lines do not imply morphological stasis. Figured specimens drawn with a camera lucida to the same scale (10 μm bar shown below C^1). (A) *Pterocanium korotnevi* (V21–148, 20 cm: Late Pleistocene), (B) *P. praetextum* (RC 12–66, 0–2 cm:Late Pleistocene), (C_1) *P. charybdeum allium* (RC12–66, 2,555 cm:Late Miocene), (C_2) *P. charybdeum* (RC12–66, 0–2 cm), (C_3) *P. trilobum* (V19–169, 9–10 cm:Late Pleistocene), (D) *P. prismatium* (RC12–66, 1,187 cm:Mid-Pliocene), (E) *P. audax* (V19–171), 880 cm:Mid-Pliocene). (Lazarus *et al.*, 1982)

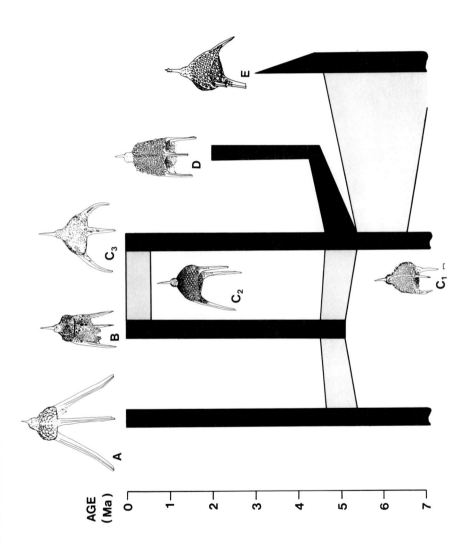

trend. The two speciation events that occurred during the interval (Fig. 4-6) were both gradual (ca. 500,000 years) and putative hybridization between lineages appears to be common, particularly in regions of up-welling. There were two extinction events, one gradual (*P. audax*) and one abrupt (*P. prismatium*). The phyletic ancestor–descendent relations and geochronological periods correlated with intergradation of form, suggesting hybridization, are presented in Fig. 4-6. An additional example of the elegant inferential lineages that can be derived by ex-amination of morphotypes and their morphological relationships is shown in Fig. 4-7 prepared by Riedel and Sanfilippo (1981). Further illustrative data on evolution of specific solitary radiolarian groups can be found in Lazarus *et al.* (1982), Moore (1982), Goll (1968, 1969, 1979), Bjørklund and Goll (1979), Goll and Bjørklund (1980); and Kellogg and Hays (1975).

Among the few reports that exist on the phylogeny of colonial ra-diolaria, Knoll and Johnson (1975) analyzed the ancestral origin of *Buccinosphaera invaginata*. This species apparently evolved during the quaternary during the interval of 125,000 to 300,000 years B.P. *B. invaginata* possesses a nearly spherical shell with small irregular pores and larger more scattered pores bearing a short, internally directed tube. Knoll and Johnson have traced the origin of *B. invaginata* to a *Collos-phaera* sp., which possessed decidedly irregular shells bearing scattered surface lumps and irregular protrusions. This collosphaerid species, named only a *Collosphaera* sp. A., progressed through transitional stages wherein the surface protrusions gradually migrated toward the large pores. As the "swellings" encircled the large pores, they induced a concavity surrounding the pore and forced it to extend inward. The ultimate effect in evolution was to produce a pore with an inwardly oriented tube and a nearly spherical shell, with the exception of the few lumps near the large pores. These data are of interest from a the-oretical viewpoint, as they illustrate gradual morphological changes that are clearly detectable within the range of biological variation pres-ent in the population at each stratum and thus provide some additional evidence from recent geological history about the tempo and mode of phylogenetic change in a colonial radiolarian lineage.

A Perspective on Phylogeny, Ecology, and Biological Variability

Given ideal conditions, a book of this kind could include a general discussion of the role of genetics in radiolarian evolution, environ-mental adaptations, and population diversity. Our knowledge of ra-diolarian genetics, however, is close to nothing. With the exception of some knowledge about chromosomal composition in the large mul-

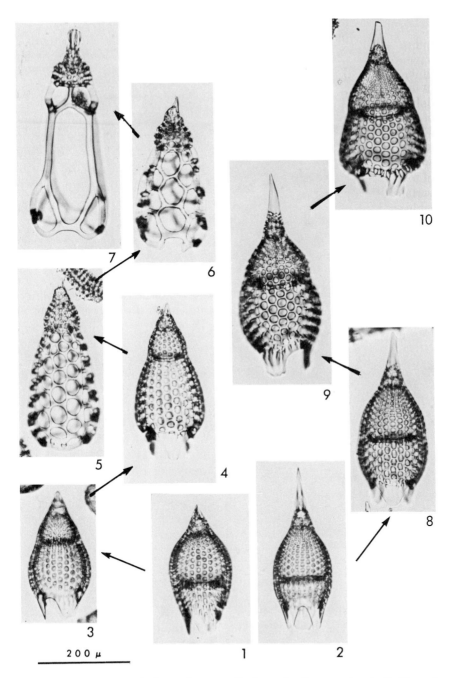

Fig. 4-7. Evolution of the subgenera *Podocyrtis* (*Lampterium*) (3–7) and *Podocyrtis* (*Podocyrtis*) (1,2,8–10). (Riedel and Sanfilippo, 1981)

tichromosomal nucleus of *Aulacantha scolymantha* (Grell, 1973), we have no knowledge of radiolarian cytogenetics or Mendelian genetics. Since we have not been able to observe sexual reproduction thus far in any species of radiolaria, it is not possible to pursue classical Mendelian-type studies, where the phenotypes of offspring, from parents of known phenotypes, are used to infer genotypes. An alternative strategy could be to examine the frequency of phenotypes in natural populations (basic inferential population genetics), with the aim of making some conclusions about genetic diversity in the populations. This kind of research, however, presumes that we have knowledge of the natural variability in populations that can be explained by environmental variables. This kind of information is essential in order to determine the extent of environmental impact on phenotypical characteristics, and thus to assess the extent to which environmental effects may confound inferences about genotypic variability. It is well documented in developmental biology and population genetics that the environment unavoidably influences morphogenesis. The radiolarian morphotype, for example, is the resultant of the interaction of basic genetic information and the impress of environmental variables on the emerging form of the radiolarian during ontogeny. All living things develop in an environment that influences the ultimate expression of genetic information as it is translated into phenotypic form and behavior. Given carefully defined, constant, and optimal environmental conditions for growth and maturation, fairly reliable Mendelian-type studies or even inferential population genetic studies can be pursued with confidence. If, however, the environment varies in a significant, and perhaps unknown, way that alters the phenotype, it is very difficult to draw inferences about the fundamental genotype or even about basic variability in gentoypes within the populations. We clearly need to know much more about how typical variations in natural environmental variables influence radiolarian morphogenesis or otherwise influence form and behavior after maturation. This is particularly critical in relation to skeletal morphology, as much of our knowledge about radiolarian evolution is based on fossil skeletal evidence. The information presented in the fossil skeleton may not represent predominately stable genetic characteristics; however, to an unknown extent it may also represent perturbations induced by transitory environmental influences. Given the uncertainty in the range of biological responses by different species of radiolaria to environmental variables, it is unwise to generalize findings from one group to another. For example, the patterns or rate of evolution inferred for one group should not be generalized uncritically to a different group, unless there is clear evidence of similarities in their response patterns to environmental variables. This evidence is not easily inferred from the fossil record. Concurrent biological research, however, with closely related extant species, may help to elucidate the extent of these similarities in response variations.

Radiolaria are capable of making dramatic changes in skeletal morphology in response to environmental variables. I have observed spine-bearing radiolaria in laboratory culture resorb their spines by dissolution of the silica near the base and at various points along the spine, thereby fracturing it into small pieces which were shed. All of this occurred in a period of minutes. Broken spines in a healthy radiolarian (e.g., a Cubosphaerid) are regenerated within a period of hours, and are similar in morphology to the original spine, but usually are more slender (Anderson, 1981). There is additional fine structure evidence (Cachon and Cachon, 1971b) that radiolaria may be able to metabolize silica. Partially degraded biogenic particles of silica have been observed in digestive vacuoles of radiolaria. This suggests that radiolaria may be able to obtain silicate sources by ingestion and degradation of diatoms and other siliceous phytoplankton in addition to resorption and possibly recycling of their skeletal silicon. Little is known about the effect of variations in silicate nutrient concentrations on skeletal morphogenesis or alterations in morphology of the skeleton. For example, it is of interest to know if radiolarian skeletons are thinner or otherwise altered to make them less bulky when grown in a silicate-impoverished environment. These changes may include alterations in pore pattern or number, or arrangement and number of spines and other ornamentation on the skeleton. Current research on these kinds of questions has just begun in our laboratory. It is equally possible that quality or quantity of available light may influence symbiont primary productivity and indirectly influence skeletal morphology. Likewise, variations in nutrition during ontogeny may have significant skeletal morphological effects. Given this uncertainty in our knowledge, both about the occurrence of sexual reproduction in radiolaria and about environmental influences on their morphology, great care should be exercised in making inferences about phylogenetic relationships including hybridization and genetic variability in past populations of radiolaria based on fossil evidence. Although the sum total of current micropaleontological knowledge about variation in morphotypes within strata and across geological time spans strongly points toward sexual reproduction in radiolaria, definite biological evidence for syngamy and zygote formation by reproductive swarmers has not been reported.

Much additional research is needed on the range of variability in form and behavior of radiolaria during ontogeny to document better natural variations within species. This research may effectively be pursued with samples collected from plankton in the natural environment and laboratory studies of specimens maintained in culture under varying environmental conditions. The results of this research should be particularly valuable as an aid to micropaleontologists in their search for biological explanations for fossil events observed in the biostratigraphic record. As stated earlier, an earnest effort should be made to examine the organic content of living and fossil radiolarian skeletons to determine if isotopic fractionation is characteristic of symbiont-bearing species, as an aid to determining the presence or absence of symbionts in fossil forms. Clearly, the presence of symbionts may be a very significant factor in explaining phylogenetic changes and the possible success of new species.

The opportunity for a rich and productive interdisciplinary effort may now emerge, as the scientific basis for the study of living radiolaria begins to match the substantial repertoire of theory and research already developed by micropaleontologists. The union of efforts by biologists, micropaleontologists, and researchers in cognate disciplines promises to yield much useful information about the evolution, ecology, basic physiology, and biological variability in form and behavior of radiolaria throughout their long history. I trust that in some measure this book may help to encourage such an interdisciplinary effort among scholars who have much to gain intellectually and much to contribute by their cooperative endeavors to understand a most remarkable group of aesthetically pleasing mineral-secreting protists—the radiolaria.

References

Adshead, P. C. 1980. Pseudopodial variability and behavior of Globigerinids (Foraminiferida) and other planktonic Sarcodina developing in cultures. In W. V. Sliter (ed.), *Studies in Marine Micropaleontology and Paleoecology: A Memorial Volume to Orville L. Bandy*, Cushman Foundation for Foraminiferal Research, Special Publication No. 19, pp. 96–126.

Anderson, O. R. 1976a. A cytoplasmic fine-structure study of two spumellarian Radiolaria and their symbionts. *Mar. Micropaleontol.* 1:81–89.

Anderson, O. R. 1976b. Ultrastructure of a colonial radiolarian *Collozoum inerme* (Müller) and a cytochemical determination of the role of its zooxanthellae. *Tissue Cell* 8:195–208.

Anderson, O. R. 1976c. Fine structure of a collodarian radiolarian (*Sphaerozoum punctatum* Müller 1858) and cytoplasmic changes during reproduction. *Mar. Micropaleontol.* 1:287–297.

Anderson, O. R. 1977a. Cytoplasmic fine structure of nassellarian Radiolaria. *Mar. Micropaleontol.* 2:251–264.

Anderson, O. R. 1977b. Fine structure of a marine ameba associated with a blue-green alga in the Sargasso Sea. *J. Protozool.* 24:370–376.

Anderson, O. R. 1978a. Fine structure of a symbiont-bearing colonial radiolarian, *Collosphaera globularis*, and ^{14}C isotopic evidence for assimilation of organic substances from its zooxanthellae. *J. Ultrastruct. Res.* 62:181–189.

Anderson, O. R. 1978b. Light and electron microscopic observations of feeding behavior, nutrition, and reproduction in laboratory cultures of *Thalassicolla nucleata*. *Tissue Cell* 10:401–412.

Anderson, O. R. 1980. Radiolaria. In M. Levandowsky and S. Hutner (eds.), *Biochemistry and Physiology of Protozoa*. Second Edition, Vol. 3, pp. 1–40. Academic Press, New York.

Anderson, O. R. 1981. Radiolarian fine structure and silica deposition. In T. L. Simpson and B. E. Volcani (eds.), *Silicon and Siliceous Structures in Biological Systems*, pp. 347–380. Springer-Verlag, New York.

Anderson, O. R. 1982. Radiolaria/algae. In L. Goff (ed.), *Algal Symbiosis: A Continuum of Interaction Strategies*, pp. 129–160. Cambridge University Press, Cambridge.

Anderson, O. R. 1983. Cellular specialization and reproduction in planktonic foraminifera and radiolaria. In K. Steidinger and L. Walker (eds.), *Significance of Plankton Life Cycles in Population Survival and Dispersal*. Chemical Rubber Company Press, Cleveland, OH, in press.

Anderson, O. R., and A. W. H. Bé. 1976. The ultrastructure of a planktonic foraminifer *Globigerinoides sacculifer* (Brady) and its symbiotic dinoflagellates. *J. Foraminiferal Res.* 6:1–21.

Anderson, O. R., and A. W. H. Bé. 1979. Recent advances in Foraminiferal fine structure research. In R. H. Hedley and C. G. Adams (eds.), *Foraminifera*, pp. 121–202. Academic Press, London.

Anderson, O. R., and M. Botfield. 1983. Biochemical and fine structure evidence for cellular specialization in a large spumellarian radiolarian, *Thalassicolla nucleata*. *Mar. Biol.* 72:235–241.

Anderson, O. R., and W. K. Hoeffler. 1979. Fine structure of a marine proteomyxid and cytochemical changes during encystment. *J. Ultrastruct. Res.* 66:276–287.

Anderson, O. R., M. Spindler, A. W. H. Bé, and Ch. Hemleben. 1979. Trophic activity of planktonic foraminifera. *J. Mar. Biol. Assoc.* (U. K.) 59:791–799.

Anderson, O. R., and N. R. Swanberg. 1981. Skeletal morphogenesis in some living collosphaerid radiolaria. *Mar. Micropaleontol.* 6:385–396.

Anderson, O. R., and N. R. Swanberg. 1982. Selective trophic activity of some solitary radiolaria. In preparation.

Banner, F. T., W. J. Clarke, J. L. Cutbill, F. E. Eames, A. J. Lloyd, W. R. Riedel, and A. H. Smout. 1967. Protozoa. In W. B. Harland, C. H. Holland, M. R. House, N. F. Hughes, A. B. Reynolds, M. J. S. Rudwick, G. E. Satterthwaite, I. B. H. Tarlo, and E. C. Wiley (eds.), *The Fossil Record*, pp. 291–332. Geological Society of London, London.

Bardele, C. 1976. Particle movement in Heliozoan axopods associated with lateral displacement of highly ordered membrane domains. *Z. Naturforsch.* 31:189–194.

Bé, A. W. H., J. E. Damuth, L. Lott, and R. Free. 1976. Late Quaternary climatic record in Western equatorial Atlantic sediment. In R. M. Cline and J. D. Hays (eds.). *Investigations of Late Quaternary Paleoceanography and Paleoclimatology*, Memoir 145, pp. 165–200. Geological Society of America, Boulder, CO.

Bé, A. W. H., Ch. Hemleben, O. R. Anderson, M. Spindler, J. Hacunda, and S. Tuntivate-Choy. 1977. Laboratory and field observations of iving planktonic foraminifera. *Micropaleontology (N.Y.)* 25:294–307.

Bé, A. W. H., Ch. Hemleben, O. R. Anderson, M. Spindler. 1980. Pore structures in planktonic foraminifera. *J. Foraminiferal Res.* 10:117–128.

Beers, J. R., and G. L. Stewart. 1969a. The vertical distribution of micro-zooplankton and some ecological observations. *J. Cons. Int. Explor. Mer.* 33:30–44.

Beers, J. R., and G. L. Stewart. 1967. Numerical abundance and estimated biomass of microzooplankton. In J. D. H. Strickland (ed.), *The Ecology of Plankton off La Jolla California in the Period April through September 1967. Scripps Inst. Oceanogr. Bull.* 17:67–87.

Beers, J. R., and G. L. Stewart. 1969b. Micro-zooplankton and its abundance relative to the larger zooplankton and other seston components. *Mar. Biol.* 4:182–189.

Beers, J. R., and G. L. Stewart. 1971. Microzooplankters in the plankton communities of the upper waters of the eastern tropical Pacific. *Deep-Sea Res.* 18:861–883.

Belar, K. 1926. Der Formwechsel der Protistenkerne. *Ergebn. Zool. Jena* Vol. 6.

Berger, W. H. 1968. Radiolarian skeletons: Solution at depth. *Science* 159:1237–1238.

Berger, W. H. and P. H. Roth. 1975. Oceanic micropaleontology: progress and prospect. *Rev. Geophys. Space Phys.* 13:561–635.

Berggren, W. A. 1969. Rates of evolution in some cenozoic planktonic Foraminifera. *Micropaleontology (N.Y.)* 15:351–365.

Berggren, W. A. and C. D. Hollister. 1974. Currents of time. *Oceanus* 17:28–33.

Bjørklund, K. 1974. The seasonal occurrence and depth zonation of radiolarians in Korsfjorden, Western Norway. *Sarsia* 56:13–42.

Bjørklund, R., and R. M. Goll. 1979. Internal skeletal structures of *Collosphaera* and *Trisolenia*: A case of repetitive evolution in the Collosphaeridae. *J. Paleontol.* 53:1293–1326.

Blank, R. G., and S. V. Margolis. 1975. Pliocene climatic and glacial history of Antarctica as revealed by Southeast Indian Ocean deep sea cores. *Geol. Soc. Am. Bull.* 86:1058–1066.

Blueford, J. R. 1978. Radiolaria from the Navarin Basin province, Bering Sea. In P. R. Carlson, et al. (eds.), *Seafloor Geologic Hazards, Sedimentology, and Bathymetry; Navarin Basin Province, Northwest Bering Sea*, pp. 130–137. U. S. Geology Survey Open-File Service Section, Western Distribution Branch, Denver, CO.

Bolli, H. M., A. R. Loeblich, Jr., and H. Tappan. 1957. Planktonic foraminiferal families Hantkeninidae, Orbulinidae, Globorotaliidae, and Globotruncanidae. *Bull. U. S. Nat. Mus.* 215:1–50.

Boltovskoy, D. 1975. Ecological aspects of zooplankton (Foraminifera, Pteropoda, and Chaetognatha) of the southwestern Atlantic Ocean. *Veliger* 18:203–216.

Boltovskoy, D. 1979. Zooplankton of the southwestern Atlantic. *S. Afr. J. Sci.* 75:541–544.

Boltovskoy, D. 1981. Radiolaria. In D. Boltovskoy (ed.), *Atlas of the Zooplankton of the Southwestern Atlantic Ocean and Methods in Marine Zooplankton Research*, pp. 261–316. Instituto Nacional de Investigacion y Desarrollo Pesquero Mar del Plata, Argentina. [In Spanish]

Boltovskoy, D. 1970. Masas de agua (caracteristica, distribucion, movimentos) en la superficie del Atlantico Sudoeste, segun indicadores biologicos-Foraminiferos. *Serv. Hidrogr. Naval (Argentina)* Publication No. H-643. pp. 1–99.

Boltovskoy, D., and W. R. Riedel. 1979. Polycystine radiolaria from the Southwestern Atlantic Ocean plankton. *Rev. Esp. Micropaleontol.* 12:99–146.

Boltovskoy, E. 1970. Masas de agua (caracteristica, distribucion, movimentos) en la superficie del Atlantico Sudoeste, segun indicadores biologicos-Foraminiferos. *Serv. Hidrogr. Naval (Argentina)* Publication No. H-643. pp. 1–99.

Borgert, A. 1909. Untersuchungen über die Fortpflanzung der Tripyleen Radiolarien, speziell von *Aulacantha scolymantha* H. Th. II. *Arch. Protistenkd.* 14:134.

Bramlette, M.N. 1965. Massive extinctions in biota at the end of Mesozoic time. *Science* 148:1696–1699.

Brandt, K. 1881. Untersuchungen an Radiolarien. *Montasber. Kgl. Preuss. Akas. Wiss. Berlin Jahrg.* 1881:388–404.

Brandt, K. 1882. Ueber die morphologische und physiologische des Chlorophylles bei Thieren. *Mittheil. Zool. Stat. Neapel.* 4:191–302.

Brandt, K. 1885. Die koloniebildenden Radiolarien (Spherozoëen) des golfes von Neapel und der angrenzenden Meeresabschnitte. *Monogr. Fauna Flora Golfes Neapel.* XIII:1–276.

Brandt, K. 1890. On the division of *Thalassicolla Mitt. Ver. Schlesw.-Holstein. Aerzte* 5:12.

Brandt, K. 1902. Beitrage zur Kenntnis der Colliden. *Arch. Protistenkd.* 1:59–88.

Brandt, K. 1905. Zur Systematik der koloniebildenden Radiolarien. *Zool. Jahrb.* (Suppl.) 8:311–352.

Breckner, A. 1906. *Beiträge zur Kenntnis der koloniebildenden Radiolarien mit Nadel (Sphaerozoida).* Inaugural dissertation. Doktorwürde hohen philosph. Fak. Christian-Albrechts Univ., Kiel.

Bretsky, S. S. 1979. Recognition of ancestor-descendent relationships in invertebrate paleontology. In J. Cracraft and N. Eldredge (eds.), *Phylogenetic Analysis and Paleontology*, pp. 113–163. Columbia University Press, New York.

Broecker, W. S. 1971a. Calcite accumulation rates and glacial to interglacial changes in oceanic mixing. In K. K. Turekian (ed.), *The Late Cenozoic Glacial Ages*, pp. 239–265. Yale University Press, New Haven, CT.

Broecker, W. S. 1971b. A kinetic model for the chemical composition of sea water. *Quat. Res. (N. Y.)* 1:188–207.

Broecker, W. S. 1974. *Chemical Oceanography*, p. 214. Harcourt, Brace, Jovanovich, New York.

Brown, W. L., and E. O. Wilson. 1956. Character displacement. *Syst. Zool.* 5:49–64.

Cachon, J., and M. Cachon. 1965. L'infrastructure des axopodes chez les Radiolaires Sphaerellaires Periaxoplastidies. *C. R. Acad. Sci. (Paris)* 261:1388–1391.

Cachon, J., and M. Cachon. 1969. Revision Systématique des Nassellaires *Plectoidea* à propos de la description d'un nouveau represéntant, *Plectagonidium deflandrei* nov. gen. nov. sp. *Arch. Protistenkd.* 111:236–251.

Cachon, J., and M. Cachon. 1971a. Le système axopodial des Radiolaires Nassellaires. *Arch. Protistenkd.* 113:80–97.

Cachon, J., and M. Cachon. 1971b. Recherches sur le métabolisme de la silice chez les Radiolaires. Absorption et excretion. *C. R. Acad. Sci. (Paris)* 272:1652–1654.

Cachon, J., and M. Cachon. 1972a. Le système axopodial des Radiolaires Sphaerodiés I. Centroaxoplastidiés. *Arch. Protistenkd.* 114:51–64.

Cachon, J., and M. Cachon. 1972b. Le système axopodial des Radiolaries Sphaeroidés II. Les periaxoplastidiés III. Les cryptoaxoplastidies (anaxoplastidiés) IV. Les fusules et le système rhéoplasmique. *Arch. Protistenkd.* 114: 291–307.

Cachon, J., and M. Cachon. 1972c. Les modalitiés du dæpôt de la Silice chez les Radiolares. *Arch. Protistenkd.* 114:1–13.

Cachon, J., and M. Cachon. 1973a. *Microtubular Systems of Actinopod Axopods.* British Cell Biology Society, York, September 1973 Conference Paper. See also: Cachon et al., 1973.

Cachon, J., and M. Cachon. 1973b. Systèmes microtubulaires de l'astropyle et des parapyles de Phaeodariés. *Arch. Protistenkd.* 115:324–335.

Cachon, J., and M. Cachon. 1974. Les Systémes Axopodiaux. *Ann. Biol.* 13 (11–12):523–560.

Cachon, J., and M. Cachon. 1976. Le Système axopodial des Collodaires (Radiolaires polycystines). 1. Les Exoaxoplastidies. *Arch. Protistenkd.* 118:227–234.

Cachon, J., and M. Cachon. 1977a. Cellular transfer of membranes and its relation to the microtubular system. *Biol. Cell.* 30:137–140.

Cachon, J., and M. Cachon. 1977b. Infrastructural constitution of microtubules of the axopodial system of Radiolaria. *Bol. Soc. Bot. Mex.* 36:229–231.

Cachon, J., and M. Cachon. 1978. *Sticholonche zanclea* Hertwig: A reinterpretation of its phylogenetic position based upon new observations on its ultrastructure. *Arch. Protistenkd.* 120:148–168.

Cachon, J., M. Cachon, C. Febvre-Chevalier, and J. Febvre. 1973. Determinisme de l'edification des systèmes microtubulaires steréoplasmiques d'actinopodes. *Arch. Protistenkd.* 115:137–153.

Cachon, J., M. Cachon, L. Tilney, and M. Tilney. 1977. Movement generated by interactions between the dense material at the ends of microtubules and non-actin-containing microfilaments in *Sticholonche zanclea*. *J. Cell Biol.* 72:314–338.

Cachon, M., and B. Caram. 1979. A symbiotic green alga, *Pedimonas symbiotica* sp. nov. (Prasinophyceae), in the radiolarian *Thalassolampe margarodea*. *Phycologea* 18:177–184.

Campbell, A. S. 1954. Radiolaria. In R. C. Moore (ed.), *Treatise on Invertebrate Paleontology*, Part D, *Protista 3*, pp. D11–D163. Geological Society of America/University of Kansas Press, Lawrence, KS.

Campbell, A. S., and E. A. Holm. 1957. Radiolaria. *Geol. Soc. Am. Mem.* 67:737–743.

Carley, W. M. 1969. Silicon and the division cycle of the diatoms *Navicula pelliculosa* and *Cylindrotheca fusiformis*. *N. Am. Paleont. Conv. Proc.* 994–1009.

Casey, R. 1966. *A Seasonal Study on the Distribution of Polycystine Radiolarians from Waters Overlying the Catalina Basin, Southern California*. Ph.D. thesis, University of Southern California.

Casey, R. E. 1971a. Distribution of polycystine radiolaria in the oceans in relation to physical and chemical conditions. In B. M. Funnell and W. R. Riedel (eds.), *The Micropaleontology of Oceans*, pp. 151–159. Cambridge University Press, Cambridge.

Casey, R. E. 1971b. Radiolarians as indicators of past and present water-masses. In B. M. Funnell and W. R. Riedel (eds.), *The Micropaleontology of Oceans*, pp. 331–349. Cambridge University Press, Cambridge.

Casey, R. E. 1977. The ecology and distribution of recent radiolaria. In A. T. S. Ramsay (ed.), *Oceanic Micropaleontology*, Vol. 2, pp. 809–845. Academic Press, London.

Casey, R., L. Gust, A. Leavesley, D. Williams, R. Reynolds, T. Duis, and J. M. Spaw. 1979a. Ecological Niches of radiolarians, planktonic foraminiferans and pteropods inferred from studies on living forms in the Gulf of Mexico and Adjacent Waters. *Trans.-Gulf Coast Assoc. Geol. Soc.* 29:216–223.

Casey, R. E., and K. J. McMillen. 1977. Cenozoic radiolarians of the Atlantic Basin and margins. In F. M. Swaim (ed.), *Stratigraphic Micropaleontology of Atlantic Basin and Borderlands*, pp. 226–238. Elsevier, Amsterdam.

Casey, R. E., A. Leavesley, J. M. Spaw, K. McMillen, and J. Sloan. 1981. Radiolarian species composition, density and diversity as indicators of water structure and circulation on the south Texas shelf. *Trans.-Gulf Coast Assoc. Geol. Soc.* 31:257–263.

Casey, R., K. McMillen, R. Reynolds, J. M. Spaw, R. Schwarzer, H. Gevirtz, and M. Bauer. 1979b. Relict and expatriated radiolarian fauna in the Gulf of Mexico and its implications. *Trans.-Gulf Coast Assoc. Geol. Soc.* 29:224–227.

Casey, R. E., T. M. Partridge, and J. R. Sloan. 1970. Radiolarian life spans mortality rates, and seasonality gained from recent sediment and plankton samples. *Proceedings of the II Planktonic Conference Roma 1970*, Vol. I, pp. 159–165. Edizioni Tecnoscienza, Rome.

Casey, R. E., J. M. Spaw, and F. R. Kunze. 1982a. Polycystine radiolarian distributions and enhancements related to oceanographic conditions in a hypothetical ocean. *Trans.-Gulf Coast Assoc. Geol. Soc. 32*, in press.

Casey, R. E., J. M. Spaw, and F. R. Kunze. 1982b. Polycystine radiolarian distributions and enhancements related to oceanographic conditions in a hypothetical ocean. *Trans.-Gulf Coast Assoc. Geol. Soc. 32*, in press.

Casey, R. E., T. M. Partridge, and J. R. Sloan. 1970. Radiolarian life spans mortality rates, and seasonality gained from recent sediment and plankton samples. In *Proceedings of the II Planktonic Conference Roma 1970*, Vol. I, pp. 159–165. Edizioni Tecnoscienza, Rome.

Chatton, E. 1920. Les Peridiniens parasites. Morphologie, reproduction, éthologie. *Arch. Zool. Exp. Géné.* 59:1—475.

Chatton, E. 1934. L'origine péridinienne des Radiolaires et l'interpretation parasitaire de l'anisosporogénèse. *C.R. Acad. Sci. (Paris)* 198:309.

Cheissin, E. M. 1965. The significance of ultrastructure in the taxonomy of protozoa. In *Progress in Protozoology: II. International Conference on Protozoology, London, 1965*, Excerpta Medica International Congress Series No. 91, p. 22, Amsterdam.

Chen, P. 1975. Antarctic radiolaria. In D. E. Hayes, L. A. Frakes, *et al.* (eds.), *Initial Reports of the Deep Sea Drilling Project* 28:437–513. U. S. Government Printing Office, Washington, D. C.

Cienkowski, L. 1871. Ueber Schwärmer-bildung bei Radiolarien. *Arch. Mikrosk. Anat.* 7:372–381.

Cifelli, R. 1969. Radiation of Cenozoic planktonic Foraminifera. *Syst. Zool.* 18:154–168.

Cifelli, R., and K. N. Sachs, Jr. 1966. Abundance relationships of planktonic Foraminifera and Radiolaria. *Deep-Sea Res.* 13:751–753.

Cleve, P. T. 1899. Plankton collected by the Swedish Expedition to Spitzbergen in 1898. *K. Svenska. Vetens.-Akad. Handl.* 32:1–51.

Cleve, P. T. 1900. Notes on some Atlantic planktonic-organisms. *K. Svenska Vetensk.-Akad. Handl.* 34:1–22.

Cleve, P. T. 1901. Plankton from the Indian Ocean and the Malay Archipelago. *Handl. Kgl. Svenska Vetensk.-Akad.* 35:1–58.

Cline, R. M. and J. D. Hays (eds.). 1976. *Investigations of Late Quaternary Paleoceanography and Palaeoclimatology*, Memoir 145. Geological Society of America, Boulder, CO.

Corliss, J. O. 1979. Electron microscopy and ciliate systematics. *Am. Zool.* 19:573–587.

Cracraft, J. 1974. Phylogenetic models and classification. *Syst. Zool. 23*:71–90.

Cummings, C. E. and H. B. McCarty. 1982. Stable carbon isotope in *Astrangia danae*: Evidence for algal modification of carbon pools used in calcification. *Geochim. Cosmochim. Acta* 46:1125–1129.

Cutbill, J. L. and B. M. Funnell. 1967. Numerical analysis of the fossil record. In W. B. Harland, C. H. Holland, M. R. House, N. F. Hughes, A. B. Reynolds, M. J. S. Rudwick, G. E. Satterthwaite, I. B. H. Tarlo, and E. C. Wiley (eds.), *The Fossil Record*, pp. 791–820. Geological Society of London, London.

Deflandre, G. 1952. *Albaillella* nov. gen. Radiolaire fossile de Carbonifere inferieur, type d'une lignée aberrante éteinte. *C. R. Acad. Sci.* 234:872–874.

Deflandre, G. 1953. Radiolaries fossiles. In P. P. Grassé (ed.), *Traité de Zoologie*, Vol. 1, fasc. 2, pp. 389–433. Masson, Paris.

Desrosieres, R. 1969. Surface macroplankton of the Pacific Ocean along the equator. *Limnol. Oceanogr.* 14:626–632.

DeWever, P., A. Sanfilippo, W. R. Riedel, and B. Gruber. 1979. Triassic radiolarians from Greece, Sicily and Turkey. *Micropaleontology (N.Y.)* 25:75–110.

Dodge, J. D. 1973. *The Fine Structure of Algal Cells*, pp. 8–11, 143–146, and 209–213. Academic Press, London.

Douglas, R., and W. V. Sliter. 1966. Regional distribution of some Cretaceous Rotaliporidae and Globotruncanidae (Foraminiferida) within North America. *Tulane Stud. Geol.* 4:89–130.

Dow, R. L. 1978. Radiolarian distribution and late Pleistocene history of the Indian Ocean. *Mar. Micropaleontol.* 3:203–207.

Downie, C. 1967. The geological history of the microplankton. *Rev. Palaeobot. Palynol.* 1:269–281.

Dumitrica, P. 1970. Cryptocephalic and cryptothoracic Nassellaria in some Mesozoic deposits of Romania. *Rev. Roum. Geol. Geophys. Geogr. Geol. Ser.* 14:45–124.

Ehrenberg, C. G. 1847a. Über eine halibiolithische, von Herrn R. Schomburgk entdeckte, vorherrschend aus mikroskopischen Polycystinen gebildete, Gebirgsmasse von Barbados. *Monatsber. Preuss. Akad. Wiss. Jahrg.* 1846:382–385.

Ehrenberg, C. G. (1847b). Über die mikroskopischen kieselschaligen Polycystinen als mächtige Gebirgsmasse von Barbados und über das Verhältnis der aus mehr als 300 neuen Arten bestehenden ganz eigenthümlichen Formengruppe jener Felsmasse zu den lebenden Thieren und zur Kreidebildung. Eine neue Anregung zur Enforschung des Erdlebens. *Monatsber. Preuss. Akad. Wiss. Jahrg.* 1847:40–60.

Ehrenberg, C. G. 1872. Mikrogeologische Studien als zussamenfassung seiner Beobachtungen des kleinsten Lebens der Meeres—Tiefgründe aller Zonen und dessen geologischen Einfluss. *Monatsber. Preuss. Akad. Wiss. Jahrg.* 1872:265–322.

Ehrenberg, C. G. 1873. Grössere Felsproben des Polycystinen-Mergels von Barbados mit weiteren Erläuterungen. *Monatsber. Preuss. Akad. Wiss. Jahrg.* 1873:213–263.

Ehrenberg, C. G. 1875. Fortsetzung der mikrogeologischen Studien als Gesammt-Uebersicht der mikroskopischen Paläontologie gleichartig analysierter Gebirgsarten der Erde, mit specieller Rücksicht auf den Polycystinen-Mergel von Barbados. *Abh. Preuss. Akad. Wiss. Jahrg.* 1875:1–226.

Emery, K. O., and S. Honjo. 1979. Surface suspended matter off Western Africa: Relations of organic matter, skeletal debris, and detrital minerals. *Sedimentology* 26:775–794.

Engleman, G. F. and E. O. Wiley. 1977. The place of ancestor-descendent relationships in phylogeny reconstruction. *Syst. Zool.* 26:1–11.

Enriques, P. 1919. Richerche sui Radiolari. *Comit. Talassogr. Ital. Mem.* 71:1–55.

Enriques, P. 1931. Formazione e sviluppo dello scheletro siliceo nei Radiolari. *Boll. Sci. Ital. Biologia Sperimentale* 6:1–7.

Erez, J. 1978. Vital effect of stable-isotope composition seen in foraminifera and coral skeletons. *Nature (London)* 273:199–202.

Ericson, D. B., M. Ewing, and G. Wollin. 1964. The Pleistocene epoch in deep-sea sediments. *Science* 146:723–732.

Ericson, D. B., and G. Wollin. 1968. Pleistocene climates and chronology in deep-sea sediments. *Science* 162:1127–1234.

Esaias, W. E., and H. C. Curl. 1973. Effect of dinoflagellate bioluminescence on copepod ingestion rates. *Limnol. Oceanogr.* 17:901–906.

Fager, E. W. 1957. Determination and analysis of recurrent groups. *Ecology* 38:586–595.

Fairbanks, R. G., and R. E. Dodge. 1979. Annual periodicity of the $^{18}O/^{16}O$ and $^{13}C/^{12}C$ ratios in the coral Montastrea annularis. Geochim. Cosmochim. Acta 43:1009–1020.

Febvre, J. 1971. Le myoneme d'Acantaire: Essai d'interpretation ultrastructurale et cinetique. Protistologica 7:379–381.

Febvre, J. 1972. Le cortex ectoplasmique des acanthaires I. Les systèmes maillés. Protistologica 8:169–178.

Febvre, J. 1973. Le cortex des acanthaires II. Ultrastructure des zones de jonction entre les pièces corticales. Protistologica 9:87–94.

Febvre, J. 1974. Relations morphologiques entre les constituants de l'enveloppe, les myonèmes, le squelette et le plasmalemme chez les Arthracantha Schew [Acantharia]. Protistologica 10:141–158.

Febvre, J., and C. Febvre-Chevalier. 1979. Ultrastructural study of zooxanthellae of three species of acantharia (protozoa: actinopoda), with details of their taxonomic position in the Prymnesiales (Prymnesiophyceae, Hibberd, 1976). J. Mar. Biol. Assoc. (U. K.) 59:215–226.

Fillon, R. H. 1973. Radiolarian evidence of later Cenozoic ocean paleotemperatures: Ross Sea Antarctica. Palaeogr. Palaeoclimatol. Palaeoecol. 14:171–185.

Fol, H. 1883. Sur le Sticholonche zanclea et un novel ordre des Rhizopodes. Mem. Inst. Nat. Geneve 15:1–35.

Foreman, H. P. 1973. Radiolaria of Leg X, with systematics and ranges for the families Amphipyndacidae, Artostrobiidae and theoperidae. In J. L. Worzel, W. Bryant, et al. (eds.), Initial Rep. Deep Sea Drilling Project 10. U.S. Government Printing Office, Washington, D. C.

Foreman, H. P. and W. R. Riedel. 1972. Catalogue of Polycystine Radiolaria. Vols. 1–2. Special Publication Series 1 (1834–1900), American Museum of Natural History, New York.

Franke, W. W., J. Kartenbeck, and H. Spring. 1976a. Involvement of bristle coat structures in surface formations and membrane interactions during coenocytotomic cleavage in caps of Acetabularia mediterranea. J. Cell Biol. 71:196–206.

Franke, W. W., M. Luder, J. Kartenbeck, H. Zerban, and T. Keenan. 1976b. Involvement of vesicle coat material in casein secretion and surface regeneration. J. Cell Biol. 69:173–195.

Frerichs, W. E. 1970. Paleobathymetry, paleotemperature, and tectonism. Geol. Soc. Am. Bull. 81:3445–3452.

Gamble, F.W. 1909. The Radiolaria. In E. R. Lankester (ed.), A Treatise on Zoology, pp. 94–153. Adam and Charles Black, London.

Gardner, J. V., and J. D. Hays. 1976. Response of sea-surface temperature and circulation to global climatic change during the past 200,000 years in the eastern equatorial Atlantic. In R. M. Cline and J. D. Hays (eds.). Investigations of Late Quaternary Paleoceanography and Paleoclimatology, Memoir 145, pp. 221–246. Geological Society of America, Boulder, CO.

Garrone, R., T. L. Simpson, and J. Pottu-Boumendil. 1981. Ultrastructure and deposition of silica in sponges. In T. L. Simpson and B. E. Volcani (eds.), Silicon and Siliceous Structures in Biological Systems, pp. 495–526. Springer-Verlag, New York.

Geddes, P. 1882. On the nature and functions of the "Yellow Cells" of radiolarians and coelenterates. Proc. Roy. Soc. Edin. 11:377–397.

Geptner, V. G. 1936. Oyashchaya Zoogeografiya. Bimedgiz, Leningrad.

Goll, R. M. 1968. Classification and phylogeny of Cenozoic Trissocyclidae (Radiolaria) in the Pacific and Caribbean Basins, Part I. J. Paleontol. 42:1409–1432.

Goll, R. M. 1969. Classification and phylogeny of Cenozoic Trissocyclidae (Radiolaria) in the Pacific and Caribbean Basins, Part II. *J. Paleontol.* 43:322–339.

Goll, R. M. 1972a. Systematics of eight *Tholospyris* taxa (Trissocyclidae: Radiolaria). *Micropaleontology (N. Y.)* 18:443–475.

Goll, R.M. 1972b. Leg 9 synthesis, Radiolaria. In J. D. Hays, *et al.* (eds.), *Initial Reports of the Deep Sea Drilling Project* 9:947–1058. U.S. Government Printing Office, Washington, D. C.

Goll, R. M. 1979. The Neogene evolution of *Zygocircus*, *Neosemantis* and *Callimitra*: Their bearing on nassellarian classification. A revision of the Plagiacanthoidea. *Micropaleontology (N. Y.)* 25:365–396.

Goll, R. M., and K. R. Bjørklund. 1971. Radiolaria in surface sediments of the North Atlantic Ocean. *Micropaleontology (N. Y.)* 17:434–454.

Goll, R. M., and K. R. Bjørklund. 1974. Radiolaria in surface sediments of the South Atlantic. *Micropaleontology (N. Y.)* 20:38–75.

Goll, R. M., and G. Merinfeld. 1979. Radiolaria. In R. W. Fairbridge and D. Jablonsky (eds.), *The Encyclopedia of Earth Sciences*, Vol. VII, pp. 673–684. Dowden, Hutchison and Ross, Stroudsburg, PA.

Goll, R. M., and K. R. Bjørklund. 1980. The evolution of *Eucornis fridtjofnanseni* and its application to the Neogene biostratigraphy of the Norwegian-Greenland Sea. *Micropaleontology (N. Y.)* 26:356–371.

Gordon, A. L. 1971. Circulation of Antarctic bottom waters south of Australia and New Zealand. In R. J. Adie (ed.), *Antarctic Geology and Geophysics*, p. 773. Oslo, Universitetsforlaget.

Gould, S. J., and N. Eldredge. 1977. Punctuated equilibria: The tempo and mode of evolution reconsidered. *Paleobiology* 3:115–151.

Gran, H. H. 1902. Das Plankton des norwegischen Nordmeeres von biologischen und hydrographischen Geischspunkt behandelt. *Rep. Norw. Fishery Mar. Invest.* 2:1–222.

Grell, K. G. 1973. *Protozoology*. Springer-Verlag Berlin, Heidelberg.

Grimstone, A. V. 1959. Cytology, homology, and phylogeny—A note on "organic design." *Am. Nat.* 93:273–282.

Haeckel, E. 1860a. Über neue lebende Radiolarien des Mittelmeeres. *Monatsber. Kgl. Preuss. Akad. Wiss. Berlin Jahrg.* 1860:794–834.

Haeckel, E. 1860b. Abbildungen und diagnosen neuer Gattungen und Arten von lebenden Radiolarien des Mittelmeeres. *Monatsber. Kgl. Preuss. Akad. Wiss. Berlin Jahrg.* 1860:835–845.

Haeckel, E. 1862. *Die Radiolarien (Rhizopoda Radiaria)*. Eine Monographie, Reimer, Berlin.

Haeckel, E. 1881. Prodromus Systematic radiolarium, entwurf eines radiolarien—Systems auf Grund von studien der Challenger–Radiolarien. *Jenaische Zeitschr. fur Naturw.* 15:418–472.

Haeckel, E. 1887. Report on Radiolaria collected by H. M. S. Challenger during the years 1873–1876. In C. W. Thompson and J. Murray (eds.), *The Voyage of H. M. S. Challenger*, Vol. 18, pp. 1–1760 Her Majesty's Stationery Office, London.

Haeckel, E. 1881. Entwurf eines Radiolarien-Systems auf Grund von Studien der Challenger–Radiolarien. *Jena. Z. Naturw.* 15 (New Ser. 8):418–472.

Haecker, V. 1908a. Tiefsee-Radolareen. Spezieller Teil, Lfg. 2. Die Tripyleen, Collodarien und Mikroradiolarien der Tiefsee. *Wissenschaftl. Ergebn. Deutsch. Tiefsee Exped.* 14:337–476.

Haecker, V. 1908b. Tiefsee-Radiolarien. Allgemeiner Teil. Form und Formbildung bei den Radiolarien. *Wissenschaftl. Ergebn. Deutsch. Tiefsee Exped.* 14:477–706.

Haecker, V. 1908c. Tiefsee-Radiolarien. Allgemeiner Teil. Form. und Formbildung bei den Radiolarien. *Wiss. Ergebn. Deutsch. Tiefsee-Exp. Dampfer (Valdivia) 1898–1899.* 14:477–706.

Haldane, J. B. S. 1956. Can a species concept be justified? In *The Species Concept in Paleontology*, pp. 95–96. Systematics Assoc. Publication No. 2.

Harbison, G. R., L. P. Madin, and N. R. Swanberg. 1978. On the natural history and distribution of oceanic ctenophores. *Deep-Sea Res.* 25:233–256.

Hardin, G. 1960. The competitive exclusion principle. *Science* 131:1292–1298.

Harper, C. W. 1976. Phylogenetic inference in paleontology. *J. Paleontol.* 50:180–193.

Harper, H. E., Jr., and A. H. Knoll. 1975. Silica, diatoms, and Cenozoic radiolarian evolution. *Geology* 3:175–177.

Harvey, E. N. 1926. Oxygen and luminescence with a description of methods for removing oxygen from cells and fluids. *Biol. Bull. Mar. Biol. Lab. (Woods Hole)* 51:89–97.

Hausmann, K., and D. J. Patterson. 1982. Pseudopod formation and membrane production during prey capture by a heliozoon (Feeding by Actinophrys, II). *Cell Motil.* 2:9–24.

Hays, J.D. 1965. Radiolaria and late Tertiary and Quaternary history of Antarctic Seas. In G. A. Llano (ed.), *Biology of Antarctic Seas II. Amer. Geophys. Union, Antarct. Res. Ser.* 5:125–184.

Hays, J.D. 1970. The stratigraphy and evolutionary trends of radiolaria in North Pacific deep-sea sediments. In J. D. Hays (ed.). *Geol. Soc. Am. Mem.* 126:185–218.

Hays, J. D., and J. G. Donahue. 1972. Antarctic Quaternary record and radiolarian and diatom extinctions. In R. J. Adie (ed.), *Antarctic Geology and Geophysics*, Ser. B:733–738. International Union of Geological Sciences.

Hays, J. D., J. Imbrie, and N. J. Shackleton. 1976a. Variations in the Earth's orbit: Pacemaker of the ice ages. *Science* 194:1121–1132.

Hays, J. D., J. A. Lozano, N. Shackleton, and G. Irving. 1976b. Reconstruction of the Atlantic and western Indian Ocean sectors of the 18,000 B. P. Antarctic Ocean. In R. M. Cline and J.D. Hays (eds.). *Investigations of Late Quaternary Paleoceanography and Paleoclimatology*, Memoir 145, pp. 337–374. Geological Society of America, Boulder, CO.

Hays, J. D., and N. Opdyke. 1967. Antarctic radiolaria, magnetic reversals, and climatic change. *Science* 158:1001–1011.

Hays, J. D., T. Saito, N. D. Opdyke, and L. H. Burckle. 1969. Pliocene-Pleistocene sediments of the equatorial Pacific—their paleomagnetic, biostratigraphic and climatic record. *Geol. Soc. Am. Bull.* 80:1481–1514.

Herring, P. J. 1979. Some features of the bioluminescence of the radiolarian *Thalassicolla* sp. *Mar. Biol.* 53:213-216.

Hertwig, R. 1879. Der Organismus der Radiolarien. *Jenaische Denkschr.* 2, Taf. vi–xvi:129–277.

Hertwig, R., and E. Lesser. 1874. Ueber Rhizopoden und denselben nahestenden Organismen. *Arch. Mikr. Anat.* 10 (Suppl.:35).

Hibberd, D. J. 1977. Observations on the ultrastructure of the cryptomonad endosymbiont of the red-water ciliate *Mesodinium rubrum*. *J. Mar. Biol. Assoc. (U. K.)* 57:45-61.

Hilmers, K. 1906. *Zur Kenntnis der Collosphaeriden.* Inaugural Dissertation. Kiel, pp. 5–93.

Holdsworth, B. K. 1969. The relationship between the genus *Albaillella* Deflandre and the ceratoikiscid Radiolaria. *Micropaleontology (N. Y.)* 15:230–236.

Holdsworth, B. K. 1977. Paleozoic radiolaria: Stratigraphic distribution in Atlantic borderlands. In F. M. Swaim (ed.), *Stratigratigraphic Micropaleontology of Atlantic Basins and Borderlands*, p. 167. Elsevier, Amsterdam.

Hollande, A. 1974. Données ultrastructurales sur les isospores des Radiolaires. *Protistologica* 10:567–572.

Hollande, A., J. Cachon, and M. Cachon. 1970. La signification de la membrane capsulaire des Radiolaires et ses rapports avec le plasmalemme et les membranes du réticulum endoplasmique. Affinités entre Radiolaires, Heliozaires et Peridiniens. *Protistologica* 6:311–318.

Hollande, A., M. Cachon, and J. Valentin. 1967. Infrastructure des axopodes et organisation générale de *Sticholonche zanclea* Hertwig (Radiolaire Sticholonchidea). *Protistologica* 3:155–166.

Hollande, A., and D. Carre. 1974. Les xanthelles des Radiolaires Sphaerocollides, des Acanthaires et de *Vellela vellela*: Infrastructure-cytochimie-taxonomie. *Protistologica* 10:573–601.

Hollande, A., and M. Enjumet. 1953. Contribution à l'étude biologique des Sphaerocollides (Radiolaires Collodaires et Radiolaires polycyttaires) et de leurs parasites. *Ann. Sci. Nat. Zool.* 15:99–183.

Hollande, A., and M. Enjumet. 1954. Morphologie et affinities due Radiolaire *Sticholonche zanclea* Hertwig. *Ann. Sci. Nat.* (11e Serie) 16:337–342.

Hollande, A., and M. Enjumet. 1955. Parasites et cycle evolutif des Radiolaires et des Acanthares. *Bull. Stn. Agric. Peche Castiglione* (Nouv. Ser.) 7:153–176.

Hollande, A., and M. Enjumet. 1960. Cytologie, evolution et systematique des Sphaeroides (Radiolaires). *Arch. Mus. Nat. Hist. Natur.* (7e Serie) 7:1–134.

Hollande, A., and R. Martoja. 1974. Identification du cristalloide des isospopres de Radiolaires à un cristal de celestite ($SrSO_4$). Détermination de la constitution du cristalloide per voie cytochimique et à l'aide de la microsonde électronique et du microanalyseur à émission ionique secondaire. *Protistologica* 10:603–609.

Honjo, S. 1978. Sedimentation of materials in the Sargasso Sea at a 5,367 m deep station. *J. Mar. Res.* 36:469–492.

Hovasse, R. 1924. Sur les Perdiniens parasites des Radiolaires coloniaux. *Bull. Soc. Zool. (France)* 48:337.

Hurd, D. C., H. S. Prankratz, V. Asper, J. Fugate, and H. Morrow. 1981. Changes in the physical and chemical properties of biogenic silica from the central equatorial Pacific: Part III, specific pore volume, mean pore size, and skeletal ultrastructure of acid-cleaned samples. *Amer. J. Sci.* 281:833–895.

Huth, W. 1913. Zur Entwicklungsgeschichte der Thalassicollen. *Arch. Protistenkd.* 30:1.

Imbrie, J., and N. G. Kipp. 1971. A new micropaleontological method for a quantitative paleoclimatology: Application to a late Pleistocene Caribbean core. In K. K. Turekian (ed.), *The Late Cenozoic Glacial Ages*, pp. 71–81. Yale University Press, New Haven, CT.

Johnson, D. A., and A. H. Knoll. 1974. Radiolaria as paleoclimatic indicators: Pleistocene climatic fluctuations in the equatorial Pacific Ocean. *Quat. Res. (N. Y.)* 4:206–216.

Johnson, D. A., and C. Nigrini. 1980. Radiolarian biogeography in surface sediments of the western Indian Ocean. *Mar. Micropaleontol.* 5:111–152.

Johnson, D. A., and C. Nigrini. 1982. Radiolarian biogeography in surface sediments of the eastern Indian Ocean. *Mar. Micropaleontol.* 7:237–281.

Jörgensen, E. 1900. Protophyten and protozoen im Plankton aus der norwegischen Westküste. *Bergens Mus. Arb. 1899* 6:51–95.

Jörgensen, E. 1905. The protist plankton and the diatoms in bottom samples. *Bergens Mus. Skr.* (Ser. 1) 7:49–151.

Kamatani, A. 1971. Physical and chemical characteristics of biogenous silica. *Mar. Biol.* 8:89–95.

Kanaya, T., and I. Koizumi. 1966. Interpretation of diatom thanatocoenoses from the North pacific applied to a study of core V20-130 (studies of deep-sea core V20–130, Part IV). *Tohoku Univ. Sci. Rep. Sendai 2d Ser. (Geol.)* 37:89–130.

Keany, J. 1973. *Antarctissa strelkovi* as a paleoclimatic index in the southern ocean. *Antarct. J. U. S.* 8:287–288.

Keany, J. 1976. Early Pliocene paleoclimatology and radiolarian biostratigraphy of the Southern Ocean. *Antarct. J. U. S.* 11:171–173.

Keany, J. 1977. *Late Cenozoic Antarctic Radiolarian Distributions and Paleoceanographic Implications.* Ph.D. thesis. University Rhode Island, Kingston, RI.

Kellogg, D. E. 1976a. Character displacement in the radiolarian genus, *Eucyrtidium. Evolution* 29:736–749.

Kellogg, T. B. 1976b. Late Quaternary climatic changes: Evidence from deep-sea cores of Norwegian and Greenland Seas. In R. M. Cline and J. D. Hays (eds.). *Investigations of Late Quaternary Paleoceanography and Paleoclimatology*, Memoir 145, pp. 77–110. Geological Society of America, Boulder, CO.

Kellogg, D. E. and J. D. Hays. 1975. Microevolutionary patterns in late Cenozoic Radiolaria. *Paleobiology* 1:150–160.

Kennett, J. P. 1970. Pleistocene paleoclimates and foraminiferal biostratigraphy in Subantarctic deep-sea cores. *Deep-Sea Res.* 17:125–140.

Kennett, J. P. 1979. Zoogeography of Antarctic microfossils. In S. Van Der Spoel and A. C. Pierrot-Bults (eds.). *Zoogeography and Diversity of Plankton*, pp. 328–355. Wiley, New York.

Kent, D., N. D. Opdyke, and M. Ewing. 1971. Climate change in the North Pacific using ice-rafted detritus as a climatic indicator. *Geol. Soc. Am. Bull.* 82:2741–2754.

Kerkut, G. A. 1961. *Implications of Evolution.* Pergamon Press, Oxford.

Khmeleva, N. N. 1967. Rol'radiolyarii pri otsenke pervichnoi produktsii v krasnom More i adenskom zalive. *Doklady Akademii Nauk SSSR* 172:1430–1433.

King, K. 1974. Preserved amino acids from silicified protein in fossil Radiolaria. *Nature (London)* 252:690–692.

King, K. 1975. Amino acids composition of the silicified organic matrix in fossil polycystine Radiolaria. *Micropaleontology (N. Y.)* 21:215–226.

King, K. 1977. Amino acid survey of recent calcareous and siliceous deep-sea microfossils. *Micropaleontology (N. Y.)* 23:180–193.

Kling, S. A. 1966. Castanellid and Circoporid Radiolarians: Systematics and zoogeography in the Eastern North Pacific. Ph.D. thesis. University of California, San Diego, CA.

Kling, S. A. 1971. Dimorphism in radiolaria. In A. Farinacci (ed.), *Proceedings of the II Planktonic Conference, Roma 1970*, pp. 663–672. Edizioni Tecnoscienza, Rome.

Kling, S. A. 1973. Radiolaria from the eastern North Pacific, Deep Sea Drilling Project 18. In L. D. Kulm, R. von Huene, et al. (eds.). *Initial Reports of the Deep-Sea Drilling Project* 18:617–671.

Kling, S. A. 1976. Relation of radiolarian distributions to subsurface hydrography in the North Pacific. *Deep-Sea Res.* 23:1043–1058.

Kling, S. A. 1978. Radiolaria. In B. U. Haq and A. Boersma (eds.), *Introduction to Marine Micropaleontology*, pp. 203–244. American Elsevier, New York.

Kling, S. A. 1979. Vertical distribution of polycystine radiolarians in the central North Pacific. *Mar. Micropaleontol.* 4:295–318.

Knoll, A. W., and D. A. Johnson. 1975. Late Pleistocene evolution of the collosphaerid radiolarian *Buccinosphaera invaginata* Haeckel. *Micropaleontology (N. Y.)* 21:60–68.

Kruglikova, S.B. 1976. Radiolarians in the upper Pleistocene sediments of the boreal and northern subtropical zones of the Pacific Ocean. *Oceanology* 16:61–64.

Labrachérie, M. 1980. Les Radiolaires temoins de l'evolution hydrologique depuis le dernier maximum glaciaire au large du Cap Blanc (Afrique du Nord-Ouest). *Palaeogr. Palaeoclimatol. Palaeoecol.* 32:163–184.

Lazarus, D., J. D. Hays, and D. R. Prothero. 1982. Evolution of the radiolarian species-complex *Pterocanium*, a preliminary survey. (MS available from authors)

Lee, J. J. 1980. Nutrition and physiology of the Foraminifera. In M. Levandowsky and S. Hutner (eds.), *Biochemistry and Physiology of Protozoa*, Second Edition, Vol. 3, pp. 43–66. Academic Press, New York.

Levine, N. D., J. O. Corliss, F. E. G. Cox, G. Deroux, J. Grain, B. M. Honigberg, G. F. Leedale, A. R. Loeblich III, J. Lom, D. Lynn, E. G. Merinfeld, F. C. Page, G. Poljansky, V. Sprague, J. Vavra, and F. G. Wallace. 1980. A newly revised classification of the protozoa. *J. Protozool.* 27:37–58.

Ling, H. Y. 1966. The radiolarian *Protocystis thomsoni* (murray) in the northeast Pacific Ocean. *Micropaleontology (N.Y.)* 12:203–214.

Ling, H. Y., and K. Takahashi. 1977. Observation on microstructure of selected phaeodarian Radiolaria. *Mem. Geol. Soc. China* 2:207–212.

Lipps, J. 1970. Plankton evolution. *Evolution* 24:1–22.

Lisitzin, A. P. 1971. Distribution of siliceous microfossils in suspension and in bottom sediments. In B. M. Funnell and W. R. Riedel (eds.), *The Micropaleontology of Oceans*, pp. 173–195. Cambridge University Press, Cambridge.

Lo-Bianco, S. 1903. La pesche abissale eseguite da F. A. Krupp col Yacht *Puritan* nelle adiacenze di Capri ed in altre localita del Mediterraneo. *Mitt. Zool. Sta. Neapol.* 16:109–279.

Lombari, G. A. and G. Boden. 1982. Paleobiogeography and diversity of radiolaria: Recent vs. Miocene. Geological Society of America Abstracts—95th Annual Meeting. pp. 548–549.

Lozano, J. A., and J. D. Hays. 1976. Relationship of radiolarian assemblages to sediment types and physical oceanography in the Atlantic and Western Indian Ocean sectors of the Antarctic Ocean. *Geol. Soc. Am. Mem.* 145:303–336.

Mague, T. H., N. M. Weare, and O. Holm-Hansen. 1974. Nitrogen fixation in the North Pacific Ocean. *Mar. Biol.* 24:109–119.

Malmgren, B. A., and B. U. Haq. 1982. Assessment of quantitative techniques in paleobiogeography. *Mar. Micropaleontol.* 7:213–236.

Manton, I., and M. Parke. 1965. Observations on the fine structure of two species of *Platymonas* with special reference to flagellar scales and the mode of origin of the theca. *J. Mar. Biol. Assoc.* (U. K.) 45:743–754.

Manton, I., K. Oates, and M. Parke. 1963. Observations on the fine structure of the *Pyramimonas* stage of *Halosphaera* and preliminary observations on three species of *Pyramimonas. J. Mar. Biol. Assoc.* (U.K.) 43:225–238.

Margalef, R. 1968. *Perspectives in Ecological Theory*. University of Chicago Press, Chicago, IL.

Mast, H. 1910. Die Astrosphaeriden. *Wiss. Ergebn. de Deutschen Tiefsee-Expedition auf dem Dampfer "Valdivia" 1898–1899 (Jena)*, Vol. 19(4), pp. 125–190.

Matthews, J. B. L., and N. J. Sands. 1973. Ecological studies on the deep-water pelagic community of Korsfjorden, Western Norway. The topography of the area and its hydrography in 1968–1972, with a summary of the sampling programmes. *Sarsia* 52:29–52.

Maurasse, F. J. 1976. Paleoecologic and paleoclimatic implications of radiolarian facies in Caribbean Paleogene deep-sea sediments. *Caribb. Geol. Conf. Trans.* 7:185–204.

Maurasse, F. J. 1979. Cenozoic radiolarian paleobiogeography; Implications concerning plate tectonics and climatic cycles. *Palaeogr. Palaeoclimatol. Palaeoecol.* 26:253–289.

McGowan, J. A 1971. Oceanic biogeography of the Pacific. In B. M. Funnell and W. R. Riedel (eds.), *The Micropaleontology of Oceans.* pp. 3–74. Cambridge University Press, Cambridge.

McIntyre, A., and A. W. H. Bé. 1967. Modern Coccolithiphoridae of the Atlantic Ocean. I. Placoliths and cyroliths. *Deep-Sea Res.* 14:561–597.

McIntyre, A., N. Kipp, *et al.* 1976. Glacial North Atlantic 18,000 years ago: A CLIMAP reconstruction. In R. M. Cline and J. D. Hays (eds.). *Investigations of Late Quaternary Paleoceanography and Paleoclimatology*, Memoir 145, pp. 43–76. Geological Society of America, Boulder, CO.

McMillen, K. J., and R. E. Casey. 1978. Distribution of living polycystine radiolarians in the Gulf of Mexico, and Caribbean Sea, and comparison with the sedimentary record. *Mar. Micropaleontol.* 3:121–145.

Meyen, F. 1834. Über das Leuchten des Meeres und Beschreibung einiger Polypen und anderer niederer Thiere. *Verh. Kaiserl. Leopoldin.–Carolin. Aka. Naturwiss.* 16:125–216.

Mielck, W. 1913. Resumé des observations sur le plankton des meres explorées par le Conseil pendant les années 1902–1908; Radiolaria. *Result. Crois. Period Cons. Perm. Inst. Explor. Mer. Bull. Trimestr.* 3:303–402.

Molinari, R. L., J. F. Festa, and D. W. Behringer. 1978. The circulation in the Gulf of Mexico derived from estimated dynamic height fields. *J. Physic. Oceanogr.* 8:987–996.

Moore, T. C., Jr. 1969. Radiolaria: Change in skeletal weight and resistance to solution. *Geol. Soc. Am. Bull.* 80:2103–2108.

Moore, T. 1971. Radiolaria. In J. I. Tracey, Jr., *et al.* (eds.), *Initial Rep. Deep Sea Drilling Project* 8:727–775. U. S. Government Printing Office, Washington, D. C.

Moore, T. C., Jr. 1972. Mid-Tertiary evolution to the radiolarian genus *Calocycletta. Micropaleontol. (N. Y.)* 18:144–152.

Moore, T. C. Jr. 1973. Late Pleistocene-Holocene oceanographic changes in the northeastern Pacific. *Quat. Res. (N. Y.)* 3:99–109.

Moore, T. C., Jr. 1978. The distribution of radiolarian assemblages in the modern and Ice-age Pacific. *Mar. Micropaleontol.* 3:229–266.

Moore, T. C., Jr. 1982. Mid tertiary evolution of the radiolarian genus *Calocycletta Micropaleontology (N. Y.)* 18:144–152.

Morley, J. J. 1977. *Upper Pleistocene Climatic Variations in the South Atlantic Derived from a Quantitative Analysis: Accent on the Last 18,000 Years.* Ph.D. dissertation. Columbia University, New York.

Morley, J. J. 1979. A transfer function for estimating paleoceanographic conditions based on deep-sea surface sediment distribution of radiolarian assemblages in the South Atlantic. *Quat. Res. (N. Y.)* 12:381–395.

Morley, J. J., and J. D. Hays. 1979. Comparison of glacial and interglacial ocean-ographic conditions in the South Atlantic from variations in calcium carbonate and radiolarian distributions. *Quat. Res. (N. Y.)* 12:396–408.

Müller, J. 1855. Über *Sphaerozoum* und *Thalassicolla*. *Monastber. Kgl. Preuss. Akad. Wiss. Berlin Jahrg.* 1855:229–253.

Müller, J. 1858a. Ueber die Thalassicollen, Polycystinen and Acanthometren des Mittelmeeres. *Akad. Wiss.* (zu Berlin).

Müller, J. 1858. Über die Thalassicollen, Polycystinen und Acanthometren des Mittelmeeres. *Abh. Preuss. Akad. Wiss. Jahrg.* 1858:1–62.

Neagu, T. 1970. Microbiostratigraphy of the Cenomanian deposits from the southern part of eastern Carpathians (with some evolutionary-phylogenetic considerations regarding the planktonic Foraminifera). *Rev. Roum. Geol. Geophys. Geogr. Ser. Geol.* 14:171–188.

Newell, N. D. 1966. Mass extinctions at the end of the Cretaceous period. *Science* 149:922–924.

Nicol, J. A. C. 1958. Observations on luminescence in pelagic animals. *J. Mar. Biol. Assoc. (U. K.)* 37:705–752.

Nigrini, C. 1967. Radiolaria in pelagic sediments from the Indian and Atlantic Oceans. *Bull. Scripps Inst. Oceanogr.* 11:1-106.

Nigrini, C. 1968. Radiolaria from eastern tropical Pacific sediments. *Micropaleontology (N. Y.)* 14:51–63.

Nigrini, C. 1970. Radiolarian assemblages in the North Pacific and their applications to a study of Quaternary sediments in Core V20-130. In J. D. Hays (ed.), *Geological Investigations of the North Pacific. Geol. Soc. Am. Mem.* 126:139–183.

Nigrini, C. 1971. Radiolarian zones in the Quaternary of the equatorial Pacific Ocean. In B. M. Funnell and W. R. Riedel (eds.), *The Micropaleontology of Oceans,* pp. 443–461. Cambridge University Press, Cambridge.

Nigrini, C., and T. C. Moore. 1979. *The Guide to Modern Radiolaria.* Special Publication No. 16, Cushman Foundation for Foraminiferal Research.

Parke, M., and I. Manton. 1965. Preliminary observations on the fine structure of *Prasinocladus marinus. J. Mar. Biol. Assoc. (U. K.)* 45:525–536.

Parke, M., and I. Manton. 1967. The specific identity of the algal symbiont in *Convoluta roscoffensis. J. Mar. Biol. Assoc. (U. K.)* 47:445–464.

Pätau, K. 1937a. Ueber die Natur der Anisosporen der Radiolarien. *Verh. Deutsch. Zool. Ges. (Leipzig)* 39:93.

Pätau, K. 1937b. Sat Chromosome und spiral Struktur der Chromosomen der extra-capsularen Körpern (Meriodinium sp.) von *Collozoum inerme* Müller. *Cytologia Fujii Jubil,* p. 667.

Patterson, D. J. 1979. On the organization and classification of the protozoon, *Actinophrys sol* Ehrenberg, 1830. *Microbios* 26:165–208.

Pavshtiks, E. A. 1956. Sezonnye ismeneniya v planktone i kormovye migratsii sel'di. *Trudy Polyarnogo Nauchnoissledovatel' skogo Instituta morskogo rybnogo khozyaistva i okeanografii.* Vol. 9, pp. 93–123.

Pavshtiks, E. A., and L. A. Pan'Kova. 1966. O pitanii pelagicheskoi molodi morskikh okunei roda *Sebastes* planktonon v Devisovom prolive. *Materialy nauchnoi sessii Polyarnogo nauchnoissledovatel'skogo Instituta morskogo rybnogo khozyaistva i okeanografii,* Vol. 6, p. 87.

Pearse, B. 1975. Coated vesicles from pig brain purification and biochemical characterization. *J. Mol. Biol.* 97:93–98.

Pessagno, E. A. 1977. Radiolaria in Mesozoic stratigraphy. In A. T. S. Ramsay (ed.), *Oceanic Micropaleontology,* pp. 913–950. Academic Press, New York.

Petrushevskaya, M. G. 1965. Osobennosti konstruktsii skeleta radiolyarii Botryoidae (otr. Nassellaria). *Trudy Zool. Inst. Leningr.* 35:79–118.

Petrushevskaya, M. G. 1966. Radiolyarii v planktone i v donnykh osadkakh. In *Geokhimiya Kremnesema*, pp. 219–245. Nauka, Moscow.

Petrushevskaya, M. G. 1971a. On the natural system of polycystine Radiolaria (Class Sarcodina). In A. Farinacci (ed.), *Proceedings of the II Planktonic Conference Roma 1970*, pp. 981–992. Edizioni Tecnoscienza, Rome.

Petrushevskaya, M. H. 1971b. Radiolaria in the plankton and recent sediments from the Indian Ocean and Antarctic. In B. M. Funnell and W. R. Riedel (eds.), *The Micropaleontology of Oceans*, pp. 319–329. Cambridge University Press, Cambridge.

Petrushevskaya, M. G. 1975a. Cenozoic radiolarians of the Antrctic, Leg 29, DSDP. In J. P. Kennett, R. E. Houtz, *et al.* (eds.), *Initial Reports of the Deep Sea Drilling Project*, Vol. 29, pp. 541–675. U. S. Government Printing Office, Washington, D. C.

Petrushevskaya, M. G. 1975b. The structure of skeletons in radiolaria. *Cytology (U. S. S. R.)* 12:1436–1437. [in Russian]

Petrushevskaya, M. G. 1977. On the origin of radiolaria. *Zool. J. (U. S. S. R.)* 56:1448–1458. [in Russian]

Petrushevskaya, M. G. 1981. *Radiolaria, Order Nassellaria*. Academy of Sciences of the Soviet Union, Zoological Institute. Leningrad Science Publishers, Leningrad Division. [in Russian].

Petrushevskaya, M. G., and K. R. Bjørklund. 1974. Radiolarians in Holocene sediments of the Norwegian-Greenland seas. *Sarsia* 57:33–46.

Petrushevsaya, M. G., and G. E. Kozlova. 1972. Radiolaria. In J. D. Hays, *et al.* (eds.), *Initial Reports of the Deep Sea Drilling Project 14*. U. S. Government Printing Office, Washington, D. C.

Pielou, E. C. 1969. *An Introduction to Mathematical Ecology*. Wiley-Interscience.

Pisias, N. G. 1976. Late Quaternary sediment of the Panama Basin: Sedimentation rates periodicities and controls of carbonate and opal accumulation. In R. M. Cline and J. D. Hays (eds.). *Investigations of Late Quaternary Paleoceanography and Paleoclimatology*, Memoir 145, pp. 375–392. Geological Society of America, Boulder, CO.

Pisias, N. G. 1978. Paleoceanography of the Santa Barbara Basin during the last 8000 years. *Quat. Res. (N. Y.)* 10:366–384.

Pitelka, D. R. 1963. *Electron-microscopic Structure of Protozoa*. Pergamon Press, Oxford.

Popofsky, A. 1913. Die Sphaerellarien des Warmwassergebietes. *Deutsche Südpolar-Exp. 1901–1903 Berlin Zool. Bd. 13*, 5:75–159.

Popofsky, A. 1917. Die Collosphaereden der Deutschen Südpolar-Expedition, 1901–1903. Mit Nachtrag zu den Spumellarien und Nassellarien. *Deutsche Südpolar-Exped. 1901–1903, Berlin. Zool. Bd. 8, H. 3*:235–278.

Prell, W. L., and W. B. Curry. 1981. Faunal and isotopic indices of monsoonal upwelling: Western Arabian Sea. *Oceanol. Acta* 4:91–98.

Prezelin, B. B., and B. M. Sweeney. 1978. Photoadaptation of photosynthesis in *Gonyaulax polyedra*. *Mar. Biol.* 48:27–35.

Prothero, D. R., and D. B. Lazarus. 1980. Planktonic microfossils and the recognition of ancestors. *Syst. Zool.* 29:119–129.

Reeve, M. R. 1981. Large cod-end reservoirs as an aid to the live collection of delicate zooplankton. *Limnol. Oceanogr.* 26:577–579.

Renz, G. W. 1976. The distribution and ecology of Radiolaria in the Central Pacific plankton and surface sediments. *Bull. Scripps Inst. Oceanogr.* 22:1–267.

Reshetnyak, V. V. 1955. Vertikalnoe raspredelenie radiolayarii Kurilo-Kamchatskoi vpadiny. *Trudy Zool. Inst. Akad. Nauk U. S. S. R.* 21:94–101.

Reynolds, R. A. 1979. *Neogene Radiolarian Biostratigraphy and Paleocean-ography of the Northwest Pacific.* Ph.D. thesis, Rice University, Houston, TX.

Riedel, W. R. 1951a. Number of Radiolaria in sediments. *Nature (London)* 167:75.

Riedel, W. R. 1951b. Sedimentation in the Tropical Indian Ocean. *Nature (London)* 168:737.

Riedel, W. R. 1958. Radiolaria in Antarctic Sediments. *Repts. B. A. N. Z. Antarct. Res. Exped. Ser. B.* 6:217–255.

Riedel, W. R. 1959. Oligocene and lower Miocene Radiolaria in tropical Pacific sediments. *Micropaleontology (N. Y.)* 5: 285–302.

Riedel, W. R. 1967a. Some new families of Radiolaria. *Proc. Geol. Soc. Lond.* 1640:148–149.

Riedel, W. R. 1967b. Subclass Radiolaria. In W. B. Harland, *et al.* (eds.), *The Fossil Record*, pp. 291–298. Geological Society of London, London.

Riedel, W. R. 1967c. An annotated and indexed bibliography of polycystine radiolaria. Manuscript available from W. R. Riedel, Scripp's Institution of Oceanography, La Jolla, CA.

Riedel, W. R. 1971a. Systematic classification of polycystine Radiolaria. In B. M. Funnell and W. R. Riedel (eds.), *The Micropalaeontology of Oceans*, pp. 649–661. Cambridge University Press, Cambridge.

Riedel, W. R. 1971b. Radiolarians from Atlantic deep-sea drilling. In A. Farinacci (ed.), *Proceedings of the II Planktonic Conference Roma*, pp. 1057–1967 (erratum sheet distributed subsequently by author). Edizioni Tecnoscienza, Rome.

Riedel, W. R. 1971c. Systematic classification of polycystine radiolaria. In B. M. Funnell and W. R. Riedel (eds.), *The Micropaleontology of Oceans*, pp. 649–661. Cambridge University Press, Cambridge.

Riedel, W. R. 1978. Systems of morphologic descriptors in paleontology. *J. Paleontol.* 52:1–7.

Riedel, W. R., and E. A. Holm. 1957. In J. Hedgepeth (ed.), Treatise on marine ecology and paleoecology. *Geol. Soc. Am. Mem.* 67:1069–1072.

Riedel, W. R., and A. Sanfilippo. 1970. Radiolaria, Leg 4, deep sea drilling project. In R. G. Bader, *et al.* (eds.), *Initial Rep. Deep Sea Drilling Project* 4:503–575. U. S. Government Printing Office, Washington, D. C.

Riedel, W. R., and A. Sanfilippo. 1971. Cenozoic Radiolaria from the western tropical Pacific, Leg. 7. In E. L. Winterer, *et al.* (eds.), *Initial Rep. Deep Sea Drilling Project* 7:1529–1672. U. S. Government Printing Office, Washington, D. C.

Riedel, W. R., and A. Sanfilippo. 1977. Cainozoic Radiolaria. In A. T. S. Ramsay (ed.), *Oceanic Micropaleontology*, Vol. 2, pp. 847–912. Academic Press, London.

Riedel, W. R., and A. Sanfilippo. 1978. Stratigraphy and evolution of tropical Cenozoic radiolarians. *Micropaleontology (N. Y.)* 24:61–96.

Riedel, W. R., and A. Sanfilippo. 1981. Evolution and diversity of form in radiolaria. In T. L. Simpson and B. E. Volcani (eds.), *Silicon and Siliceous Structures in Biological Systems*, pp. 323–346. Springer-Verlag, New York.

Robinow, C. F., and J. Marak. 1966. A fiber apparatus in the nucleus of the yeast cell. *J. Cell. Biol.* 29:129–151.

Robinson, W. J., and R. M. Goll. 1978. Fine skeletal structure of the radiolarian *Callimitra carolotae* Haeckel. *Micropaleontology (N. Y.)* 24:432–438.

Ruddiman, W. F., and A. McIntyre. 1976. Northeast paleoclimatic changes over the past 600,000 years. In R. M. Cline and J. D. Hays (Eds.). *Investigations of Late Quaternary Paleoceanography and Paleoclimatology*, Memoir 145, pp. 111–146. Geological Society of America, Boulder, CO.

Sachs, H. M. 1973. North Pacific Radiolarian assemblages and their relationship to oceanographic parameters. *Quat. Res. (N. Y.)* 3:73–88.

Sachs, H. M. 1975. Radiolarian-based estimate of North Pacific summer sea-surface temperature regime during the latest glacial maximum. *Alaska Sci. Conf. Proc.* 24:37–42.

Saito, T. 1976. Geologic significance of coiling direction in the planktonic foraminifera *Pulleniatina. Geology* 4:305–309.

Sancetta, C. 1978. Neogene Pacific microfossils and paleoceanography. *Mar. Micropaleontol.* 3:347–376.

Sancetta, C. 1979a. Paleogene Pacific microfossils and paleoceanography. *Mar. Micropaleontol.* 4:363–398.

Sancetta, C. 1979b. Use of semiquantitative microfossil data for paleoceanography. *Geology* 7:88–92.

Sanfilippo, A., and W. R. Riedel. 1970. Post-Eocene "closed" theoperid radiolarians. *Micropaleontology (N. Y.)* 16:446–462.

Sanfilippo, A., and W. R. Riedel. 1973. Cenozoic Radiolaria (exclusive of theoperids, artostrobiids and amphipyndacids) from the Gulf of Mexico, DSDP Leg. X. In J. L. Worzel, et al. (eds.), *Initial Rep. Deep Sea Drilling Project 10.* U. S. Government Printing Office, Washington, D. C.

Sanfilippo, A., M. J. Westberg, and W. R. Riedel. 1981. Cenozoic radiolarians at site 462, Deep Sea Drilling Project Leg. 61, Western tropical Pacific. In R.L. Larson, et al. (eds.), *Initial Reports of the Deep Sea Drilling Project* 61:495–505. U. S. Government Printing Office, Washington, D. C.

Sarjeant, W. A. S. 1967. The stratigraphical distribution of fossil dinoflagellates. *Rev. Paleobot. Palynol.* 1:323–343.

Schaaf, A. 1981. Late Early Cretaveous radiolaria from deep sea drilling project Leg. 62. In J. Thiede, T.L. Vallier, et al. (eds.), *Initial Reports of the Deep Sea Drilling Project* 62:419–470. U. S. Government Printing Office, Washington, D. C.

Schuster, F. L. 1975. Ultrastructure of mitosis in the amoeboflagellate *Naegleria gruberi. Tissue Cell* 7:1–12.

Schwab, D. 1975. *The Nuclear Fine Structure of Some Monothalamous Foraminifera during Interphase and Nuclear Division: Conference Paper.* Benthonics 1975, Dalhousie University, Halifax.

Schwartz, A. 1931. Ueber den körperbau der Radiolarien (Erg. paläontologischer arbeitsmethoden). *Abh. Senckenberg. Naturforsch. Ges.* 43:1–17.

Shaw, A. B. 1969. Adam and Eve, paleontology and the non-objective arts. *J. Paleontol.* 43:1085–1098.

Sheehan, R., and F. T. Banner. 1972. The pseudopodia of *Elphidium incertum. Rev. Esp. Micropaleontol.* 4:31–63.

Silver, M. W., and A. L. Alldredge. 1981. Bathypelagic marine snow: deep-sea algal and detrital community. *J. Mar. Res.* 39:501–530.

Simpson, T. L. 1981. Effects of germanium on silica deposition in sponges. In T. L. Simpson and B. E. Volcani (eds.), *Silicon and Siliceous Structures in Biological Systems*, pp. 527–550. Springer-Verlag, New York.

Sloan, J. R. 1981. *Radiolarians of the North Philippine Sea: Their Biostratigraphy, Preservation, and Paleoecology.* Unpublished Ph.D. thesis. University of California, Davis, CA.

Smayda, T. J. 1958. Biogeographical studies of marine phytoplankton. *Oikos* 9:158–191.

Spaw, J. M. 1979. *Vertical Distribution, Ecology and Preservation of Recent Polycystine Radiolaria of the North Atlantic Ocean (Southern Sargasso Sea Region).* Ph.D. thesis, Rice University, Houston, TX.

Spooner, B. L. 1975. Microfilaments, microtubules and extracellular materials in morphogenesis. *Bio. Sci. 25*:440–451.

Stanley, E. M. 1973. Minor element abundance in radiolarian tests (abstr.). In *Cordilleran Section, Geol. Soc. Am. Abstr. 69th Annual Meeting,* Vol. 5, p. 110.

Stanley, E. M. 1981. *Biogeography and Evolution of "Bipolar" Radiolaria.* Ph.D. thesis, University of California, Davis, CA.

Stewart, W. D. P. 1971. In J. D. Costlow (ed.), *Nitrogen Fixation in the Sea,* pp. 537–564. Gordon Beach, New York.

Strelkov, A. A. and V. V. Reshetnyak. 1971. Colonial Spumellarian. Radiolarians of the World Ocean. In *Explorations of the Fauna of the Seas IX (XVII), Radiolarians of the Ocean, Reports on the Soviet Expeditions,* Academy of Sciences of the U. S. S. R., pp. 295–417. [English translation by W. R. Riedel]

Strickland, J. D. H., and T. Parsons. 1972. A Practical handbook for seawater analysis. *Fish. Res. Board Can. Bull.* 125.

Sullivan, C. W., and B. E. Volcani. 1981. Silicon in the cellular metabolism of diatoms. In T. L. Simpson and B. E. Volcani (eds.), *Silicon and Siliceous Structures in Biological Systems,* pp. 15–42. Springer-Verlag, New York.

Swain, F. M. (ed.). 1977. *Stratigraphic Micropaleontology of Atlantic Basins and Borderlands.* Elsevier, Amsterdam.

Swanberg, N. R. 1979. *The Ecology of Colonial Radiolarians: Their Colony Morphology, Trophic Interactions and Associations, Behavior Distribution and the Photosynthesis of Their Symbionts.* Ph.D. thesis. Woods Hole Oceanographic Institution and Massachusetts Institute of Technology.

Swanberg, N. R. 1974. The feeding behavior of *Beroe ovata. Mar. Biol. 24*:69–76.

Swanberg, N. R. and O. R. Anderson. 1981. *Collozoum caudatum* sp. nov.: A giant colonial radiolarian from equatorial and Gulf Stream waters. *Deep-Sea Res. 28A:* 1033–1047.

Swanberg, N. R., and G. R. Harbison. 1980. The ecology of *Collozoum longiforme, sp. nov.* a new colonial radiolarian from the equatorial Atlantic Ocean. *Deep Sea Research 27:*715–731.

Szalay, F. S. 1977. Ancestors, descendents, sister groups, and testing of phylogenetic hypotheses. *Syst. Zool. 26:*12–18.

Takahashi, K. 1981. *Vertical Flux, Ecology and Dissolution of Radiolaria in Tropical Oceans: Implications for the Silica Cycle.* Ph.D. thesis. Woods Hole Oceanographic Institution and Massachusetts Institute of Technology.

Takahashi, K. 1983. Radiolaria. In S. Honjo (ed.), *Ocean Biocoenosis,* in press, Micropaleontology Press, New York.

Takahashi, K., and S. Honjo. 1981. Vertical flux of Radiolaria: A taxon-quantitative sediment trap study from the western tropical Atlantic. *Micropaleontology (N. Y.) 27:*140–190.

Takahashi, K., and H. Y. Ling. 1980. Distribution of *Sticholonche* (Radiolaria) in the upper 800 M of the waters in the Equatorial Pacific. *Mar. Micropaleontol. 5:*311–319.

Tan, Z., H. Gao, and X. Su. 1978. The quantitative distribution of *Sticholonche zanclea* in the Western part of the East China Sea. *Oceanol. Limnol. Sin. 9:*59–66. [In Chinese]

Tan, Z., and T. Tchang. 1976. Studies on the radiolaria of the East China Sea, II. Spumellaria, Nassellaria, Phaeodaria, Sticholonchea. *Studia Marina Sin.*

11:217–313. [In Chinese]

Tappan, H. 1968. Primary production, isotopes, extinctions and the atmosphere. *Palaeogeogr. Palaeoclimatol. Palaeoecol.* 4:187–210.

Tappan, H., and A. R. Loeblich, Jr. 1972. Fluctuating rates of protistan evolution, diversification and extinction. *International Geological of Congress 24, Montreal, 1972, Sect. 7, Paleontology*, pp. 205–213.

Tappan, H., and A. R. Loeblich, Jr. 1973. Evolution of the oceanic plankton. *Earth-Sci. Rev.* 9:207–240.

Taylor, D. L. 1974. Taxonomy and biological fitness. In W. B. Vernberg (ed.), *Symbiosis in the Sea*, pp. 245–262. University of South Carolina Press, Columbia, SC.

Theyer, F., and S. R. Hammond. 1974. Paleomagnetic polarity sequence and radiolarian zones, Brunhes to Polarity Epoch 20. *Earth Planetary Sci. Lett.* 22:307–319.

Theyer, F., C. Y. Mato, and S. R. Hammond. 1978. Paleomagnetic and geochronologic calibration of latest Oligocene to Pliocene radiolarian events, equatorial Pacific. *Mar. Micropaleontol.* 3:377–395.

Thomas, W. H. 1964. An experimental evaluation of the C^{14} method for measuring phytoplankton production using cultures of *Dunaliella primolecta* Butcher. *Fishery Bull.* 63:273–292.

Thompson, D. W. 1942. *On Growth and Form*. Macmillan, New York.

Tregouboff, G. 1953. Classe des Radiolaires. In P. P. Grassé (ed.), *Traité de Zoologie Anatomie, Systematique, Biologie*, Vol. 1., pp. 322-436. Masson, Paris.

Trench, R. K. 1980. Integrative mechanisms in mutualistic symbioses. In C. B. Cook, et al. (eds.), *Cellular Interactions in Symbiosis and Parasitism*, pp. 275–297. Ohio State University Press, Columbus, OH.

Tyler, S. 1979. Contributions of electorn microscopy to systematics and phylogeny. *Amer. Zool.* 19:541–543.

Valentine, J. W., and E. M. Moores. 1970. Plate-tectonic regulation of faunal diversity and sea level: A model. *Nature (London)* 228:657–659.

Volcani, B. E. 1981. Cell wall formation in diatoms: Morphogenesis and biochemistry. In T. L. Simpson and B. E. Volcani (eds.), *Silicon and Siliceous Structures in Biological Systems*, pp. 157–200. Springer-Verlag, New York.

Weaver, F. M., R. E. Casey, and A. M. Perez. 1981. Stratigraphic and paleoceanographic significance of early Pliocene to middle Miocene radiolarian assemblages from northern to Baja California. In R. E. Garrison (ed.), *The Monterey Formation and Related Siliceous Rocks of California*, pp. 71–86. Society of Economic Paleontologists and Mineralogists.

Withers, N. W., W. C. M. C. Kokke, M. Rohmer, W. H. Fenical, and C. Djerassi. 1979. Isolation of sterols with cyclopropyl-containing side chains from the cultured marine alga *Peridinium foliaceum*. *Tetrahedron Lett.* 38:3605–3608.

Zickler, D. 1970. Division spindle and centrosomal plaques during mitosis and meiosis in some Ascomycetes. *Chromosoma (Berlin)* 30:287–304.

Appendix
A Synopsis of Major Protozoan Taxonomic Groups

This summary of protozoan taxonomy* provides a context for the place of radiolaria among other major protozoan groups. It is categorized to only the phylum level with citations to some commonly observed genera within each phylum.

Kingdom. Protista

Phylum 1. Sarcomastigophora

The flagellated and/or pseudopod-bearing protozoa. Nucleus of a single type, except in heterokaryotic (bearing more than one nucleus and of different shape and/or size) Foraminiferida. Sexuality, when present, essentially syngamy (fusion of gametes to form a zygote giving rise to a new individual). The flagellated organisms (e.g., *Euglena* sp., dino-flagellates and colorless flagellates) are in the subphylum Mastigophora. The pseudopod-bearing organisms and those that move by protoplasmic streaming (e.g., *Amoeba* sp.) are placed in the subphylum Sarcodina. Among the Sarcodina, the rhizopod-bearing organisms (e.g., amoeba and foraminifera) are in the superclass Rhizopoda. The axopod-bearing organisms, however, are included within the superclass Actinopoda. The Actinopoda embrace the classes of Acantharea (Acantharia), and Polycistinea and Phaeodaria (the latter two are equivalent to the classical "Radiolaria"). The class Heliozoea (sun-ray animals; e.g., *Actinophrys* sp., *Sticholonche* sp.) are also included here.

* Based on the system of Levine *et al.*, 1980.

Phylum 2. Labyrinthomorpha
Feeding stage of growth characterized by a sheath-like network with spindle-shaped or spherical, nonameboid cells; in some genera, amoeboid cells move within the network by gliding. Saprobic (feeding by absorbing organics from the environment) and parasitic on algae, mostly in marine and estuarine waters. The genus *Labyrinthula* and its relatives are included in the phylum.

Phylum 3. Apicomplexa
A group of parasitic species bearing an apical complex (visible with electron microscope) consisting of polar ring(s), subpellicular microtubules, etc. present at some stage; micropores generally present at some stage; cilia absent. The "gregarine" parasites, Coccidia, and Babesia-like parasites occur in the phylum. The malaria parasite and other blood-parasitic species are included here.

Phylum 4. Microspora
Obligatory intracellular parasites in nearly all major animal groups; producing unicellular spores, each with an imperforate wall, containing one uninucleate or dinucleate sporoplasm and simple or complex extrusion apparatus. Some of these parasites cause diseases of economic importance including diseases of the honey bee (*Nosema* sp.) and infections in some fish.

Phylum 5. Ascetospora
All parasitic; spore multicellular (or unicellular), with one or more sporoplasms; without polar capsules or polar filaments.

Phylum 6. Myxozoa
All species are parasitic; spores of multicellular origin, with one or more polar capsules and sporoplasms; bearing 1, 2, or 3 (rarely more) valves.

Phylum 7. Ciliophora
Most species free-living, a few are parasitic, but many are commensals or living upon the surface of other organisms. Locomotion is by simple cilia or compound ciliary organelles with a subpellicular infraciliature present even in species lacking cilia; two types of nuclei typically (a macronucleus and a micronucleus); sexual and/or asexual reproduction (binary fission transverse, but budding and multiple fission also occur). Among the numerous ciliated species included here are some commonly recognized ciliates such as *Paramecium* sp., *Tetrahymena* sp., *Vorticella* sp., *Stentor* sp., tintinnida, etc.

Glossary

abdomen The third segment or joint in the shell of Nassellaria (see Fig. 1-8I); more generally, the third body segment of an organism exhibiting head, thorax, and abdomen regions.

acid phosphatase A digestive enzyme with maximal activity (pH optimum) in the acid range; it occurs in digestive vacuoles and catalyzes the hydrolytic cleavage of phosphate from organic molecules; a marker enzyme for lysosomes.

albuminoid spherules A cytoplasmic deposit or granule containing protein.

alveoli Bubble-like cytoplasmic compartments filled either with fluid or gas (see Fig. 1-1), often forming a soap-bubble-like frothy layer around some spumellarian radiolaria.

amino acid An organic compound composed of carbon, nitrogen, oxygen, hydrogen and sometimes sulfur; the building blocks (monomers) for protein polymers.

anabolism The process of building-up of organic compounds and other biologically significant molecules; the process of synthesis as opposed to degradation (catabolism).

Animalia A kingdom (animals) in the five kingdom classification including Monera, Protista, Fungi, Plantae, and Animalia; multicellular organisms consuming other organisms as prey.

apophyses Branched or forked lateral processes arising from radial spines in the skeleton of some radiolaria; sometimes becoming much branched and fused at their ends forming a lattice shell.

asexual reproduction Reproduction by some form of division of the parent organism without formation of gametes; producing offspring almost identical to the parent.

axoflagellum A thickened, whisker-like projection formed by lateral fusion of many axopodia arising from a specialized pore area in the central capsular wall of some spumellaria.

axoneme   A central shaft or bundle of intracytoplasmic microtubules within an axopodium which provides structural support for and stiffness of the axopodium.

axoplast   A specialized region within the cytoplasm of radiolaria where the microtubules of the axopodia originate; occurring juxtanuclear or embedded within a membrane-lined pocket surrounded by the nucleus.

axopodia   Stiffened, ray-like pseudopodia (cytoplasmic processes) radiating outward from the cell body of radiolaria and other Actinopoda.

bar termination response   A step in skeletal pore production whereby an existing large pore is subdivided into two or more smaller pores by a cross-bridge growth that flares or expands at its distal end upon contact with the opposite rim, thus terminating the new pore production and yielding two smaller slightly ovoid pores (see Fig. 2-26).

capsular wall   An organic, usually perforated, wall surrounding the central capsule of radiolaria (see Fig. 1-6).

carnivore   An organism that preys upon other animals for food.

catabolism   A metabolic process of breaking-down molecules to obtain energy as may occur in intracellular digestion.

cephalis   The helmet-shaped shell, either the entire shell or uppermost dome-like segment, of nassellarian skeletons (see Fig. 1-8I).

chromatin   The genetic material (DNA and associated macromolecules) within the nucleus composing the chromosomes.

cisterna   A membrane-lined canal or space within the cytoplasm of a cell.

coccoid alga   An altered state compared to the free-living alga, often characterized by loss of or changes in the thecal wall (when present), and absence of flagella in motile species; characteristic of some symbiotic algae.

coelopodia   A hollow pseudopodial structure enclosing prey or other apprehended objects trapped by radiolaria (see Figs. 3-2 and 3-5).

cortical shell   An outermost shell enclosing one or more concentrically located inner shells.

crista(ae)   A pouch-like or tubular inwardly-directed fold of the inner membrane of a mitochondrion; the site of ATP production during aerobic metabolism.

cyanophyte   An alga or related organism with "blue-green" pigmentation composed of chlorophylls, carotenoids, c-phycocyanin, and c-phycoerythrin.

cytochalasin B   A compound used in cellular biology to disrupt intracellular microfilaments as a means of investigating the role of microfilaments in cellular activity such as motility.

cytochrome c oxidase   An enzyme catalyzing the reduction of oxygen and the coupled oxidation of the respiratory pigment cytochrome c concurrent with ATP production in mitochondria of protista and other higher organisms.

cytokalymma   A specialized, cytoplasmic sheath in radiolaria that establishes the architecture of the skeleton and secretes the silica during skeletal morphogenesis within its cisterna (see Figs. 2-18 and 2-19).

dinoflagellate   An alga often flagellated and bearing a complex, ornate theca (wall) composed of closely intercalated segments (thecal plates); the theca is constricted at the midline to form a conical epicone and a lower conical or nearly conical, opposed segment called a hypocone; coccoid forms are frequent symbionts in radiolaria.

DNA (deoxyribonucleic acid)   A nitrogen-containing, sugar-phosphate compound that is the building-block of the chromatin polymers in the chromosomes.

endoplasmic reticulum (ER)   An intracellular canal-like, membranous network penetrating deeply into the cytoplasm and possibly providing cisternae for transport of substances throughout the cell (see Fig. 2-6).

exocytosis   A cellular process of eliminating substances from the cell by emptying them from a vacuole or vesicle that fuses with the plasma membrane to form a cup-like depression from which the substances are expelled.

extracapsulum   The mass of cytoplasm connected with and surrounding the central capsular cytoplasm.

filopodia   Fine, thread-like pseudopodia, lacking a stiffened central rod or axoneme, radiating outward from the cell body.

fission   A process of asexual cellular reproduction accompanied by division of the mother cell nucleus into daughter nuclei followed by segregation of the nuclei into fragments of cytoplasm produced by a furrow-like cleavage of the mother cell.

food vacuole   A vacuole containing food particles that have not been digested.

Fungi   A group of lower organisms lacking chlorophyll, and possessing a cell wall sometimes chitinous in quality; spreading by mycelia and reproducing by spores; includes the "molds," mildews, and "mushrooms;" compare to Animalia.

fusule   A complex structure of the capsular wall membranes in radiolaria composed of a strand of cytoplasm connecting intracapsulum
with extracapsulum and often passing through a unique collar
structure in the capsular membrane (see Fig. 2-14).

gelatinous envelope   A viscous or sometimes turgid coat, usually hyaline, surrounding the cytoplasm of radiolaria and often extending
outward to distances well over twice the diameter of the central
capsule.

genotype   The genetic characteristics of an organism expressed as the
gene composition as opposed to the phenotype or physical characteristics of the organism.

Golgi body   An intracellular organelle forming a horseshoe-shaped or
fan-shaped stack of cisternae that bud-off secretory vesicles near
the periphery; origin of some lysosomes and secretory vesicles that
become distributed throughout the cell (see Fig. 2-6).

herbivore   An organism preying on plants or other primary producers.

holoplankton   Planktonic organisms (floating organisms or those mainly
transported by ocean currents) that dwell in open ocean water of
great depth.

intracapsulum   The cytoplasm within the central capsule of radiolaria
including the nucleus or nuclei, subcellular organelles, food reserve
bodies, vacuoles, etc.

karyokinesis   Division of the nucleus to form two or more daughter
nuclei.

lipid   An oily substance forming food reserves within the cytoplasm
and also a significant component of cellular membranes.

macrosphere   An innermost shell (50 μm dia. or larger) within a concentric set of shells in radiolaria.

microbody   A subcellular organelle surrounded by a single membrane
and containing a slightly granular matrix, frequently with a crystalline inclusion of enzymes, apparently catalase; its function is to
mediate nitrogen metabolism and in some protista to link with the
glyoxylate cycle, an important process in conversion of fats to sugar.

microsphere   An innermost shell (less than 50 μm dia.) within a concentric set of shells.

microtubule   A slender intracellular tubule (ca. 300 Å dia.) composed
of protein and forming a cytoskeletal framework within the cell
(see Fig. 3-6).

mitochondrion A subcellular organelle surrounded by a double membranous envelope and enclosing enzyme systems mediating aerobic metabolism including glucose and fat metabolism resulting in production of high energy compounds (e.g., ATP) utilized to drive cellular processes (see Fig. 2-6).

Monera One-celled organisms that do not possess a "true nucleus" and lack mitochondria; includes the bacteria; compare to Animalia.

morphogenesis The origin of form in a living system and the pattern of development of an organism during ontogeny.

morphology The study of form and structure of an organism.

nucleolus An organelle within the nucleus which often appears more densely stained than the surrounding nucleoplasm; a possible site of ribosomal RNA and protein synthesis (see Figs. 2-1 and 2-8).

nucleoplasm The living substance of the nucleus enclosed within the nuclear membranous envelope.

nucleus An organelle within a cell containing much of the genetic information (chromosomes) and the center for coordination of cellular activity.

omnivore An organism preying on plants *and* animals as sources of food.

ontogeny The process of growth and development of an organism from inception of growth to maturity.

organelle A structure within a cell that serves a particular function; e.g., mitochondrion, lysosome, Golgi body.

perialgal vacuole A vacuole enclosing an algal symbiont surrounded by a cellular membrane that may serve as a specialized barrier to permit appropriate isolation of the alga from the host cytoplasm.

perinuclear cisterna The space within the membranous envelope surrounding the nucleus; it is sometimes dilated and surrounds the plastids in a common envelope with the nucleus.

phenotype The physical characteristics of an organism resulting from the combined influence of genetic and environmental factors during development of the organism; contrasted with the genotype or genetic characteristics of an organism.

Plantae A group of photosynthesizing organisms ("plants") in a five kingdom classification; compare to Animalia.

plasma membrane The outer membrane surrounding a cell and regulating exchange of material between cell and environment.

plastid A light-trapping organelle (e.g., chloroplast) containing chlorophyll and/or other light-absorbing pigments that transform light energy into chemical energy to drive photosynthesis; pigment-containing membranes (thylakoids) within the plastid are arranged in thin lamellae or more complex organized stacks called grana (see Fig. 2-16, thylakoid lamellae).

pore termination response A stage in pore development in the radiolarian skeleton characterized by rounded apertures produced by growth of the cytokalymma in a circular pattern resulting in its final stages (termination response) as a rounded pore in the siliceous skeleton (see Fig. 2-26F).

porochora (pore field) A specialized portion of the capsular wall in Nassellaria containing closely spaced pores where the fusules occur.

prasinophyte An alga belonging to the class Prasinophyceae characterized by tiny scales covering the flagella and bearing a nucleus that protrudes into the pyrenoid in some species (see Fig. 2-16C).

primary lysosome A vesicle containing digestive enzymes that are destined to catalyze the hydrolytic decomposition of food in secondary lysosomes produced by fusion of the primary lysosome with a food vacuole.

promitochondrion A precursor of mictochondria, usually much smaller than mature mitochondria and appearing as a small vesicle with slightly denser matrix than the surrounding cytoplasm.

Protista A group of organisms encompassing the classical group "protozoa" possessing a true nucleus enclosed within a membranous envelope; compare to Animalia.

Pseudopodia Cytoplasmic projections that serve specialized roles of locomotion, feeding and other physiological functions.

pyrenoid A subcellular organelle associated with the plastid and a site of starch accumulation during photosynthesis (see Fig. 2-16).

reserve substance Food substances stored within the cell to provide sources of energy during periods of environmental deprivation or when external food cannot be utilized, for example, during quiescent states of the organism.

rhizopodia Fine pseudopodia with a branching or reticulate pattern.

sagittal ring A ring-like component of radiolarian skeletons that lies in a medial sagittal plane separating the skeleton into segments.

sarcomatrix The vacuolated, rich layer of cytoplasm immediately surrounding the capsular wall where, among other functions, digestion occurs extensively within digestive vacuoles.

sexual reproduction Reproduction by gamete formation and their fusion (syngamy) to form a zygote, the earliest stage of a new individual.

spicule A stout or needle-like spine incorporated within or emanating from the radiolarian skeleton.

spine   A skeletal projection; defined differently by various authors as either a major rod-like projection from the skeleton or a minor barb-like enation on the skeleton; in the latter case a major projection is a spicule.

symbiont   An organism living in close association with another organism (the host) resulting in some degree of mutual benefit.

symbiotroph   An organism deriving some or all of its nutritional requirements from its symbionts either by assimilating organic products from the symbiont or by occasionally ingesting a limited number of the symbionts as food.

thecal vesicle   A membranous, sometimes flattened vesicle on the surface of a cell wherein the thecal plate of the surrounding cell wall is secreted, as for example in dinoflagellates.

thorax   The second shell segment or joint in the nassellarian skeleton.

thylakoid   A plastid, internal membrane containing photosynthetic pigments.

tintinnid   A marine ciliate typically forming a conical or trumpet-shaped lorica either formed by a secreted organic substance or by cementing together small particles (e.g., sand grains) gathered from the environment.

tripod   A component of or the major skeletal structure in Nassellaria; composed of three basal feet and a vertical apical spine (see Fig. 1-9E).

trophic activity   The mode or pattern of feeding activity including the quantity, kind and range of prey consumed; and the physiological mechanisms for prey apprehension, ingestion and digestion.

zooxanthella(e)   An alga associated with a host, usually as a symbiont, and possessing a yellow-green pigment as opposed to a clearly green pigment when viewed with light optics; the latter are zoochlorellae.

Index

Note: Italicized numbers indicate figure references.

bioluminescence of, 198
fine structure, 90–94, 112–114, 167
fusule of, 114
longevity (life span), 212
parasites of, 123
in symbiosis research, 199
Thalassicollida, 23
Thalassiosira (diatom) as prey, 175
Thalassolampe, 11, 179, 220
fine structure, 168
margarodes, 121
Thalassophysa, 156, 179
pelagica, 156
sanguinolenta, 156
spiculosa, 156
vesicle production, 189
Thalassophysidae, 156
Thalassopila, 179
Thalassoplegma, 66
Thalassosphaera, 179
Thalassosphaerida, 22, 23
Thalassoxanthium, 179
Thecosphaera bulbosa, 65
Thecosphaera radians, 65
Thecosphaeridae, 51, 64
Theocalyptra, 18, 232
bicornis, 18, 242, 245
davisiana, 18, 219, 242
Theoconus, 18, 241
minythorax, 245
zancleus, 18, 239, 249, 250
longevity (life span) of, 211
Theocorys, 42, 223
veneris, 242
Theocorythium dianae, 18
Theocorythium trachelium, 229
Theocosphaera, 2, 230
inermis, 2
Theocotyle, 42
Theocyrtida, 29
Theocyrtis, 43
Theoperidae, 40, 293
Theophormis, 241
Tholocubus tesseralis, 235
Tholonida, 24
Tholoniidae, 39
Tholospira, 222, 230
Tholospyrida, 29

Tholospyris, 219, 226, 234
devexa, 219, 226
procera, 234, 235
scaphipes, 219, 234
Thyrsocyrtis, 42
Toctopyle stenzona, 256
Transition Faunal Zone, 215, 217, 220–225
Transitional Assemblage, 248
Transitional specialization (protozoa), 86, 88
Tribonosphaera, 69, 75, 261
centripetalis, 75, 263
Triceraspyris antarctica, 229
Trilobatum, 230
Triopyle hexagona, 235
Triospyris antarctica, 234
Triplecta triactis, 26
Tripocalpida, 29
Tripocyrtida, 29
Tripocyrtis pteides, 238
Tripod, 13, 21
Trissocyclidae, 293
Tristylospyris, 234, 241
palmipes, 234, 235
Trochophore larvae as prey, 175
Trophic activity, 174–189
algal predation, 93, 175, 177, 180–183, 187, 203–204, 268–270
animal predation, 14, 98, 175, 177, 182–188, 269
assessment of, 176–178, 186–188, 264–267
compared to foraminifera, 177, 188, 197, 270
and ecological niche, 268–270
infusorial predation, 14, 98, 180–182
predation behavior, 14, 98, 177–178, 180–189
prey diversity, 174–178
role of symbionts, 120–121, 187, 199–206, 266–270
Tropical Assemblage, 247
Tropical complex, 234
Tropical Convergence, 232, 233, 287
Trypanosphaera, 69
Tuscarora, 224
Tuscarorida, 31, 33